Marine Pollutant Transfer

Marine Pollutant Transfer

H. L. Windom
Skidaway Institute of
Oceanography

R. A. Duce
University of Rhode Island

Lexington Books
D. C. Heath and Company
Lexington, Massachusetts
Toronto

Library of Congress Cataloging in Publication Data

Main entry under title:

Marine pollutant transfer.

1. Marine pollution—Congresses. I. Windom, H. L. II. Duce, Robert
A.
GC1081.M37   628.1'686'162   76-17507
ISBN 0-669-00855-9

Published simultaneously in Canada

Printed in the United States of America

International Standard Book Number: 0-669-00855-9

Library of Congress Catalog Card Number: 76-17507

Contents

Preface

The ocean environment is the final repository of waste resulting from man's continental and marine activities. Scientific evidence suggests that the rate of pollutants such as heavy metals, petroleum and chlorinated hydrocarbons entering the ocean is increasing. A better understanding of transfer pathways to and within the ocean environment is of paramount importance if we are to understand the potential impact of this increase and react in a rational manner to preserve this system.

Realizing the importance of this subject, the National Science Foundation's Office for the International Decade of Ocean Exploration initiated a research program on pollutant transfer in the marine environment. Under this NSF/IDOE Pollutant Transfer Program, research projects have been conducted by a number of scientists in the United States over the past four years. In January 1976 a Pollutant Transfer Workshop was held at the Skidaway Institute of Oceanography in Savannah, Georgia, to evaluate their research progress in this field and to make recommendations for future research. During this workshop particular emphasis was given to evaluating our present knowledge relative to the quantitative fluxes of the various pollutants to the marine environment. In addition to individuals specifically involved in this NSF sponsored program, other scientists concerned with problems of marine pollution, both from the United States and other nations, participated in this workshop. The deliberations and recommendations of the workshop participants are presented in the initial section of this volume. As a further outgrowth of this workshop, the investigators participating in the NSF/IDOE Pollutant Transfer Program agreed to summarize their findings to date in a number of papers which form the major part of this volume. Although this volume is not meant to be a complete evaluation of pollutant transfer in the marine environment, it nonetheless serves to bring together the results of these complementary studies.

The initial paper in this volume summarizes the deliberations and recommendations of the Pollutant Transfer Workshop. The paper is organized into five sections, which attempt to summarize briefly our state of knowledge about the transfer of matter across the five primary interfaces in the marine environment: (1) Air/Sea (2) River/Sea (3) Biosphere/Sea (4) Particle/Sea (5) Sediment/Sea. The contributed papers are grouped generally according to pollutant category: heavy metals, petroleum hydrocarbons and chlorinated hydrocarbons.

The Pollutant Transfer Workshop that initiated this volume was supported by a grant from the National Science Foundation's Office for the International Decade of Ocean Exploration (OCE76-09731). We would like to thank all of the participants at this workshop for their time and efforts.

H. L. Windom
R. A. Duce

1

Pollutant Transfer: Workshop Deliberations
Workshop Participants

The subject of pollutant transfer in the marine environment was discussed in a NSF sponsored Workshop at the Skidaway Institute of Oceanography in January 1976. The participants, listed in Appendix 1A, included experts from the United States and other nations. The Workshop was charged with evaluating the state of knowledge of pollutant transfer across the five primary interfaces of the marine environment and making recommedations for future research. Although the Workshop broke into five working panels, individuals were free to contribute to each. In this way the interdisciplinary nature of the problem could be considered. The deliberations of each panel are given in the following sections.

Air/Sea Interface Exchange Processes

James Butler: Panel Chairman

The boundary region between the lower atmosphere and the surface of the ocean is an active zone of interaction. Exchange of pollutant and naturally occurring gases and particles occurs in both directions across the air/sea interface. Concentrated efforts to understand chemical exchange processes across this interface have been relatively recent, however. The past decade has seen significant development in the area of air/sea gas exchange models as well as laboratory and field investigations of the chemistry and physics of processes occurring when bubbles burst at the air/sea interface. There has also been considerable interest in the chemistry and dynamics of the sea surface microlayer, particularly as a concentration zone for various pollutants. Although we are beginning to get some insight into the importance of the surface microlayer and bubble scavenging processes relative to the chemistry of atmospheric particles produced by bursting bubbles, we still know relatively little about the downward fluxes and mechanisms of deposition of atmospheric particles to the ocean.

"State of Knowledge" Matrix

The air/sea interface panel has attempted to evaluate the present state of knowledge relative to pollutant transfer across this interface by contructing the matrix shown in Table 1-1. Columns represent three classes of pollutants: halocarbons, heavy metals and hydrocarbons. Each of these has been divided into a more volatile and less volatile group. For example, freons are much more volatile than PCB or DDT residues and thus behave differently with respect to general transport processes. Rows represent areas of understanding we seek about each of the classes of pollutants. In addition to the concentrations of pollutants and rates of transfer associated with either side of the air/sea interface, we have added "origin," since there is considerable controversy over whether the source of some materials is anthropogenic or natural. We have also added "ultimate fate" since this is one of the fundamental questions asked about any pollutant. In the matrix, 0 represents lack of knowledge, 1 represents some but not adequate knowledge, and 2 represents substantial knowledge.

As an example of the use of this matrix, concentrations in the surface microlayer have been measured for all materials (except freons) but only in a few locations, and with devices that fall far short of the ideal of sampling the outermost molecular layer of the sea surface. Thus the first row of the matrix is mostly 1's. Similarly, freons and light hydrocarbons are almost certainly in

Table 1-1.

State of Knowledge Matrix for Air/Sea Interface Exchange Processes

| | Halocarbons | | Metals | | Hydrocarbons | |
	Freons	Others	Volatile (Hg, Pb, etc.)	Others	Light ($<C_{12}$)	Heavy
Concentration in Surface Microlayer	0	1	1	1	1	1
Concentration in Air	2	1	1	1	1	0
Air/Sea Partition at Equilibrium	1	1	0	1	2	1
State of Aggregation in Air	2	1	1	2	2	1
Rate of Fallout From Air	0	1	0	1	1	0
State of Aggregation in Water	0	1	0	0	1	1
Rate of Reemission From Water	0	0	1	1	1	0
Origin	2	2	1	2	1	1
Ultimate Fate	1	1	1	1	1	1

0 = We know nothing
1 = We know something
2 = We know a lot

the gas phase in the atmosphere and so these are designated 2 in the horizontal row "State of Aggregation in Air."

Depending on the objectives of any particular research program, more or less emphasis should be given to the items marked 1 and 0. For example, determining the precise physical and chemical state of freons in seawater is a very difficult fundamental problem, and would require much more effort than measuring the distribution of a freon compound between the sea and air, which involves only determining total concentration in the water and air, not the detailed state of aggregation. However, if the objective of a program were a detailed understanding of the chemical and physical state of pollutants in the environment, the former might be given priority.

Under the present NSF/IDOE Pollutant Transfer Program, the objectives are to trace the pathways followed by pollutants introduced into the marine environment and to provide quantitative assessment of rates of exchange and quantities in each reservoir. Crudely speaking, we are dealing conceptually with a box-and-pipeline model. With this in mind, it is clear that what is desired is direct, phenomenological measurements of the vertical transport of materials within and into the sea with attention to the physical-chemical properties of the materials (solubility, vapor pressure, diffusivity, adsorption energy) and attention to relevant oceanographic and meteorological conditions (air and sea stability and turbulence, presence and type of surface films, temperature, insolation, humidity, etc.).

Future Research

Some priority areas of future research are clearly suggested by Table 1-1. Sampling of the surface microlayer should be more extensively carried out, with equipment (e.g., hydrophilic teflon, the BIMS device developed at the University of Rhode Island) which more closely approximates the ideal of collecting the topmost monolayer. This surface microlayer, although minute in volume compared to the atmosphere or ocean, has an influence on the marine biota far out of proportion to its volume, and represents the interface through which all air/sea interchange must pass. In connection with this work, the further development of the theory of transport across the interface, particularly under conditions of turbulence and rapid bubble formation and breaking, is essential. Measurement of the bubble size and number distribution on the open ocean as a function of wind speed is urgently needed. It would be desirable to know the aerial coverage of given portions of the sea by natural and pollutant surface films, and how these form and disperse. Air surveys using side-looking X-band radar, which measures capillary wave intensity, would be useful for such studies.

Closely related, but almost unexplored, is the measurement of rates of reemission of materials from the sea back into the air. Only for the heavy metals and the light hydrocarbons is there any information and much of this is speculative.

Investigation of all the transport processes for freons is necessary. Although it is known that freons are anthropogenic, that they exist as gases in the atmosphere, and that they probably find their ultimate sink in the stratosphere, virtually nothing is known about what quantities of the freons find their way into the oceans, or what the rates of exchange are.

Similarly, little or nothing is known of the rates at which atmospheric trace metals, and particularly volatile metals (like Hg) or those forming volatile compounds during high temperature anthropogenic processes (like lead) are exchanged at the air/sea interface. There is virtually no reliable data available on the concentration of trace metals (and most other pollutants) in precipitation in open ocean areas, nor is there information on the dry deposition rate of these materials to the open ocean surface. There is some evidence that aeolian transport of particulate matter to the ocean is significant to marine sedimentation on a global scale, but this requires considerable additional quantitative data. The state of aggregation in air is rather uncertain for volatile metals because the relative amount in the gas phase compared with that associated with particles has not been adequately studied for most of them. In water, this is further complicated by the lack of information on chemical speciation.

The heavy fraction of hydrocarbons should receive more attention than the light fraction (typically methane and lower alkanes) because of the potentially greater impact of heavy aromatics and heterocyclics (some of them carcinogens) on the biosphere and on human health. No estimate, let alone measurement, has been made of their rate of fallout from the air or their rate of reemission from the sea surface into the air. Evaporation *per se* is not an important mechanism, and because of the complex character of petroleum, the distinction between true solubility, micelle (emulsion) formation, particle formation and adsorption on particulates of other origins is difficult to accomplish and is poorly understood. This is further complicated by the difficulty of distinguishing between hydrocarbons of petroleum and biogenic origin. Ideally, this work should not be done with "crude oil" or its residues but with compounds of known biological impact found in crude oils. However, pinpointing the optimum study compound and developing adequate analytical methods for isolating it is a research project in itself.

River/Sea Interface Exchange Processes

Roy Carpenter: Panel Chairman

Rivers deliver the largest quantities of many anthropogenic chemical compounds

—both organic and inorganic—to the estuarine and coastal zones of the ocean. Since these are the areas with the greatest biological productivity and are most intensively utilized by man, an understanding of river/sea exchange processes is particularly important. Any program investigating pollutant transport in this region must be concerned with (1) characterizing the river inputs of organic and inorganic chemicals, and (2) determining the interactions that occur when the chemicals supplied by the river enter the sea and what fractions of the river input are transported through the estuarine zone or immediate area of the river mouth to the more open ocean environments.

Generally we know much more about the river input of inorganic chemicals to the ocean than organics. There have been, however, two recent examples of the importance of river input of organic materials. One is the input of Kepone coming from a chemical plant on the James River in Hopewell, Virginia. This synthetic organic killed bacteria in a sewage treatment plant and the river is now closed to sportfishing, shellfish harvesting, etc. The other example is the large quantity of PCBs in the Hudson River system which are or will be transferred to the ocean at an unknown rate.

Characterization of River Input

In characterizing river inputs one faces the important problem of designing a sampling program to acquire representative samples. Significant vertical and horizontal inhomogeneities in the amounts of suspended matter, its size distribution, mineralogy, chemical composition and reactivity occur in most rivers. Predicatable seasonal changes in many rivers lead to variations in amounts and types of suspended and dissolved phases transported. Varying seasonal flow rates change the total quantity of material and possibly change the relative importance of various inorganic fractions. The ratio of biogenic/inorganic solids is also seasonally dependent. Such seasonal variations are especially significant in arid and semiarid regions. Sampling programs can be designed to account for these changes, but not for the effect of unpredictable episodic pulses—those due to floods, storms, spills, etc. Occasional storms, such as Tropical Storm Agnes, may transport more material to the ocean in a few days than occurred in many years immediately preceding, and we are not now in a position to estimate reliably the effect of such events. The transport of most chemicals with the coarser bed load of rivers is probably less significant than transport with finer suspended matter, so the bed load tends to be overlooked in most sampling programs. The bed load characteristics and behavior depend on the flow of the river and estuarine hydrodynamics. There are two known cases where the bed load contains major quantities of PCBs and must be considered— the Hudson and Duwamish Rivers. Because of the dynamics of these two estuaries, the bed load of these rivers and their associated chemical compounds have been transported *upstream* under normal river flow rather than to the sea. Eventually river floods will carry the material to the sea, however. In general, the in-station variability of bottom sediment is much greater in rivers than on continental shelves and in the deep sea. This variability must be reliably established before variations and real differences between stations can be considered.

We generally feel that adequate analytical techniques are now available for determining the total quantity of substances in most dissolved and solid phases in a particular river sample. It is more difficult to determine the chemical forms and phases with which various organic pollutants and inorganic chemicals are associated. This information is most important to obtain since the phase and chemical speciation will control the biological availability, toxicity, and the chemical reactivity when the material enters the sea. The fate, behavior, and effects of each chemical form will be different. We urge better characterization of the chemical forms and phases with which organic and inorganic chemical compounds are associated in a few environmentally-typed rivers rather than determinations of only total quantities in many rivers.

Several chemicals which are potential pollutants are associated primarily with the solids rather than the dissolved phase in rivers, so it is important to characterize the solids as well. We now have a rather complete characterization of the solid phase of Fe, Mn, Cu, Co, Ni, and Cr carried by the Amazon and Yukon Rivers. Equivalent data for these elements in other rivers and for other elements in these two rivers are not yet available and would be most valuable.

One complication in understanding the reactions of river basin derived solids in the sea is that they may acquire an organic coating which dominates and controls reactivity with the dissolved species. The chemical reactivity of such coated materials may be quite different from the uncoated bulk material. Analytical techniques to characterize such surface coatings (such as ESCA) are just now being developed and have not yet been applied to river or marine solids. The partitioning behavior between dissolved forms and such coated solids is predictable for many organic chemicals whose chemical reactivity with surfaces is dominated by London-dispersive interactions for the range of ionic strengths commonly encountered in estuaries and rivers. Evidence of this behavior and its prediction has been provided to date only for the Puget Sound estuary.

Reactions of River Derived Materials in the Ocean

Once the river-borne materials enter the ocean they undergo several types of reactions with time scales varying from hours to millenia. Examples of such reaction which must be considered are

1. Inorganic sorption/desorption in response to changes in pH, ionic strength or concentration of a particular species.
2. Coagulation/flocculation and settling out.

3. Inorganic precipitation or dissolution—e.g., some of the river soluble iron may precipitate out when contacting the higher pH environment of seawater.
4. Biological uptake, excretion and alteration.

It should be stressed that some of these reactions occur quickly in the zone of river/seawater mixing. It is important to know the fraction of the dissolved, solid and solid-associated chemical elements and compounds which are actually transported away from the estuarine area at the river mouth to the open ocean. This fraction will vary for different chemicals and may be dominated by episodic pulses. It is clearly dependent on a complex combination of biological, chemical and physical oceanographic processes. If the great majority of river derived materials are trapped within a few kilometers of the river mouth, the amount of river input to the more open ocean is greatly overestimated by determining just the amount carried by the river.

Efforts to determine the percentage of river-transported Hg and Cd brought to the southeastern U. S. coast that is trapped in estuaries and the percentage that is transported to the more open continental shelf are now underway as part of the present NSF/IDOE Pollutant Transfer Program. This panel stresses the need for more such studies, both of other rivers and other chemical substances. The PCB transfer through the Hudson River and its estuary would be especially significant to determine.

Biosphere/Sea Interface Exchange Processes

Donald Button: Panel Chairman

The general area of biosphere/sea exchange involves processes central to current biochemical research, chemical speciation theory and population dynamics. It also taxes the lower limits of chemical sensitivity. Two general types of pollutants—heavy metals and hydrocarbons (both chlorinated and nonchlorinated) —were considered by this panel. Each of these types of pollutants may be accumulated by active or passive biological processes, then retained, metabolized or eliminated.

Incorporation of Metals

It is recognized that metal ions are actively taken up (e.g., by cell walls, gill tissue, gut tissue) by organisms within many given trophic schemes. However, such phenomena within the marine biosphere may not represent a major pathway for metal uptake since such recognition is based on laboratory experiments dealing with only a few types of organisms, selected types of metals, and without adequate knowledge of the environmental forms of metals available for incorporation.

A major route of required metal uptake by marine organisms may be via selected metal complexes, including food. Incorporation by lower organisms (e. g., zooplankton) can be primarily from food sources, especially when food is ingested in relatively high concentrations (e.g., organisms feeding on phytoplankton blooms).

A major route for metal transfer among trophic levels is via metals adsorbed to the surfaces of organisms. The phenomenon is particularly significant among dead organisms since it represents a mechanism for making metals available for transfer through the detrital food web.

Incorporation of Hydrocarbons

Specific active transport systems for alkane and aromatic hydrocarbons exist and kinetic information is available for some of these systems. Chlorinated hydrocarbons are probably handled by the same mechanism. Both may or may not be metabolized, depending on structure.

Direct partitioning with the environment based on partition coefficients occurs. There is substantive information for direct accumulation but relatively little for incorporation from solid food. Direct accumulation across respiratory membranes in those organisms having gill structures seems to be the dominant mechanism. There are reports describing how high concentrations of hydrocarbons, including chlorinated hydrocarbons, alter the composition of biological communities—e.g., in the vicinity of a major oil spill. The ecological mechanisms are still unresolved, and this appears to be a fruitful area of further study.

Losses of Heavy Metals and Hydrocarbons

Loss of heavy metals and hydrocarbons by animals may occur either before or after incorporation (assimilation). Loss before assimilation will be in the form of fecal material (excrements). It is differentiated between unmodified and modified compounds. Modification may be caused by digestive processes in the gut of the animal. Loss after the pollutant has been assimilated occurs by excretion (dissolved). Heavy metals may be discharged in a detoxified form as bound metals. Some hydrocarbons are excreted as water soluble derivatives. Pollutants may reenter the water in an unmodified form through excretion.

Inhibition Mechanisms

Inhibitory effects of pollutants can modify the extent to which organisms affect fluxes. General pertinent mechanisms include interference of catalysts and transport rates by associations with proteins or lipids. Similar mechanisms are possibly involved in affecting changes in reproductive or chemotactic behavior. The low level effects remain unresolved.

Biomagnification in Marine Food Chains

The tendency for any pollutant to be biomagnified in marine food chains is dependent on the chemical characteristics of the pollutant and its interaction with the tissues of marine organisms. Recently, attempts have been made to relate such properties as octanol-water partition coefficients in the case of hydrocarbons and protein binding properties in the case of several heavy metals to the tendency of these materials to be biomagnified. This type of approach could be very fruitful. If sufficient correlative data of this sort are obtained it may be possible to construct a predictive model relating one or a few chemical and physical properties of a pollutant to its tendency to be biomagnified in marine food chains.

Ultimately biomagnification depends on the rates of uptake and release of the pollutant by marine organisms, the nature of the tissue "binding" of the pollutant, and finally the availability of ingested tissue-bound pollutants to consumers at higher trophic levels. There is a growing volume of published data on uptake and release rates of various pollutants by many species of marine organisms. However, there is relatively little information about the behavior of pollutants once they are absorbed or adsorbed by the organism. If we are to predict the potential for a given pollutant to biomagnify in the marine food chain much more information is needed about such subjects as the distribution of pollutants in the tissues of contaminated organisms, the nature of the tissue-pollutant binding, and the availability of tissue-bound pollutants to consumers of contaminated organisms.

Future Research

Biological processes may represent a major mechanism in the transport of pollutants through the mixed layer of productive areas of the oceans. The magnitude of this flux depends on a number of factors, the importance of which need evaluation. Uptake at the primary producer level is the most important mechanism for metal transfer into the biosphere and therefore requires considerable further research. The relationship between pollutant uptake and primary production should also be better established. In the open ocean the relative importance of pollutant removal for the mixed layer by settling dead phytoplankton cells and fecal pellets should be evaluated in relation to variations in primary and secondary production rates.

Particle/Sea Interface Exchange Processes

Peter Betzer: Panel Chairman

Genesis of Particles in the Ocean System

External sources. Both the atmosphere and rivers/runoff are significant external sources for particles in the ocean. These have been discussed by the Air/Sea and River/Sea Exchange panels.

Mid-ocean ridges and spreading centers are areas of considerable geologic activity: large volumes of rock are emplaced or extruded here, the epicenters of earthquakes are concentrated in these regions and substantial shifting of crustal material takes place along fracture zones. Metal-rich sedimentary deposits are found in the Pacific, over the East Pacific Rise and in the Atlantic on the Mid-Atlantic Ridge. These deposits appear to have formed from the precipitation of iron and manganese hydroxides from seawater. Other transition metals are associated with these deposits and may also be added to the deep ocean at active spreading centers. There is a need for studies from such active areas which could be used (1) to identify which elements are added to the ocean system in these areas, and (2) to determine how these species react once they are added to the ocean system (i.e., are the materials being added to these areas removing substantial amounts of material from seawater). Iron, manganese, mercury, arsenic, copper, lead, cadmium, nickel and cobalt are elements which may be added in substantial quantities over active spreading centers by submarine volcanism.

The input of glacial flour in polar areas amounts to about 30×10^{14} grams/year. This is approximately 10% of the suspended materials carried annually by rivers.

Ocean dumping results in the release of substantial quantities (2×10^{12} grams) of petroleum and petrochemical wastes to waters of the open ocean as well as coastal waters of the United States. In addition, in selected areas such as waters of the New York Bight, the dumping of sewage sludge, acid mine waste, cellar dirt and dredge spoil exceeds the particle inputs for all rivers between the Canadian border and Chesapeake Bay.

Generation within the ocean. It has been demonstrated that particle growth or aggregation can be facilitated by bubbles rising through a column of water. There is also convincing evidence that surface active organic components play an important role in the aggregation of these organic particulates. Trace metals can be associated chemically or physically with these aggregated particles. There is also evidence that some trace metals that are soluble in seawater may accumulate in particulate aggregates during bubbling. It is not known if these soluble species are inorganic in nature or are chemically chelated by soluble organic components.

Another type of mechanism for the generation of particles in marine systems is the nucleation and growth of solid phases from dissolved components. Particle sizes may subsequently increase by flocculation. Knowledge of these processes has been obtained primarily through relatively simple chemical studies under laboratory conditions and in a few instances their occurrence has been documented in marine systems. Nucleation and particle growth require that seawater be supersaturated with respect to the particular solid phase. Calcium and carbonate in surface ocean waters provide one of the best known examples of a supersaturated condition, but in this case nucleation and growth do not occur because of the inhibitory effects of other seawater constituents such as magnesium and organic matter. Flocculation is likely to be a significant process at the riverine/ocean interface. The behavior of colloidal particles in marine systems is poorly known. One would expect that electrostatically dispersed colloids will coagulate during transport through a salinity gradient. On the other hand, it has been argued that marine colloids may be protected from flocculation by organic substances. Generation of particulates by the biomass has been considered by the panel on Biosphere/Sea Exchange.

Chemical Exchange Between Particles and Seawater

Particles are thought to be vehicles for the transport of pollutants into and within the marine ecosystem. There are two primary regions where particles are generated: near the sea surface and near the sea floor. As described previously, particles near the sea surface may either be introduced from external sources such as the atmosphere or river runoff, or they may be produced in the water itself by a variety of mechanisms such as the biological production of plankton cells and the generation of particles from dissolved organic matter. Particles near the sea floor are generated by the resuspension of sediments.

Depending upon their size, their shape, their chemical composition, and whether they are alive or dead, these different types of particles may have different environments. Inorganic particles may be more important carriers of pollutants in the vicinity of river mouths whereas organic detritus may be the predominant carrier in biologically productive areas such as the high latitude cyclonic gyres of the continental shelf upwelling regions.

In order to describe quantitatively, as well as qualitatively, the flux of pollutants into and within the marine ecosystem a number of questions must be answered:

1. Are particles quantitatively important carriers of pollutants?
2. What types of pollutants are preferentially adsorbed onto what type of particle? For instance, partition coefficients on the order of 10^5 have been found for chlorinated hydrocarbons between seawater and particles regardless of the particle organic content.

3. Is the uptake by organisms of pollutants associated with particles an impor-
    tant mechanism for their introduction into and transfer through marine
    food chains? It should be noted that many of these questions are likely to
    be answered by comparatively simple laboratory experiments.

It has generally been believed that the chemical exchange of substances between seawater and solid interfaces plays a significant role in regulating the
composition of marine waters. In the open ocean environment it has been possible to uniquely show that a given chemical distribution is controlled by these
surface processes. The significant environmental variables influencing adsorption,
desorption and ion exchange are surface interfacial area, pH, salinity, and solution phase complexation. The chemical properties of the interface are of secondary importance. These surface processes are likely to be most significant in
estuarine, coastal and productive surface waters as well as in the interstitial
waters of sediments.

Solubility equilibria can in principle provide a mechanism for particle-seawater exchange. Biogenic production of solid phases is more prevalent than
chemical precipitation, but this latter process may occur when chemically concentrated fluids enter the ocean—e.g., during submarine volcanism or manmade
waste discharge. Solid phase dissolution is an important process for carbonate
and silicate phases in the deep ocean. As a first approximation solubility equilibria may be described simply by a solubility product. However, in multicom-
ponent natural solutions such as seawater, solid phase substitution reactions
rather than simple solubility equilibria may prevail. In the oceanic environment
the kinetics of solubility reactions may be the controlling factor.

Changes in clay and carbonate mineral lattices undoubtedly take place in
the marine environment. These alterations range from being fairly rapid (i.e.,
the movement of magnesium out of a high magnesium calcite lattice after the
death of an organism) to slow (i.e., exchange of interlattice cations in certain
clays). The former are probably important in determining the fate and exchange
of certain pollutants in the water column while the latter would have a larger
impact on the distribution of pollutants in sediments and pore waters. In any
case, little is known about the importance of matrix changes in the marine environment and their involvement with pollutant transfer.

Particle Transport Through the Marine Environment

Particles in marine systems will be transported in part by the physical processes
of advection and turbulent mixing. In addition, particles are subjected to gravitational settling. The relative magnitudes of these two types of processes depends on the size and density of the particles. Progress has been made in recent
years in defining the size distribution of particles in the marine environment

using Coulter Counter and optical scattering techniques. Present knowledge is limited by the lack of characterization of particles in the size range of 0.01-1.0 μm.

The resuspension of sediments by bottom currents has long been recognized as an important part of sediment dispersal in the ocean and is, therefore, an integral part of pollutant transfer in the marine environment.

As wave-produced bubbles rise through the water column they effectively transport a significant fraction of the aggregated particles to the water surface. Transport of inorganic particulates to the surface is related to the quantity of surface active organic matter in the water. This may lead to an increase in the residence time of inorganic particulate matter in a given water column. As the bubbles break at the water surface, some organic and inorganic material is ejected into the atmosphere thereby providing a removal mechanism from the water column. It is not known if particle flotation during bubbling causes a net transfer of material to the atmosphere or accelerates the sedimentation of particles to the deep ocean. Pollutant organic substances may effectively alter the normal surface active character of the water such that aggregation and bubble flotation of particulates is accelerated.

Future Research

The movement of many natural and manmade materials within the ocean system is related to and mediated by suspended materials. Given this important fact, it is surprising that so little is known of the variation in size, in surface area, in number and in composition (both surface and bulk) of particulate matter with space and time. In fact, most of the effort applied to chemical studies of particles in the ocean has been applied to simple bulk chemical analysis. While this information is a necessary starting point, it is insufficient for a meaningful understanding of seawater-particle exchange. A systematic program of oceanic particle-seawater interaction should be formulated which will establish the relationship between particle reactivity and particle characteristics. Studies of surface active organic compounds which could affect particles in the ocean in relation to growth or disintegration throughout the water column would be beneficial. Also information is needed on the possible selective association of pollutant trace metals (lead, cadmium, mercury, etc.) and organics such as petroleum hydrocarbons, PCBs, and DDT with inorganic particles, planktonic or organic detritus. The population of naturally occurring oceanic particles should be characterized in terms of the inorganic (carbonates, clays, quartz, hydroxides) and organic (aggregates, frustules, coccoliths) components, surface morphology (scanning electron microscopy), mineralogy (X-ray diffraction), surface chemical characteristics (electron microprobe and ESCA), surface area and functional groups. The modification of particle size distributions by marine processes should be considered. These efforts to understand the nature of oceanic particles and their role in chemical transport and transformation in marine systems should be undertaken relative to their chemical exchange with seawater and not merely for their own sake.

Attention should be given to the micro-environments created by chemical and biochemical processes near interfaces which may influence particle-solution exchange. Many surface exchange processes are highly sensitive to pH and the pH of the bulk solution may not be representative of that near the interface. Chemical inhomogeneities along a surface may result in large differences in chemical reactivity.

The foregoing investigations of seawater-particle exchange could be integrated with efforts to determine the degree of particle uptake, assimilation, and modification by major marine species. For example, what chemical and physical changes occur at the surface of particles as they pass through the gut of benthic organisms and zooplankton? Also, does the association of particles facilitate or depress microbial decomposition?

Is the sinking of particles associated with pollutants a quantitatively important mechanism for their vertical transport and does this provide a means of transporting organic pollutants through the pycnocline? Intensive research near a large, geochemically significant river such as the Mississippi would help us understand the fate of both natural and manmade materials which are delivered to freshwater/saltwater interfaces. Laboratory studies aimed at understanding the role that pH, salinity, complexation and particle type play in flocculation, and adsorption-desorption reactions over substantial ionic strength gradients should complement the field work.

Finally, research should be supported that is devoted to new methodology of collection, analysis and speciation.

Sediment/Sea Interface Exchange Processes

Michael Bender: Panel Chairman

Material transport across the sediment/sea interface (Figure 1-1) involves the dissolved and suspended components of the bottom water, sedimentary solids and interstitial water, and benthic and pelagic organisms. Mass transfer between these phases may occur by a number of mechanisms. Particles are removed to the sediments by gravitational settling and uptake by benthic biota. Resuspension of this sediment by bottom currents and bioturbation may temporarily return particles to the water column. Dissolved substances may accumulate in sediments by adsorption and inorganic precipitation and by incorporation into the benthic biota. Dissolved components may also migrate into sediments as a result of low concentrations of certain dissolved species in pore waters. Bacterial

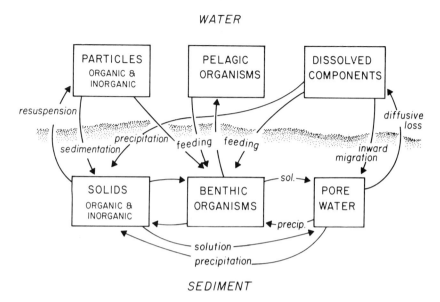

Figure 1-1. Material transport pathways across the sediment/sea interface.

decomposition of sedimentary organic detritus brings about the dissolution of nutrients and other chemicals to set up a vertical concentration gradient and a diffusive flux out of the sediments. These biologically mediated reactions change the Eh and pH conditions of pore waters and thus may change the solubility of various chemicals and allow dissolution and migration out of the sediments.

It is believed that the three key processes of interest to pollutant transfer are (1) the particulate flux to the sediments; (2) biologically mediated dissolution, precipitation, and/or degradation in the pore water; and (3) transport of the dissolved constituents out of pore waters by diffusion.

Sedimentation of Pollutants

Understanding sedimentation rates in the nearshore environment is important both for determining the ultimate fate of pollutants and for ascertaining whether pollutants entering coastal areas ever reach the open ocean. In some cases, fluxes of material to the sediments can be determined directly from concentrations of chemicals in sediments and their respective accumulation rates. However, in most nearshore areas sedimentation rates are so variable that this approach may not be fruitful. Comparison of elemental concentrations in sediments to levels in source material may provide information on whether nearshore sedimentation is a sink for various chemicals. Also, in water bodies where the water balance is well understood (as in an estuary where salinity can be used to determine relative amounts of fresh and salt water present), fluxes of dissolved and particulate chemicals across the sediment/sea interface may be estimated from the distribution of dissolved and particular components.

Pelagic sediments are apparently important sinks for pollutants having residence times between days and centuries in the ocean and atmosphere. (Materials having shorter residence times will not reach the open ocean; those having longer residence times will not reach the sediments for some time.) Sedimentation rates of pollutants that do not occur naturally (such as synthetic organics and plutonium) can be determined from integrated accumulations in pelagic sediments. Pelagic accumulation rates of other pollutants (such as petroleum hydrocarbons and trace metals) generally cannot be determined, because of our inability to sort out anthropogenic and naturally occurring components. Removal of these substances can probably be determined by studying removal rates of either natural or anthropogenic analogues and relative abundances in the overlying waters. For example, accumulation rates of petroleum hydrocarbons in sediments might be estimated from PCB accumualtion rates and the relative abundance of petroleum hydrocarbons and PCBs in the water column.

Work on pelagic sedimentation of pollutants should be carried out at a number of specific sites (each perhaps one square kilometer in area) distributed throughout the world oceans. Concentration on a limited number of areas would allow for excellent documentation of the sedimentology. It would provide a data base for each site that would be useful for all investigators, it would allow various workers to intercalibrate on collection and analysis methods, and it would provide a continuous historical record of use to present and future investigators. In addition, sites could be chosen so as to be accessible to research vessels of many nations, which would foster international cooperation.

Biologically Mediated Reactions

Biologically mediated reactions are the driving force behind the recycling of elements in natural biogenic detritus, and are likely to do the same for some pollutants. Pollutant transport across the sediment/sea interface may be dependent on biodegradation reactions in pore waters. Understanding such reactions is critical because petroleum hydrocarbons and synthetic organics may undergo biodegradation in sediments. Until we know the extent of this process, we can regard measured accumulation rates as lower limits on the rates of removal of these compounds into sediments. In addition, we should examine whether degradation products are themselves harmful.

Biologically mediated reactions in sediments change Eh and pH conditions, thereby making certain compounds more soluble than in the overlying bottom waters (Mn, for example) and others less soluble (Cd, for example).

The partitioning of pollutants between the solid and liquid phases in the sediments is dependent on the chemical nature of the pollutants, the surface layers of the sediment particles, and the pore water components. This partitioning and resulting transfer by diffusion may therefore be modified with time as degradation and diagenic processes change the character of the sediment water system.

Diffusion Processes

Dissolved components may diffuse into or out of sediments as a result of concentration gradients between pore and bottom waters. Biological decomposition may cause a buildup of components in pore water (relative to the overlying water) and cause them to diffuse out of the sediment. Rates of transfer of dissolved constituents out of pore waters can be estimated both from diffusion models along with pore water concentration gradients, and direct measurements of benthic fluxes using bell jars. Where such benthic fluxes may be significant in pollutant budgets (for example, phosphorus migration out of nearshore sediments polluted with sewage) data should be collected to evaluate the fluxes.

Within the present NSF/IDOE Pollutant Transfer Program far more effort has gone into assessing pollutant fluxes across the sea/air interface than across the sediment/sea interface. However, from this and other programs we are accumulating some information on rates of sedimentation of pollutants in the estuarine and coastal environments (for metals) and in pelagic sediments (for PCBs).

Conclusions and Recommendations

The preceding panel reports have given an indication of our present understanding of material transport across the five critical interfaces in the marine environment and have discussed in some detail areas in which additional effort is required. Major recommendations of these panels relative to future research in these areas follow.

Air/Sea Interface

1. Investigation of the physical and chemical mechanisms by which many trace substances are concentrated at the air/sea interface.
2. Determination of the fluxes of particulate and gaseous trace substances in both directions across the air/sea interface in open ocean areas.

River/Sea Interface

1. Characterization of river-transported materials in major river systems according to their dissolved/particle distribution; vertical, horizontal, and temporal variability; and chemical speciation.
2. Determination of the fate (sorption, desorption, coagulation, flocculation, precipitation, dissolution, biological uptake, etc.) of river-derived materials in estuarine areas. This is critical to evaluate the magnitude of their input to the open ocean.

Biosphere/Sea Interface

1. Investigation of pollutant uptake at the primary producer level as a function of production and water chemistry.
2. Characterization of the role of the biosphere in the transport of pollutants out of the mixed layer.
3. Characterization of pollutant transfer in food chains to establish whether or not biomagnification occurs.

Particle/Sea Interface

1. Characterization of marine suspended material relative to its role in chemical transport and transformation in marine systems. This includes studies of the particle reactivity as a function of particle type (inorganic, organic detritus, etc.), size, surface morphology, mineralogy, surface chemical characteristics, etc.
2. Investigation of the importance of the interaction of marine particles with major species in the marine biosphere.

Sediment/Sea Interface

1. Determination of the rates and primary mechanisms by which marine particles are removed to pelagic sediments at a number of specific sites throughout the world ocean.
2. Evaluation of the significance of biologically mediated reactions in sediments. Of particular importance is the possibility of biodegradation of petroleum hydrocarbons and synthetic organics.
3. Investigation of the rates of transfer of dissolved constituents out of pore waters by diffusion processes.

Appendix 1A: Pollutant Transfer Workshop Participants

Jack W. Anderson
Department of Biology
Texas A&M University
College Station, Texas 77843

Neil R. Andersen
Marine Chemistry Program
National Science Foundation
Room 317, 1800 G Street, N.W.
Washington, D.C. 20006

Richard Barber
Duke Marine Laboratory
Beaufort, North Carolina 28516

Michael L. Bender
Graduate School of Oceanography
University of Rhode Island
South Ferry Road
Kingston, Rhode Island 02881

Peter R. Betzer
Marine Science Department
University of South Florida
830 First Street South
St. Petersburg, Florida 33701

Terry F. Bidleman
Chemistry Department
University of South Carolina
Columbia, South Carolina 29208

James N. Butler
Engineering and Applied Physics
    Department
Harvard University
Pierce Hall 29 Oxford Street
Cambridge, Massachusetts 02138

D. K. Button
Marine Science Department
University of Alaska
Fairbanks, Alaska 99701

John A. Calder
Department of Oceanography
Florida State University
Tallahassee, Florida 32306

Edward J. Carpenter
Marine Sciences Research Center
State University of New York
Stony Brook, New York 11794

Roy Carpenter
Department of Oceanography
University of Washington
Seattle, Washington 98195

T. J. Chow
Ocean Research Division
Scripps Institution of Oceanography
La Jolla, California 92037

Thomas M. Church
Environmental Quality Program
    Manager
Officer for the International Decade of
    Ocean Exploration
National Science Foundation
1800 G Street, N.W.
Washington, D.C. 20006

Pedro J. Depetris
Limnologia Fisica
Instituto Nacional de Limnologio
J. Macia 1933
Santo Tome, Santa Fe
Republica Argentina

Robert A. Duce
Graduate School of Oceanography
University of Rhode Island
Kingston, Rhode Island 02881

William M. Dunstan
Skidaway Institute of Oceanography
P. O. Box 13687
Savannah, Georgia 31406

Manfred G. Ehrhardt
Marine Chemistry
Institut Für Meereskunde
22 Duesternbrooker Weg
Kiel, Germany 0431

Richard Eppley
Marine Resources Department
Scripps Institution of Oceanography
La Jolla, California 92037

William F. Fitzgerald
Marine Sciences Department
University of Connecticut
Groton, Connecticut 06340

William O. Forster
Ecological Sciences Branch
Division of Biomedical and Environ-
    mental Research
Energy Research and Development
    Administration
Washington, D.C. 20545

Wayne S. Gardner
Skidaway Institute of Oceanography
P. O. Box 13687
Savannah, Georgia 31406

William D. Garrett
Ocean Sciences Division
Naval Research Laboratory
Washington, D.C. 20375

C. S. Giam
Departments of Chemistry and
    Oceanography
Texas A&M University
College Station, Texas 77843

Marcia C. Glendening
Oceanography Section
National Science Foundation
1800 G Street, N.W.
Washington, D.C. 20550

Ronald Gibbs
College of Marine Studies
University of Delaware
Lewes, Delaware 19958

Edward D. Goldberg
Geological Research Division
Scripps Institution of Oceanography
La Jolla, California 92093

M. Grant Gross
Chesapeake Institute
Johns Hopkins University
Baltimore, Maryland 21218

Robert Harriss
Department of Oceanography
Florida State University
Tallahassee, Florida 32306

George R. Harvey
Chemistry Department
Woods Hole Oceanographic Institution
Woods Hole, Massachusetts 02543

Gerald L. Hoffman
Graduate School of Oceanography
University of Rhode Island
Kingston, Rhode Island 02881

Alan V. Holden
Department of Agriculture and
    Fisheries for Scotland
Freshwater Fisheries Laboratory
Faskally, Pitlochry, Scotland

Feenan D. Jennings
Office for the International Decade of
    Ocean Exploration
National Science Foundation
1800 G Street, N.W.
Washington, D.C. 20550

Dana Kester
Graduate School of Oceanography
University of Rhode Island
Kingston, Rhode Island 02881

George A. Knauer
Department of Oceanography
Florida State University
Tallahassee, Florida 32306

Richard Lee
Skidaway Institute of Oceanography
P. O. Box 13687
Savannah, Georgia 31406

Victor J. Linnenbom
Superintendent, Ocean Sciences
    Division
Naval Research Laboratory
4555 Overlook Avenue, S.W.
Washington, D.C. 20375

Vance McClure
Fisheries Bureau
Department of Interior
8604 La Jolla Shores Drive
La Jolla, California 92037

John H. Martin
California State University,
    San Francisco
Moss Landing Marine Laboratory
Moss Landing, California 95039

David Menzel
Skidaway Institute of Oceanography
P. O. Box 13687
Savannah, Georgia 31406

Joan Mitchell
Office for the International Decade of
    Ocean Exploration
National Science Foundation
1800 G Street, N.W.
Washington, D.C. 20550

Byron F. Morris
Bermuda Biological Station
St. George, Bermuda

Jerry M. Neff
Department of Biology
Texas A&M University
College Station, Texas 77843

J. A. Nicol
Institute of Marine Science
University of Texas
Port Aransas, Texas 78373

Charles E. Olney
Food and Resource Chemistry Depart-
    ment
University of Rhode Island
Kingston, Rhode Island 02881

Michael J. Orren
Chemistry Department
Woods Hole Oceanographic Institution
Woods Hole, Massachusetts 02543

Gustav-Adolf Paffenhofer
Skidaway Institute of Oceanography
P.O. Box 13687
Savannah, Georgia 31406

Partrick L. Parker
Institute of Marine Science
University of Texas
Port Aransas, Texas 78373

Norman J. Pattenden
Environment and Medical Sciences
    Division
Atomic Energy Research Establish-
    ment
Harwell, OXON, OX11, ORA, England

Clair C. Patterson
Division of Geological Sciences
California Institute of Technology
1201 East California Street
Pasadena, California 91125

Spyros Pavlou
Department of Oceanography
University of Washington
Seattle, Washington 98195

Bob Presley
Department of Oceanography
Texas A&M University
College Station, Texas 77843

Robert W. Risebrough
Bodega Marine Laboratory
University of California
P.O. Box 247
Bodega Bay, California 94923

William M. Sackett
Department of Oceanography
Texas A&M University
College Station, Texas 77843

William A. Sheppard
Central Research & Development Department
E. I. DuPont deNemours and
Company
Experimental Station Building 328
Wilmington, Delaware 19898

Lowell V. Sick
College of Marine Studies
University of Delaware
Lewes, Delaware 19958

David L. Stalling
Fish and Wildlife Service
United States Department of Interior
Fish Pesticide Research Laboratory
Route 1
Columbia, Missouri 65201

Graham Topping
Department of Agriculture and
Fisheries for Scotland
Marine Laboratory, Aberdeen
P.O. Box 101
Torry, Aberdeen, Scotland

John Trefrey
Department of Oceanography
Texas A&M University
College Station, Texas 77843

Shizuo Tsunogai
Chemistry Department
Woods Hole Oceanographic Institution
Woods Hole, Massachusetts 02543

Karl K. Turekian
Geology and Geophysics Department
Yale University
New Haven, Connecticut 06520

Herbert L. Windom
Skidaway Institute of Oceanography
P.O. Box 13687
Savannah, Georgia 31406

2

Transport of Pollutant Lead to the Oceans and Within Ocean Ecosystems

C. Patterson, D. Settle, B. Schaule, and M. Burnett

Extent of Industrial Lead Pollution

Concentrations of the toxic metal lead are about 12 μg/g within the earth's crust (Chow and Patterson, 1962), and were $\sim$.004 μg/g wet weight or less within the earth's biomass during neolithic times (Appendix 2A, Note 1). The mining of lead from the earth and dispersal of it as aerosols in smelter fume and auto exhausts has been carried out on such a massive scale during the past century ($\sim$130,000,000 tons lead smelted, $\sim$10,000,000 tons lead alkyls burned) that neolithic natural reservoirs and fluxes of lead ($\sim$10,000 tons lead in biosphere, $\sim$10,000 tons/yr soluble lead added to oceans, 1000 tons/yr aerosol Pb transferred to oceans) in the biosphere have been completely overwhelmed (Appendix 2A, Note 2). Today the amount of lead mined each year in the world and dispersed by industrial activity (3,000,000 tons/yr) is about 500 times larger than the total amount of lead now circulating annually within the growth cycle of the earth's biosphere ($\sim$5000 tons lead/yr). This has elevated by contamination the average concentration of lead in the biosphere by an estimated factor of $>$20 to $\sim$0.1 μg/g wet weight (Appendix 2A, Note 1). The lead pollution factor is greater for humans ($>$10^2) than for most other organisms in the biosphere because their food chains are contaminated by a large number of additional inputs from industrial sources other than smelter fume and gasoline exhausts (pesticides, coatings, bearings and alloys in food machinery, paints, glazes, food can solder, paper and cloth pigments, pipe cutings, etc.) (Appendix 2A, Note 3). The health hazard to virtually all humans by environmental lead pollution in the form of central nervous system disfunctions, although not yet widely appreciated, is a serious matter (Elias, Hirao and Patterson, 1975), and merits considerable effort in the evaluation of the problem.

Knowledge of the extent and effects of industrial lead pollution of the oceans should have a high priority, even though humans receive much of their lead pollution from interactions with their environment on land. The lead pollution factor for seafood probably ranges from an extimated minimum of ten to an observed ten thousand (Appendix 2A, Note 4). It is important to know both the extent of lead pollution in marine organisms that provide food for humans and the manner in which the pollution was transferred to the organisms. It is perhaps even more important to know what the effects of this massive injection of industrial lead into the oceans at rates far above natural levels has upon the existence of marine life. The oceans are a fundamental unit of the world ecosystem in which humans live, and we had better find out the ramifications of dumping prodigious quantities of toxic metals such as lead into them. Deleterious effects may be prolonged by release of lead from polluted nearshore sediments and terrestrial soils.

Atmospheric Transport of Industrial Lead to the Seas

The atmosphere is a major route for the transport of heavy metals to the open oceans. Aerosols account for about one third of the industrial lead added to the oceans, as shown by the estimates listed in Table 2-1. About 40,000 tons of these aerosols are added annually by dry deposition and washout from the atmosphere, while about 60,000 tons are added by rivers and sewers as storm runoff from paved surfaces which collected the aerosols on land.

The amount of lead aerosols introduced to the oceans by dry deposition can be estimated from two types of data. The observed atmospheric concentration of lead over the open ocean can be combined with an estimated transfer velocity to yield a deposition rate. A 1 ng Pb/m^3 concentration in the mid-Pacific atmosphere has been observed by Chow, Earl and Bennett (1969). A 3.4 ng Pb/m^3 concentration in the North Atlantic has been observed by Duce, Hoffman, Ray, Fletcher, Wallace, Fasching, Piotrowicz, Walsh and Hoffman (1976). The North Atlantic is much more polluted than oceans in other regions of the earth and an average value over all the oceans of 1.8 ng Pb/m^3 can be found by giving the mid-Pacific value of Chow, Earl and Bennett (1969) a weight of 2 and the North Atlantic value of Duce, Hoffman, Ray, Fletcher, Wallace, Fasching, Piotrowicz, Walsh and Hoffman (1976) a weight of 1. Elias, Hirao, Hinkley, and Patterson (1975) measured a transfer velocity of 0.10 cm/sec for lead aerosols on horizontally oriented plastic surfaces in the Yosemite subalpine region. They also

Table 2-1
Approximate Lead Input for Total Oceans
(In Tons/Year)

Industrial Inputs	
Aerosols (gasoline)	37,000
Aerosols (smelters and forest fires)	3,000
Rivers and Sewers (soluble, mainly from aerosols)	60,000
Rivers and Sewers (solids)	200,000
Neolithic Inputs	
Aerosols	1,000
Rivers (soluble)	13,000
Rivers (solids)	100,000

measured transfer velocities of 0.023 cm/sec for lead aerosols on vertically oriented surfaces in this same region. These investigators believe that the observed large deposition velocities of lead-containing aerosols on horizontal surfaces at this remote site result from on-site agglomeration of Aitkin sized lead-rich particles with much larger non-lead particles. These investigators as well as other (Duce, Turekian, private communication) believe that seaspray-generated micron-sized salt particles may serve to scavenge Aitkin-sized lead-rich particles from the marine air to the sea surface by gravitational settling. It is therefore likely that the measured transfer velocity for lead aerosols on horizontal surfaces in the remote alpine region can be related to mid-ocean regions because the lead particles had traveled hundreds of kilometers through the atmosphere before deposition. However, one is forced to assume, according to experiments of Vittori (1975), that the particle deposition flux on water is about one fourth that on dry surfaces. Therefore, the resulting transfer velocity for lead aerosols of 0.025 cm/sec can be combined with an atmospheric lead concentration of 1.8 ng Pb/m^3 to give a mid-ocean lead aerosol deposition rate of 4.5×10^{-17} g/cm^2 sec. This calculates to 5000 tons Pb/yr added to the world's oceans by dry deposition.

Duce, Hoffman, Ray, Fletcher, Wallace, Fasching, Piotrowicz, Walsh and Hoffman (1976) used a lead dry deposition velocity of 0.25 cm/sec calculated from the data of Cambray, Jeffries and Topping (1975) and multiplied this by their observed 3.4 ng Pb/m^3 concentration at Bermuda and obtained a deposition rate of 8.5×10^{-16} ng Pb/cm^2 sec (part of their model 3). If this deposition is extended to the entire oceans, a value of 100,000 tons Pb/yr results, which is considerably higher than the first calculated value mainly because a larger deposition velocity was used.

Additional inputs of Pb aerosols by dry deposition to the oceans can be estimated by considering the concentrated inputs from near-urban coastal waters as a special case. Studies of atmospheric input to the Souther California Bight by Huntzicker, Friedlander, and Davidson (1975), and by Patterson and Settle (1973), may be summarized as follows. Depositions on horizontal teflon plates were about 45 ng Pb/cm^2/day at urban locations, and about 1.4 ng Pb/cm^2/day at an island location at the outer periphery of the Bight. These and additional data can be combined to yield an average deposition rate of 7 ng Pb/cm^2/day by dry deposition on areas of the Bight. If this rate on plastic surfaces is reduced by a factor of one fourth to account for deposition on water, the rate becomes 1.7 ng Pb/cm^2/day. If we enlarge this strip of urban coastal water to include all such waters 40 km wide surrounding North America, Europe and within the Mediterranean Sea (about 1×10^6 km^2), a deposition of 6000 tons Pb/yr is estimated.

These two estimates of the input of lead aerosols by dry deposition to the oceans sum to a total magnitude of about 10,000 tons Pb/yr. To this input, contributions by precipitation must be added. There are no reliable published measurements of lead in marine rain at present, and this includes the data of Cambray, Jeffries, and Topping (1975). We can estimate a value from the observed reliable concentration measurements made in recent firn in northern Greenland (Murozumi, Chow, and Patterson, 1969). Snow is at least as efficient a scavenger of lead from the atmosphere as rain in remote regions because lead is contained in Aitkin-sized particles and the outer portions of the snowflake contain large numbers of such particles, while raindrops, because of smaller surface-to-mass ratios, may contain smaller amounts of these particles. Recent preliminary observations show that reliable lead concentrations are approximately equal in snow and rain precipitated at the same time of the year in high mountains (Elias, Hirao, Hinkley and Patterson, 1975). Northern Greenland firns today average about 0.2 μg Pb/kg by weight and if this concentration is assigned to ocean rain, an input of about 60,000 tons Pb/yr is obtained. The input by rain is probably considerably smaller, say half, or 30,000 tons Pb/yr because the observed lead in Greenland firn is the sum total of equal contributions by precipitation and by dry deposition on fallen snow. The combined input of lead to the oceans by dry deposition and precipitation is therefore probably about 40,000 tons Pb/yr.

Another method of estimating total lead aerosol inputs to the oceans is to consider the mass transport of particles from land to sea estimated by Hidy and Brock (1971), Robinson (1976), Prospero (1976) and Goldberg (1971). These estimates vary from 500×10^6 tons of mostly silicate particles (Goldberg, 1971) for deposition in the Pleistocene to 100×10^6 tons of mostly nonsilicate anthropogenic particles (Robinson, 1976) at present times. The Pb/silicate ratio is observed to be about 100 times above natural levels in air particles at remote locations in North America (Elias, Hirao, Hinkley, and Patterson, 1975), in northern Greenland (Murozumi, Chow, and Patterson, 1969), and over the Atlantic ocean (Duce, Hoffman, Ray, Fletcher, Wallace, Fasching, Piotrowicz, Walsh and Hoffman, 1976). This lead/silicate enrichment can be applied to about 100×10^6 tons silicate deposition to the oceans per year. From this flux one can estimate an input of 120,000 tons of lead to the oceans each year by both dry deposition and washout. Of the two estimates for total lead input to the oceans, the 120,000 tons calculated from silicate dust is the least accurate and is probably high according to the more accurate 40,000 tons figure.

Duce, Hoffman, Ray, Fletcher, Wallace, Fasching, Piotrowicz, Walsh and Hoffman (1976) have used the data of Cambray, Jeffries and Topping (1975) in their model 2 to estimate the total input of lead by rain plus dry fallout at their Bermuda site. If we use their Bermuda calculations to apply to all the oceans (model 2 type) an input of 260,000 tons Pb/yr is obtained. This estimate is probably much too high because it requires a high mid ocean concentration of lead in rain of 0.9 μgPb/kg. Even if the model 1 Bermuda calculations of Duce,

Hoffman, Ray, Fletcher, Wallace, Fasching, Piotrowicz, Walsh and Hoffman (1976) are used to apply to all the oceans, an input of 130,000 tons Pb/yr is obtained, which still seems high since it requires a high mid-ocean concentration of 0.4 μg Pb/kg rain. Patterson and Settle (1973) observed an average reliable concentration of 5 μg/kg in four rains falling in the Los Angeles, California, basin where the yearly atmospheric concentration of lead averaged about 3000 ng Pb/m^3 throughout the 5000 km^2 area of the basin. However, it was observed (Elias, Hirao, Hinkley and Patterson, 1975) that rain falling in the Yosemite subalpine study area contained a reliable concentration of 10 μg Pb/kg where atmospheric lead concentrations average 25 ng Pb/m^3. It is believed that high lead concentrations in this type of remote region rain result from downstream precipitation of urban lead generated upstream. Los Angeles basin rain may contain less than average urban amounts of lead because the rain is precipitated from low-lead Pacific air masses before the air becomes heavily polluted. It seems unlikely that the concentration of lead in rain over the oceans would be as high as 1/10th of that in urban rain when atmospheric concentrations of lead over the oceans are 1/1000th of those in urban areas. It therefore seems rather unlikely that atmospheric inputs of lead to the oceans of 130,000 tons/yr or 260,000 tons/yr are realistic because the inputs are calculated by models that require high concentrations of lead in rain.

From the standpoint of mass balance, industrial sources can probably provide an aerosol input of 40,000 tons Pb/yr. About 400,000 tons of lead are burned as alkyls in automotive fuels each year (*Minerals Yearbook,* 1973, plus estimated non-U.S. production) of which approximately one third is widely dispersed in the atmosphere as submicron-sized aerosols. To this must be added contributions of lead aerosols from base metal smelting which amount to about 10,000 tons Pb in aerosols each year. The reentrainment of industrial lead in the form of aerosols by forest fires is also significant. Robinson (1976) estimates that the world mass of forest fire aerosols is about half the world mass of anthropogenic aerosols generated yearly. Elias, Hirao, HInkley and Patterson (1975) estimate that about one kg/km^2/yr industrial lead aerosol input to their study area is deposited on conifer needles and bark. About 75% of the lead on needles is washed off by rain, but most of the lead on bark and surface litter is retained. Their study area was only 15% forested, so it was not anomalous in this regard to rural North America. One can estimate that between 1000 and 10,000 tons Pb/yr are introduced into the atmosphere by forest fires and all of this is reentrained industrial lead originally deposited on foliage by dry deposition.

The potential source of industrial lead aerosols for ocean transport amounts to about 150,000 tons Pb/yr. At the present time reliable mass balance inputs of such industrial aerosols by dry deposition and by precipitation to a defined basin have been made only for one remote terrestrial site (Elias, Hirao, Hinkley and Patterson, 1975) and those studies indicate the following lead inputs

$(kg/km^2/yr)$: 1.3 by dry deposition; 1.5 by snow; and 1.2 by rain. The total industrial origin of these aerosols is indicated by isotopic relationships, chemical composition, and mass balance considerations. This rate of deposition, extended to the land surface of North America, Europe and Asia, would require 400,000 tons of industrial aerosol lead per year, a figure 2.7 times larger than the potential source. This suggests that the deposition rate at the Yosemite subalpine study area is about 3.6 times higher than the average for the total land mass in the northern hemisphere.

It is highly probable that industrial lead emissions have brought about at least a tenfold enrichment of lead in tropospheric aerosols in remote land, polar, and mid-ocean regions. A 2500-fold enrichment of lead above crustal silicate values observed at the South Pole in tropospheric aerosols is believed to be partially anthropogenic (Zoller, Gladney and Duce, 1974). Volcanic emanations are also believed responsible for part of the enrichment (Zoller, Gladney and Duce, 1974). Hundredfold enrichments of Pb above crustal silicate values have been observed in mid-Atlantic marine aerosols, and variations in these enrichments correlate with downstream air trajectories from urban regions generating large amounts of industrial lead aerosols, indicating anthropogenic causes of part of the enrichments (Zoller, Gladney and Duce, 1974). Wave-generated aerosols enriched in microlayer heavy metal constituents, are held responsible for some of the observed lead enrichment (Zoller, Gladney and Duce, 1974). Hirao and Patterson (1974) observed 100-fold enrichments of Pb above crustal silicate values in aerosols and snow at the Yosemite subalpine study area in 1973. 90% of this enrichment was conclusively shown to be anthropogenic on the basis of isotopic tracers, because, fortunately, the isotopic composition of lead in California mountain snow collected in 1962 had been measured by Patterson and colleagues (Chow and Johnstone, 1965). The Pb^{206}/Pb^{207} ratio in California mountain snow changed from 1.144 in 1962 to 1.183 in 1973. This change coincides with the change observed by Chow, Snyder and Earl (1975) in both gasolines and lead-rich urban aerosols derived from gasoline in San Diego during this same time period. The isotopic change was brought about by the introduction of increased amounts of Missouri lead into the U.S. pool of industrial leads. These isotopic tracers are insensitive to more than an order-of-magnitude effect, so that only part of the observed 100-fold enrichment can be ascribed by the isotopic method to anthropogenic causes.

A greater than 100-fold enrichment of lead above crustal silicate values in present-day firn in Greenland near 80° north latitude has been demonstrated by Murozumi, Chow and Patterson (1969). Here the entire two-order-of-magnitude change has been definitely ascribed to antropogenic causes, since an increase of this size with time was observed in dated layers of ice. This interpretation rests on the significance of an extremely low lead concentration ($< 1 \times 10^{-12}$ g Pb/g ice) observed by Murozumi, Chow and Patterson (1969) in a block of 3000-year-old ice cut out under ultra-clean conditions from the edge of the ice sheet and dated by C^{14}. This snow was deposited in the interior, beyond the ablation zone, 3000 yrs ago. The concentration of silicate dust in this ice was about the same as that in the interior firn sampled for more recent lead concentrations, where the 1965 layer contained 210×10^{-12} g Pb/g ice. This can be regarded as conclusive evidence that the entire ~100-fold lead enrichment over crustal silicate values in tropospheric aerosols is anthropogenic.

It is improbable that volcanic emanations can account for the ten-fold enrichment of lead above crustal silicate values in tropospheric aerosols, which some investigators suppose is a natural enrichment underlying a further anthropogenic ten-fold enrichment. The amount of lead contributed to the atmosphere by volcanic gases seems to be less than a 100-ton standing crop out of a 5000-ton total troposphere standing crop of lead. The latter figure is obtained from the 2 ng Pb/m^3 open ocean atmosphere concentration. The 100-ton upper limit number can be estimated in two ways.

One is to consider the total volume of lava emitted by volcanoes above sea level each year and to consider that the fraction of the lead in the lava that is volatilized may be about 10%. Figures which can be used are 0.5 km^3 lava/yr (Verhoogan, 1946; Sapper, 1927), 12 μg Pb/g concentration, 10% volatilization of lead, and an atmospheric residence time of two weeks. The amount of lead which can be introduced into the atmosphere by this process is limited by mass imbalances that would be created between the sedimentary cycle and the igneous cycle after a prolonged time of operation if the fraction of lead volatilized is too great.

The second method is based upon recent determinations of the concentrations of lead in two samples of volcanic fume from Hawaii (Cadle, Wartburg, Pollock, Gandrud, and Shedlovsky, 1973). These also yield an average crop of about 100 tons. In one instance the lead/sulphate ratio was 3.2×10^{-5} and in the other instance it was 7.8×10^{-4}. The following figures were used: a half km^3 of lava issued above sea level world-wide per year, the fraction of lava which is gaseous fume was set equal to half percent (G. Macdonald, private communication), 80% of the fume was assigned to sulphate (Anderson, 1975), and a two week atmospheric residence time was assumed. Cadle's lead values are upper limits because of possible high errors due to lead contamination effects.

At the present time we cannot estimate the mass balance input of anthropogenic and natural lead aerosols to the ocean from the difference between the total generated production of lead aerosols and the total deposited on land because the rates of generation of natural lead aerosols from volcanic and soil sources, and rates of land deposition are not known with sufficient accuracy. Considering the greater plane projected surface areas of the oceans compared to land, the apparently great significance of precipitation removal mechanisms, and the large dispersal range of industrial lead aerosols, it seems highly probable that an appreciable fraction of the annual 150,000 tons of industrial lead aerosol production is added to the seas. That is, a yearly input to the seas of

40,000 tons of industrial lead associated with aerosols seems possible on a mass balance basis.

As shown in Table 2-1, it is estimated that the present annual rate of atmospheric lead addition to the oceans exceeds the former neolithic rate by a factor of about 40, that the increase is probably due to industrial additions, and that the major impact is on the concentration of freely available lead in the mixed zone.

Dissolution of Anthropogenic Particle Lead in Seawater

In order to measure the dissolution fluxes of particle lead, knowledge of the speciation, distributions, and rates of dissolution must be obtained. Until 1975 knowledge of the speciation and concentration of lead in seawater was not reliable despite decades of measurement (Participants of the Lead in Seawater Workshop, 1974). Contamination of seawater with industrial artifact lead has been widespread in the past during collection, handling, and analysis, yielding data that were wrong on the high side. The work by Tatsumoto and Patterson (1963a, 1963b) and Chow and Patterson (1966) which gave the proper perspective of the occurrence of lead in seawater may have been adversely affected by artifact contamination during collection. True concentrations of lead in mid-ocean waters are probably less than those present in laboratory distilled waters (the latter range from 5 to 50 ng Pb/kg.

Investigators have published theoretical estimates of the speciation of lead in seawater based on stabilities of common possible soluble complexes; however, the interaction of dissolved lead with both living and inanimate particles may be the most important factor determining the practical aspects of lead speciation in seawater.

Lead concentrations in coastal waters near urban regions range from 25 ng Pb/kg in surface samples of ordinary coastal water to 150 ng Pb/kg in waters highly polluted with sewage (Patterson, Settle and Glover, 1976). Studies of lead in sewage effluent from Los Angeles show that virtually all the lead in sewage is contained in the particle phase before it enters the ocean but that about 11% is made freely available within a day by cation exchange when the sewage is mixed with seawater. Essentially no more dissolved lead is released from the sewage particles even after weeks of exposure (Patterson, Settle and Glover, 1976). Observers have studied the displacement of other heavy metals in particles by exchange with cations in seawater (Johnson, Cutshall and Osterberb, 1967; Kharkar, Turekian and Bertine, 1968; Evans and Cutshall, 1973). Investigations carried out by European workers on the dissolution of lead from particles into seawater along the shores of the North Sea are unreliable because of analytical error. Lead in the filtrate of seawater passed through a 0.4μ cellulose acetate filter or extracted by dithizone in chloroform from untreated seawater has been designated freely available lead. The amounts of lead measured by these two methods are nearly identical for parallel aliquots of seawater (Patterson, Settle and Glover, 1976). As the proportion of sewage lead in seawater increases, the fraction of freely available lead decreases. For example, surface seawater containing ~200 ng Pb/kg total was found to contain 29% freely available lead, while surface seawater containing 110 ng Pb/kg was found to contain 42% freely available lead (Patterson, Settle and Glover, 1976). It is believed that in areas of high sewage pollution (>50 ng Pb/kg) most of the particle lead in the waters is associated with sewage, not plankton.

Total lead concentrations appear to decline to 30 ng Pb/kg levels before contributions of plankton particle lead become significant. Single measurements of total lead (determined after evaporation and dissolution with aqua regia) in surface waters from the Straits of San Juan de Fuca and from outside the Southern California Bight showed 24 ng total Pb/kg at the first location and 25 ng total Pb/kg at the second. However, the proportion of particle lead ranged from 90% at the first location to less than 3% at the second (Patterson, Settle and Glover, 1976; Schaule, unpublished). Surface waters off La Jolla showed 36 ng total Pb/kg during a dinoflagellate phytoplankton bloom and 16 ng total Pb/kg when the water was unusually clear. About one third of the lead associated with the phytoplankton was contained in chiton (Patterson, Settle and Glover, 1976). Some total reported lead concentrations in seawaters are listed in Table 2-2.

The isotopic compositions of total leads in coastal waters indicate in some instances that more than one kind of industrial lead was present in the waters and that they were not well mixed (Patterson, Settle and Glover, 1976). The

Table 2-2
Measured and Estimated Total Leads in Seawater,
(Aqua Regia Dissolution)

Location	Date	Depth (m)	ng Pb/kg
Los Angeles (above JWPCP outfall)	April 28, 1975	0.2	230
Los Angeles (above JWPCP outfall)	March 25, 1974	0.2	(200)[a]
Los Angeles (above JWPCP outfall)	March 25, 1974	7	(330)
Los Angeles (above JWPCP outfall)	March 25, 1974	30	(1300)
Los Angeles (above JWPCP outfall)	April 9, 1973	0.2	(110)
La Jolla (5 km west of Scripps Pier)	Jan. 13, 1976	0.2	36
La Jolla (5 km west of Scripps Pier)	Nov. 1, 1972	0.2	(16)
San Juan de Fuca Straits	July 29, 1975	0.2	24
50 miles southwest of Los Angeles[b]	Feb. 29, 1976	0.2	25

Source: Patterson, Settle and Glover (1976).

[a]Concentrations in parentheses are not observed, but are calculated by multiplying measured concentrations of aliquots poured from carboys by 1.5 to correct for wall adsorption.

[b]Schaule, unpublished.

Pb^{206}/Pb^{207} ratio of total lead in coastal surface seawater collected near Los Angeles was 1.194 (concentration $\sim$100 ng Pb/kg) and a day later was 1.188 (concentration $\sim$70 ng Pb/kg) near La Jolla. A difference of 0.2% between values of this ratio is significant. It is believed that the above difference shows a lack of mixing of contributions from two different sources of lead pollution: one was a pulse of rain storm runoff of gasoline lead (Pb^{206}/Pb^{207} observed to be $\sim$1.197) from paved surfaces added to Los Angeles waters and the other component was sewage lead (Pb^{206}/Pb^{207} estimated to be $\sim$1.188) from San Diego added to La Jolla waters.

Interaction of Seawater Lead with Marine Organisms

Contrary to common opinion, lead is not necessarily concentrated in higher organisms as it is transported to the higher ends of food chains. In marine ecosystems there are two different processes competing with each other in the transport of lead along food chains: passive absorption of lead from seawater by chelating agents on organism surfaces to form stable lead complexes (enrichment process); and inefficient active transport of lead across cell membranes (depletion process). To distinguish between these opposing effects in marine organisms, it is important for mass-balance, morphologic distributions of lead among different tissues of the various organisms in food chains to be worked out. The enrichments or depletions of lead should not be expressed in such simple terms as concentrations of lead in water or in wet, dry, or ashed tissues. The actual enrichment or depletion of lead is better expressed as a change relative to the more or less fixed bulk of calcium, which is an abundant nutritious metal that flows easily along food chains, and is biochemically similar to lead in many respects. It is revealing to study the biodiminutions and bioamplifications of both barium and lead, since barium is not yet a serious pollutant in the marine environment, and observed differences between the distributions of barium and lead in food chains give clues regarding lead pollution effects (Hirao and Patterson, 1974).

Lead, barium, strontium and calcium have been studied by isotope dilution clean-laboratory techniques in a seawater-kelp-gastropod food chain (*Macrocystis pyrifera,* and *Norrisia norrisii*) (Burnett, unpublished). The data indicate that Sr, Ba, and Pb are enriched relative to Ca in going from seawater to total kelp blade, Sr/Ca increasing by 8, Ba/Ca increasing by 20, and Pb/Ca increasing by 2000. A major fraction of kelp consists of an alginic acid-rich binding matrix, holding the kelp-blade cells together. The strengths of alginic acid-metal complexes have been determined (Haug, 1961) and the relative stabilities of those complexes are Pb$\gg$Ba>Sr>Ca. The binding matrix is in contact with seawater during and after growth, and it is believed that the relative enrichments cited above, which parallel the relative strengths of the metal complexes, show passive adsorption of Pb, Ba, Sr, and Ca from seawater by the alginic acid-rich matrix.

The gastropod that was studied feeds on kelp, cutting down into and consuming the blade material (Leighton, 1971). Using the composition of total kelp blade as gastropod food, Sr, Ba and Pb are depleted relative to Ca in going from kelp to total gastropod, Sr/Ca decreasing by 70, Ba/Ca decreasing by 600, and Pb/Ca decreasing by 150. Analyses of separate organs of the gastropod indicate differences in metal depletions. The biodiminutions cited above refer to shell material, but they also apply to total gastropod because the bulk of metals in the gastropod are in its shell. The situation is different for muscle tissue, Sr/Ca decreasing by 10, Ba/Ca decreasing by 10 and Pb/Ca decreasing by 2, in going from kelp to gastropod muscle.

Comparison of the distributions of alkaline earths and lead in tuna and terrestrial animals provide insight into lead pollution in marine animals. Ninety-five percent of the Ca and Sr, and 75% of the Ba and Pb in tuna are contained in the skeleton (Patterson and Settle, 1976). These distributions, when compared to those in a terrestrial carnivore (*Martes americana*) (Elias, Hirao, Hinkley and Patterson, 1975), are nearly the same. The surface of the fur of the terrestrial carnivore contains large amounts of dry deposition aerosol lead. A tuna was observed to contain unusually high concentrations of lead in epidermal mucus which is believed to originate from artifact contamination by fishermen (Patterson and Settle, 1976).

Although the morphological distributions of the alkaline earths and lead are nearly identical in these terrestrial and marine carnivores, the concentrations are quite different. The Ba/Ca and Pb/Ca ratios in tuna are about one tenth of those in marten because much smaller amounts of Ba and Pb are associated with the Ca in seawater than with the Ca in terrestrial soils, which is shown in Table 2-3. This does not mean that there is a greater biodiminution of Ba and Pb in albacore than in marten. The total biodiminution is a factor of 1000 for both Ba and Pb relative to Ca in going from rock to carnivore in the marten food chain (Elias, Hirao, Hinkley and Patterson, 1975). In albacore, Ba/Ca is

Table 2-3
Concentrations (μg/g) of Elements at Lowest and Highest Trophic Levels of a Marine and a Terrestrial Food Chain

	Ca	Sr	Ba	Pb[a]
Seawater[b]	400	8.1	0.03	0.00002
Albacore (body burden-wet)	8800	36.	0.1	0.008
Wall Rock[c]	12000	510	1100	22.
Marten (body burden-wet)[c]	15000	9	1	0.15

[a]Surface lead deposited on the organism has been excluded from these values.

[b]Goldberg (1963).

[c]Elias, Hirao, Hinkley and Patterson (1975).

biodiminished only by a factor of 10 in going from seawater to fish, while Pb/Ca is bioamplified by a factor of 10. This marked contrast in the food chain transport of lead between albacore and marten may be due to passive adsorption effects in the marine ecosystem.

Acknowledgments

This paper originates from work performed under NSF contract No. ID074-24362.

References

Anderson, A. T. (1975). Some basaltic and andesitic gases. *Reviews of Geophysics and Space Physics, 13*, 37-55.

Bowen, H. J. M. (1966). *Trace Elements in Biochemistry*, Academic Press, New York.

Burnett, M., Preliminary results, unpublished.

Cadle, R. D., A. F. Wartburg, W. H. Pollock, B. W. Gandrud and J. P. Shedlovsky (1973). Trace constituents emitted to the atmosphere by Hawaiian volcanoes. *Chemosphere, 2*, 231-4.

Cambray, R. S., D. F. Jeffries and G. Topping (1975). An estimate of the input of atmospheric trace elements into the North Sea and the Clyde Sea (1972-73). United Kingdom Atomic Energy Authority Harwell Report AERE-R7733, 30 pp.

Chow, T. J., J. L. Earl and C. F. Bennett (1969). Lead aerosols in marine atmosphere. *Environmental Science and Technology, 3*, 737.

Chow, T. J. and M. S. Johnstone (1965). Lead isotopes in gasoline and aerosols of Los Angeles Basin, California. *Science, 147*, 502-503.

Chow, T. J. and C. Patterson (1962). The occurrence and significance of lead isotopes in pelagic sediments. *Geochimica et Cosmochimica Acta, 26*, 263-308.

Chow, T. J. and C. C. Patterson (1966). Concentration of barium and lead in Atlantic waters off Bermuda. *Earth and Planetary Science Letters, 1*, 397-400.

Chow, T. J., C. C. Patterson and D. Settle (1974). Occurrence of lead in tuna. *Nature, 251*, 159-161.

Chow, T. J., C. B. Snyder and J. L. Earl (1975). Isotope ratios of lead as a pollutant source indicator. In *Proceedings of FAO/IAEA Symposium*, Vienna, Austria.

Duce, R. A., G. L. Hoffman, B. J. Ray, I. S. Fletcher, G. T. Wallace, J. L. Fasching, S. R. Piotrowicz, P. R. Walsh and E. J. Hoffman (1976). Trace metals
in the marine atmosphere: Sources and fluxes. This volume, chapter 4.

Elias, R. W., Y. Hirao, T. Hinkley and C. Patterson (1975). Summary report to
NSF concerning Grant GB-441132 including revisions and additional preliminary data.

Elias, R., Y. Hirao and C. Patterson (1975). Impact of present levels of aerosol
lead concentrations on both natural ecosystems and humans. *Proceedings
of International Conference on Heavy Metals in the Environment*, Toronto
(to be published).

Evans, D. and N. Cutshall (1973). Effects of ocean water on the soluble-suspended distribution of Columbia River radionuclides. In *Radioactive Contamination of the Marine Environment* (STI/PUB/313), International
Atomic Energy Agency, Vienna, Austria.

Goldberg, E. D. (1971). Atmospheric dust, the sedimentary cycle and man.
Comments on Earth Sciences: Geophysics, 1, 117-132.

Goldberg, E. (1963). In *The Sea,* 3, p. 3 (M. N. Hill, ed.), Wiley and Sons, New
York.

Haug, A. (1961). The affinity of some divalent metals to different types of al-
ginates. *Acta Chem. Scand., 15,* #8, 1794-1795

Hidy, G. M. and J. R. Brock (1971). An assessment of the global sources of
tropospheric aerosols. In *Proceedings Second International Clean Air Congress* (H. Englund, ed.), Academic Press, New York.

Hirao, Y. and C. Patterson, (1974). Lead aerosol in the High Sierra overrides
natural mechanisms which exclude lead from a food chain. *Science, 184,*
989-992.

Huntzicker, J. J., S. K. Friedlander and C. I. Davidson (1975). Material balance
for automobile emitted lead in the Los Angeles basin. *Environmental
Science and Technology, 9,* 448.

Johnson, V., N. Cutshall and C. Osterberb (1967). Retention of Zn^{65} by Columbia River sediment. *Water Resources Research, 3,* 99-102.

Kharkar, D. P., K. K. Turekian and K. K. Bertine (1968). Stream supply of dissolved silver, molybdenum, antimony, selenium, chromium, cobalt, rubidium, and cesium to the oceans. *Geochimica et Cosmochimica Acta, 32,*
285-298.

Leighton, D. L. (1971). Grazing activities of benthic invertebrates in Southern
California kelp beds. In *The Biology of Giant Kelp Beds (Macrocystis) in
California* (Wheeler J. North, ed.), Verlag Von J. Cramer, Germany.

Murozumi, M., T. J. Chow and C. Patterson (1969). Chemical concentrations of
pollutant lead aerosols, terrestrial dusts and sea salts in Greenland and Antarctic snow strata. *Geochimica et Cosmochimica Acta, 33,* 1247-1294.

Participants of the Lead in Seawater Workshop (1974). Interlaboratory lead analyses of standardized samples of seawater. *Marine Chemistry, 2,* 69–84.

Patterson, C. and D. Settle, (1973). Contribution of lead via aerosol impact to the Southern California Bight. *Journal de Recherches Atmosphériques,* juillet-decembre 1974. Numéro Spécial, "Symp. Int. sur les Echanges Ocean/Atmosphère de Matière a l'Etat Particulaire" (Int. Symp. on the Chemistry of Sea/Air Particulate Exchange Processes), Nice, France.

Patterson, C., D. Settle and B. Glover (1976). Analysis of lead in polluted coastal waters. *Marine Chemistry* (in press).

Patterson, C. and D. Settle (1976). Comparative distributions of lead in tuna. *Marine Biology* (accepted for publication).

Prospero, J. M. (1976). *Diagnostics of particular transport.* SCOR-NAS Workshop on Tropospheric Transport of Pollutants to the Ocean, Background Papers.

Robinson, E. (1976). Pollutant sources of particulate materials in marine atmospheres. SCOR-NAS Workshop on Tropospheric Transport of Pollutants to the Ocean, Background Papers.

Sapper, K. (1927). *Vulkankunde,* Engelhorns.

Schaule, B., preliminary results, unpublished.

Tatsumoto, M. and C. C. Patterson (1963a). The concentration of common lead in seawater. In *Earth Science and Meteoritics,* (J. Geiss and E. D. Goldberg, eds.), North Holland, Amsterdam, 74–89.

Tatsumoto, M. and C. C. Patterson (1963b). Concentrations of common lead in some Atlantic and Mediterranean waters and in snow. *Nature, 199,* 350–352.

Verhoogan, J. (1946). Volcanic heat. *Amer. J. Sci., 244,* 745.

Vittori, O. (1975). *Approach to the study of airborne particle deposition on sea surface.* SCOR-NAS Workshop on Tropospheric Transport of Pollutants to the Ocean, Background Papers.

Zoller, W., E. Gladney and R. Duce (1974). Atmospheric concentrations and sources of trace metals at the South Pole. *Science, 183,* 198–200.

Appendix 2A

1. Ninety-five percent of the earth's biomass resides in forests (Bowen, 1966). Most published data regarding lead in trees are either erroneous because of laboratory contamination or apply only to excessively polluted urban foliage. A careful mass distribution study of lead in a forested ecosystem has been carried out at a remote subalpine location within Yosemite National Park (Elias, Hirao, Hinkley and Patterson, 1975). Eighty-five percent of the lead in that ecosystem was contained in igneous minerals in the thin soil lying on bedrock, 15% of the lead was in soil humus, and 0.3% of the ecosystem lead was in the total biomass. Ninety-eight percent of the biomass consisted of coniferous trees which contained an average of 0.09 μg Pb/g. The average lead concentration of the remaining 2% of the biomass was about 0.3 μg/g. Ninety-five percent of the lead in the biomass of the ecosystem was therefore found to reside in trees. About 85% of the lead in the trees was located on a thin outer layer of bark while 10% of the tree lead was located in foliage (mainly on needle surfaces). The mass distribution of lead in this remote subalpine ecosystem indicates that most of the lead in the biosphere of the northern hemisphere is contained in the outer bark of trees. In the subalpine ecosystem that was studied more than 90% of the tree lead is surficial, and it originated from industrial aerosols, as was shown by aerosol deposition measurements and chemical washing experiments. A major fraction of the remaining 10% is also industrial because isotopic and mass balance data indicate that much of the lead in soil humus originated from industrial aerosols, and the deposition of industrial lead aerosols on foliage is so excessive that the magnitude of foliar uptake of industrial lead probably approaches root uptake. Therefore the neolithic concentration of lead in trees (earth's biosphere) was equal to or less than about 0.004 μg/g, or considerably less than one tenth the present concentration of 0.09 μg/g. Since half of the earth's biomass consists of tropical forests and the concentration of lead in the bark of these trees has not been measured, the concentration of lead in the earth's biomass is still somewhat uncertain.

2. The contamination of the earth's biosphere by industrial lead from smelter fumes is a process that extends about 4000 years back in time, involving the primary smelting of about 70,000,000 tons of lead before A.D. 1850 (Elias, Hirao and Patterson, 1975). The per capita production of lead within the Roman Empire was about one fifth of that within today's industrialized nations. People within ancient civilizations ingested industrial lead at rates comparable to those of today because of various common practices, such as putting lead-rich additives in green wine to inhibit souring, a process that began before 600 B.C. and continued until A.D. 1850.

3. Studies of metal distributions in food chains within the Yosemite subalpine ecosystem (Elias, Hirao, Hinkley and Patterson, 1975) indicate that lead is

naturally biodiminished relative to calcium by a factor of about 1000 in going from rock to herbivore and carnivore in the absence of lead pollution. Lead in the typical U.S. person is biodiminished relative to calcium only by a factor of about 2.5 from rocks that supply metals to human foods, indicating a typical lead contamination level about 400 times above neolithic natural levels (Elias, Hirao and Patterson, 1975).

4. The concentration of freely available lead in surface seawaters has been elevated by industrial lead pollution by about 10, according to estimates of the increased input of soluble-type lead to the oceans (10,000 tons/yr neolithic input from rivers, vs. 40,000 tons/yr atmospheric industrial input plus 60,000 tons/yr industrial lead from aerosols deposited on pavements and roofs and washed into the oceans by storm runoff and rivers). Aerosol lead from gasoline exhausts is much more reactive than lead in natural silicate clays, so that it contributes most of the freely available lead in seawater. The residence time of freely available lead in surface water is very short. According to inferences from lead isotope areal patterns in sediments (Chow and Patterson, 1962), the residence time should be less than a year. This means that freely available lead concentrations in surface waters responded quickly to increased industrial inputs.

The absolute concentration of lead in albacore muscle is exceedingly small (0.0003 μg Pb/g wet weight), and in the process of preparing this tissue for human consumption, excessive contamination factors are involved (grocery store tuna is 0.5 μg Pb/g wet weight) (Chow, Patterson and Settle, 1974). Since it is probable that lead concentrations in the upper waters of the oceans have been elevated by a factor of 10 by industrial lead pollution, the overall lead pollution factor for tuna muscle, a widely consumed human food, is about 10^4

3

Heavy Metal Transport from the Mississippi River to the Gulf of Mexico

J. H. Trefry and B. J. Presley

Introduction

Anthropogenic heavy metals, along with other materials, may be transported to the ocean from continents by atmospheric processes, direct dumping, sewage outfalls, storm runoff and river and ground-water discharge. The relative importance of these pathways for most substances is not well known; however, there is general agreement with the suggestion of Dyrssen *et al.* (1972) that atmospheric transport may be mainly responsible for addition of pollutants to the open sea, whereas the other routes are probably more important to coastal waters. River discharge is certainly dominant in total transport. Garrels and Mackenzie (1971) estimate that rivers carry 90% of the total dissolved and suspended solids added to the oceans. Rivers also carry domestic and industrial wastes away from most major communities, and thus must be responsible for much of the addition of man-derived substances to the ocean.

The major U.S. river is the Mississippi. It drains 41% of the conterminous United States, with a drainage basin stretching from New York to Montana and from Canada to the Gulf of Mexico (Figure 3-1). The Mississippi is estimated to carry about 60% of the total dissolved solids (Leifeste, 1974) and 66% of the total suspended solids (Curtis *et al.*, 1973) transported to the oceans from the conterminous U.S. This probably includes a significant percentage of the total pollutant load, considering the nature of the drainage basin.

Municipal and industrial usage of Mississippi River water and discharge of wastes into the river are known to be large. For example, Everett (1971) notes that industrial withdrawal of water from the river along the 306 kilometer stretch from Baton Rouge to Port Sulfur, Louisiana, increased from 7.6 billion liters/day in 1960 to 19 billion in 1969. This latter figure represents about 7% of the total river discharge during periods of low flow.

Nearly all the withdrawn water is, of course, returned to the river, but only after it has had heat and dissolved solids added to it. Industrial additions of dissolved solids for a 241 kilometer stretch of river above New Orleans have increased tenfold from 1958 to 1969 (Figure 3-2). The 1969 figure of 18 million kilograms added per day is about 7% of the total dissolved load at average river flow and 21% of the total at low flow. Presumably similar situations could be shown for the upper river and its tributaries where such cities as Pittsburg, St. Louis, Kansas City and Memphis are located. Everett (1971) gives little data on the chemical composition of these industrial wastes, other than to show that

39

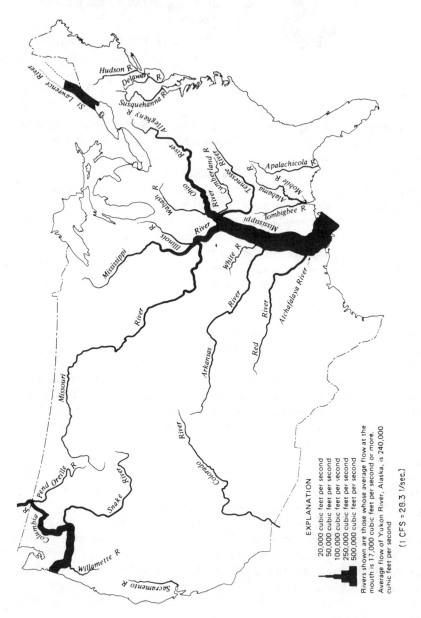

Figure 3-1. Large rivers in the United States (from Iseri and Langbein, 1974).

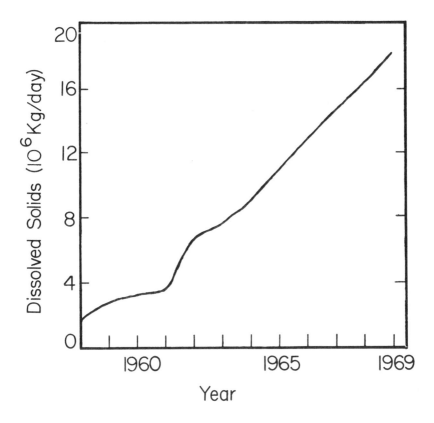

Figure 3-2. Industrial input of dissolved solids between St. Francisville and Luling Ferry, 1958–1969 (from Everett, 1971).

more than half of the dissolved solids increase is due to chloride and sulfate. Industrial wastes in general are known to contain various heavy metals (for example, NAS, 1975), but we have no estimate of the amounts added to the Missisippi by such wastes.

In addition to the inorganic substances dumped into the river, about one-half million kilograms of industrial organic wastes are discharged into the lower Mississippi River each day (Everett, 1971). Added to this industrial load is the domestic sewage from almost two million people in the area. These substances give the water a noticeably disagreeable taste and odor and are removed from municipal water supplies. Many organic wastes also use oxygen as they degrade in the river, and this causes a progressive decrease in oxygen concentrations downstream from Baton Rouge to the Gulf.

The above discussion illustrates the point that man is measurably influencing the lower Mississippi River in some of its major water quality parameters (e.g., dissolved solids and oxygen). The object of this study was to obtain estimates of the total flux of particulate and dissolved heavy metals from the river to the Gulf of Mexico and to look for evidence that this flux has been influenced by man. To accomplish this, heavy metal concentrations have been measured in Mississippi River water and suspended matter, and in Gulf of Mexico plankton and sediments. Although the study is still underway, the total river load of several heavy metals can now be estimated, as can the distribution of these metals in the Gulf plankton and sediments.

Methods

Sampling

Water, suspended matter, plankton and sediment have been collected from the Mississippi River, adjacent fresh and saltwater marshes, and the Mississippi Delta region of Gulf of Mexico during cruises in July 1973, May–June 1974, February–March 1975, July 1975, and September 1975. Many of the sampling sites have been reoccupied several times and their locations are shown in Figure 3-3.

Sediment was collected with a 2 m long, 7.5 cm diameter, plastic-lined gravity corer, or in a few cases by subsampling from a large box core using plastic tubing. The sediment cores were cut into 10 cm or smaller sections on board ship and were stored in airtight plastic containers at 5°C until analyses began.

Water for suspended matter work was collected in 30 liter Niskin bottles which were tripped immediately in order to minimize any losses of particles due to gravity settling. Gram quantities of suspended matter were obtained by vacuum filtering water through a closed system using 293 mm diameter, 0.4 μm pore size Nuclepore filters held in custom-built plastic holders. As much as 120 liters of water were put through each filter in low suspended matter areas, while in high suspended matter areas 30 liters of water were commonly used.

The filters were put into plastic bags and returned to the laboratory where they were dried at 60°C. The suspended material was then removed by crumpling the filters and scraping it off with a teflon spatula. Total suspended matter concentrations were determined at each station by filtering separate 100–1000 ml aliquots onto preweighed 47 mm, 0.4 μm Nuclepore filters and following precautions outlined by Feely (1974).

Water samples for dissolved analysis were collected separately in conventional polyethylene bottles which had been presoaked in 2N HNO$_3$ (redistilled) for several weeks. Some water samples were collected directly from the bow of a rubber boat, while others were taken by pumping water through a closed system using acid washed polyethylene tubing tied to a nylon rope and lowered

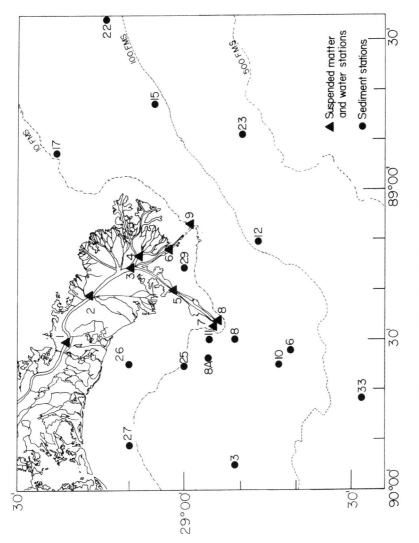

Figure 3-3. Mississippi Delta sampling stations.

from a boom extended some 10 m abeam of the ship. Most of the samples were subsequently filtered through an acid-cleaned, 47 mm, 0.4 μm Nuclepore filter arrangement and were then acidified with 1 ml of ULTREX HNO_3 per liter of sample. Many samples were stored in this manner for several months before they were analyzed, but we have not observed any effect due to this storage in our preliminary analytical work.

All plankton were collected by towing a specially constructed metal-free net from a rubber boat at some distance from the ship. The phytoplankton net was 37 μm mesh, 0.3 meters square and 1 meter long, while the zooplankton net was 308 μm mesh and 1 by 3 meters in size. Each sample collected was rinsed with deionized water on a plastic screen and, after an aliquot was taken for species identification, was frozen and stored for analysis.

At each station where water and suspended matter were collected, a vertical series of samples for total suspended matter (TSM), particulate organic carbon (POC), dissolved organic carbon (DOC), dissolved silica, pH, chloride and temperature was also taken.

Analytical Techniques

Total sediment and suspended matter metal concentrations were determined by flame atomic absorption spectrophotometry following sample dissolution in teflon beakers with an $HF\text{-}HClO_4\text{-}HNO_3$ mixture. Standards were made using appropriate matrices; and background absorbance, due to molecular band absorption and light scattering, was monitored by simultaneously measuring the absorbance of a nonspecific line and the analytical line of the element of interest. The analytical precision has been determined by analysis of numerous quadruplicates and is combined and expressed below as a coefficient of variation (CV): Fe, 3%; Al, 5%; Mn, 3%; Zn, 3%; Pb, 6%; Ni, 6%; Co, 5%; Cu, 5%; Cr, 12%; and Cd, 15%. The accuracy of our total dissolution procedures is routinely checked using numerous U.S.G.S. standard rocks and our data is consistently within 10% of the accepted values.

Dissolved heavy metal concentrations in river water samples were analyzed by direct injection into a flameless atomic absorption spectrophotometer (Perkin Elmer 306 HGA 2100) and after solvent extraction using precleaned ammonium pyrrolidine dithiocarbamate with methyl isobutyl ketone (Brooks *et al.*, 1967) and back extracting into 1N ULTREX HNO_3. Barnard and Fishman (1973) and Edmunds *et al.* (1973) have compared direct analysis versus extraction for fresh water and concluded that Fe, Cr, Cu and Mn may be successfully analyzed directly. We have shown by solvent extraction and/or method of additions analysis of raw samples that the above four metals may be analyzed directly in Mississippi River water while direct Pb and Cd values were less reliable. Nickel concentrations were not high enough in any of the samples to allow direct analysis.

Water samples with >250 mg/l dissolved solids were analyzed only after solvent extraction. The extraction method was checked by carrying out second and third extractions and by spiking raw and extracted samples. Precision (CV) for direct analyses were Fe, 15%; Mn, 15%; Cr, 5%; and Cu, 12% while precision for Pb and Cd analysis is estimated to be 20% and 30%, respectively.

Plankton samples (0.5–1g) were digested in 10 ml of a 3 : 1 HNO_3 : $HClO_4$ mixture, and after heating to near dryness were redissolved in 1N HNO_3 and filtered. Metal analyses were carried out by flame atomic absorption spectrophotometry using a nonabsorbing line to correct for matrix effects. Arsenic was determined by a modification of the colorimetric technique of Vogel (1960) following digestion in a 5 : 2 : 2 concentrated HNO_3 : $HClO_4$: H_2SO_4 mixture. Analytical precision (CV) was again calculated by analyzing several sets of quadruplicates and for samples with more than 1 ppm metal content was as follows: Al, 9%; As, 10%; Cd, 10%; Co, 9%; Cu, 5%; Fe, 6%; Pb, 12%; Mn, 3%; Ni, 12%; and Zn, 5%. For samples containing less than 1 ppm metal the coefficients of variation were often as high as 20–40%.

Particulate and dissolved organic carbon content was measured by the wet combustion infrared methods of Fredericks and Sackett (1970). Grain size distribution of sediments and suspended matter was determined by the pipet methods of Fok (1968). Suspended material was concentrated by flocculating with NaCl or by tediously centrifuging and combining residues. The flocculated material was washed by centrifugation prior to analysis. Chlorinity, dissolved silica, and pH were determined by standard titrametric, colorimetric, and electrometric methods respectively (Strickland and Parsons, 1972).

Dissolved and Suspended Matter

To determine the total amount of any material being transported from continents to oceans by rivers it is necessary to know both the dissolved and the particulate load of the rivers, as well as the concentration of the material of interest in each. We have attempted to obtain this information for several heavy metals being transported by the Mississippi River. At the present stage of this work, we have concentrated most of our efforts on determining the amounts carried as particulates, because previous work has shown this phase to be quantitatively more important for most heavy metals (Goldschmidt, 1958; Gibbs, 1973).

From the environmental quality viewpoint it is necessary to know the chemical and physical nature of the heavy metals being transported as well as the total amounts. The dissolved fractions may, for example, be complexed with various organic and inorganic ligands, while the particulate phase will include surface-held as well as lattice-held materials. These various forms are not likely to be equally available to organisms. Furthermore, the form of a metal can change with changing conditions. For example, part of the absorbed fraction might become dissolved (or vice versa) in moving from river water to seawater.

Ongoing work in our laboratory is directed toward a better understanding of the forms of the metals carried by the river and their interactions with sediments, water and organisms in river water/seawater mixing zones. This study will be described in a future publication, as dissolved and suspended matter total heavy metal concentrations and fluxes were primary concerns of the present work.

Dissolved Heavy Metals

Dissolved heavy metal concentrations (i.e., the fraction passing through whatever filter is used) have been determined for Mississippi River water by a number of investigators over the past 15 years. Table 3-1 shows that there is considerable variation in the reported results. Part of this variation can be explained by the difference in filter pore size used in the various studies. For example, Durum and Haffty in their 1961 study used 2 μm glass fiber filters, while later investigators used 0.45 or 0.4 μm membrane filters. The analytical techniques used also differ, as did the type of sampler, storage containers, etc. Considering these factors, and the analytical difficulties involved in trace analysis, the results probably agree as well as can be expected.

Our preliminary values for the few samples we have analyzed so far are somewhat lower than those previously reported (Table 3-1), but note that previous investigators found concentrations below their detection limits a large part of the time. Mississippi River dissolved metal concentrations are generally lower than those reported for average river water (Turekian, 1969) and are certainly lower than those set for drinking water standards. This is most likely due to metal adsorption on the abundant suspended matter, especially since the river pH is relatively high.

The 486 samples analyzed by the U.S.G.S. are from weekly collections at each of seven locations along the lower Mississippi River from St. Francisville, Louisiana (430 km above Head of Passes), to Venice, Louisiana (18 km above Head of Passes). The percent frequencies of detection and mean values for the observed concentrations were notably uniform along this 412 kilometer stretch and thus the data has been reduced to a single entry in Table 3-1. We find no relationship between the values found by the U.S.G.S., or the frequency of finding detectable amounts, and time of year, river flow, suspended load or other such variables, although such relationships have been found in studies of smaller rivers (for example, Carpenter, in Troup and Bricker, 1975). Suspended matter concentrations in the Mississippi decrease from over 300 mg/l during periods of normal flow to 30 mg/l or less during the three-month low flow period. The nature of the suspended matter changes and major dissolved component concentrations are also altered. However, from our limited number of analyses, we have observed no significant seasonal changes in dissolved trace metals.

Table 3-1
Mississippi River Dissolved Trace Metal Concentrations

Location	Kilometers above Head of Passes	No. of Samples		Fe (μg/l)	Mn (μg/l)	Pb (μg/l)	Cu (μg/l)	Ni (μg/l)	Cr (μg/l)	Cd (μg/l)	Zn (μg/l)	Hg (μg/l)	As (μg/l)
Baton Rouge (Durum and Haffty, 1961)	380	4		1050	81	6	10	16	5	—	—	—	—
New Orleans (Kopp and Kroner, 1967)	166	18	Mean	22	3.6	54	9	—	8	—	23	—	—
			Frequency[a] of detection (%)	(72.2)	(38.9)	(22.2)	(77.8)	—	(22.2)	(0.0)	(66.7)	—	—
Baton Rouge (Davis, 1968)	380	1		—	11	2.5	0.9	0.5	—	—	—	—	—
New Orleans (Durum et al., 1971)	166	2		—	—	<1	—	—	<1	6	<1	<0.5	—
St. Francisville to Venice (U.S.G.S., 1972, 1973, 1974)	430 to 18	486	Mean	35	—	5	8	—	2	2.5	23	0.4	3
			Frequency[b] of detection (%)	(71.5)	—	(39.3)	(21.2)	—	(<1.0)	(30.1)	(69.9)	(24.5)	(38.1)
Southwest Pass (U.S.G.S., 1975)	0 to -35	7		—	—	<1	3	3	<1	<1	10	<0.1	3
Head of Passes (This study, 1974, 1975)		10		5	10	0.2	2	1.5	0.5	0.1	—	—	—
Average River Water (Turekian, 1969)				—	7	3	7	0.3	1	—	20	0.07	2

[a]Lowest detected values were 5 μg/l for Zn and Cr, 8 μg/l for Fe, 2 μg/l for Cu, 28 μg/l for Pb, and 0.6 μg/l for Mn.

[b]Detection limits given as 0.5 μg/l for Hg; 1.0 μg/l for Cd, Cr, Co and Pb; 10 μg/l for Zn, As and Fe; however, for some analyses detection limits were higher.

An estimated flux of each element to the Gulf in dissolved form, based on flow rates of 5.7×10^{14} l/yr (Iseri and Langbein, 1974) and our best estimate of concentrations from the available data, is given in Table 3-2. It can be seen that only a small percentage of all metals are carried in solution. Similar results have been reported for the Amazon and Yukon Rivers (Gibbs, 1973).

Windom *et al.* (1971) have observed dissolved trace metal concentrations in southeastern rivers to be similar to those in adjacent saline estuaries. Andren and Harris (1975) found little difference in the Hg concentration between Mississippi River water and nearby seawater, and Slowey and Hood (1971) report Cu, Mn and Zn values from coastal Gulf of Mexico waters in extractable form to be 0.73, 1.5 and 2.5 ppb, respectively, comparable to the river values listed in Table 3-1. Davis (1968) used a chelating ion exchange resin to concentrate trace metals from 17 samples of coastal Gulf of Mexico waters. His data show considerable scatter with averages of about 0.2 ppb Mn, <0.1 ppb Ni, 0.5 ppb Cu, 5 ppb Zn and <0.1 ppb Pb for the ionic fraction. Generally higher values were obtained, especially for Ni, when samples were oxidized before the concentration step. The different techniques used by the different investigators make any comparison of dissolved concentrations in river and seawater hazardous, but it seems likely, based on the presently available data, that for many trace metals the concentrations do not differ greatly in going from the Mississippi River to the coastal Gulf of Mexico. This conclusion is consistent with our suspended matter data, which is discussed in detail below. It shows similar concentrations of most metals in the river and in the adjacent seawater.

Table 3-2
Annual Flux of Metals to the Gulf of Mexico

Element	Particulate (kg × 10⁻⁶)	Dissolved (kg × 10⁻⁶)	Particulate Total (%)	Dissolved Total (%)
Fe	13,300	2.8	>99.9	0.02
Mn	366	5.7	98.5	1.5
Zn	52	5.7	90.1	9.9
Pb	13	0.1	99.2	0.8
Cu	12	1.1	91.6	8.4
Cr	20	0.3	98.5	1.5
Ni	16	0.9	94.7	5.3
Cd	0.4	0.05	88.9	11.1
As	4	1.7	70.2	29.8

NOTE: Calculations are based on average water and suspended matter data from Tables 3-1 and 3-4, the sediment discharge data of the U.S. Army Corps of Engineers $(2.8 \times 10^{14}$ g/y; 1950-1974) and estimated water flow at the river mouths $(5.7 \times 10^{14}$ l/yr; Iseri and Langbein, 1974).

Suspended Matter

Curtis *et al.* (1973) report that the Mississippi-Atchafalaya River system carried an average of 326 million short tons (2.96×10^{14} g) of suspended matter per year for the period 1950-1969. They say this is probably 30% less than the river load in the late 1800s and early 1900s. A noticeable reduction in sediment load also occurred in the 1950s (Figure 3-4) when construction of upstream dams and farm ponds created sediment traps. Recent data by the U.S. Army Corps of Engineers (1964-1974) from below the control channel that is used to artificially divert part of the Mississippi into the Atchafalaya River show 0.86×10^{14} g/yr for the Atchafalaya and 1.5×10^{14} g/yr for the Mississippi to give a 2.36×10^{14} g/yr total suspended sediment load. Combining these figures with water flow data from the same locations gives an average suspended matter concentration of 370 mg/l for the Mississippi and 480 mg/l for the straighter, swifter Atchafalaya. This is somewhat lower than the approximately 500 mg/l calculated for the system using the 1950-1969 data.

There are, of course, significant variations from the averages on both an annual and a seasonal basis (Figures 3-4 and 3-5). These variations are related to water flow, but not in a simple way. Everett (1971) points out that suspended matter concentrations are usually higher with a rising water stage than at a corresponding flow rate on the falling stage. Thus, sediment discharge, which is a function of the availability of particles for transport, peaks before water discharge.

These trends can be seen in the total suspended matter (TSM) values we obtained on four sampling cruises to the area (Table 3-3). During May–June 1974 and July 1975 the river was near its average flow rate, but it was considerably above average for the February–March 1975 cruise. Despite this, the sediment load was actually lower in February–March, because the river had been at a high flow for some time before the cruise. During the September 1975 cruise the river was at its annual low flow stage and TSM was very much lower than on the previous cruises.

Clay (<2 μm) and silt (62.5 μm–2 μm) size particles made up virtually the entire mass of the suspended matter we collected (Table 3-3). This is in agreement with a study by the U.S. Army Corps of Engineers (1939) who analyzed 1558 samples from the three major Mississippi River passes. Their data give approximately the following particle size distribution: no material >104 μm, 1-2% between 104 and 74 μm, 1-10% between 74 and 40 μm, 30-50% between 40 and 4 μm, 20-30% between 4 and 1 μm and 20-30% less than 1 μm.

River suspended matter heavy metal concentrations vary only slightly over the first three sampling periods (Table 3-4). Differences among samples from several river locations and depths were also small as is shown by the standard deviation given for each period. The somewhat higher values for May–June 1974 may be due to a slight difference in sampling technique. This overall consistency

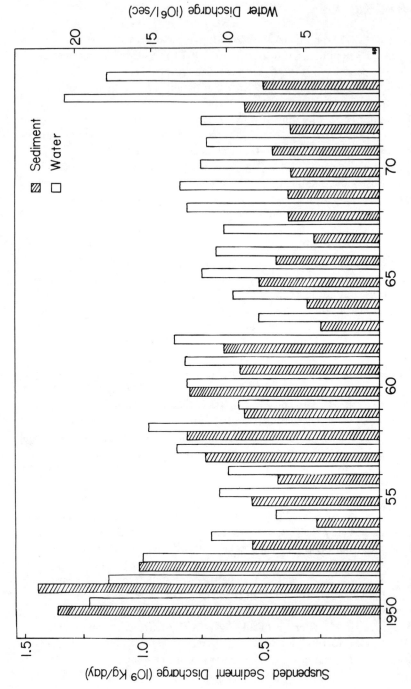

Figure 3–4. Mississippi River sediment and water discharge at Red River Landing, 1950–1962, and Tarbert Landing, 1963–1974. (Data from U.S. Army Corps of Engineers.)

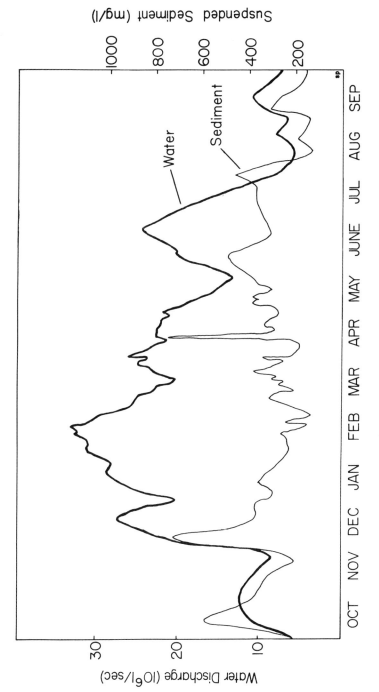

Figure 3–5. Mississippi River suspended sediment and water discharge at Tarbert Landing for water year 1974. (Data from U.S. Army Corps of Engineers 1974).

Table 3-3
Hydrological Conditions at Mississippi River Sampling Stations
(Means and Standard Deviations for Samples Analyzed)

Dates of Collection	Sample Location (Fig. 3-3)	River[a] Discharge (1/sec × 10⁻⁶)	Sediment[a] Discharge (kg/day × 10⁻⁶)	Chlorinity (mg/1)	pH	Temp (°C)	TSM (mg/1)	POC (mg/1)	POC/TSM (%)	DOC (mg/1)	Dissolved Silica (µM)	Clay (%)	Silt (%)
29 May-2 June 1974	(2,3,5,6)	16.4	6.26	21 (±1)	–	23.9 (±0.4)	224 (±54)	1.49 (0.27)	0.73 (±0.20)	–	90 (±1)	48.2 (±0.6)	51.6 (±0.6)
26 Feb-1 Mar 1975	(2-6)	21.2	4.63	19 (±1)	–	10.0 (±0.2)	156 (±32)	1.61 (0.16)	1.14 (±0.20)	3.58 (±0.54)	105 (±2)	58.6	40.9
10-14 July 1975	(1,2,3)	13.3	–	27 (±2)	8.01 (±0.04)	28.0	206 (±114)	2.10 (±0.78)	0.93 (±0.21)	3.33 (±0.20)	–	66.0 (±1.9)	33.8 (±1.9)
7-9 Sept 1975	(1,3)	6.8	–	31-240	8.00 (±0.02)	21.4	11.0 (±1.3)	0.93 (±0.14)	8.3 (±0.8)	3.14 (±0.54)	18.8 (±1.6)	–	–

[a]Data from U.S. Army Corps of Engineers (1974, 1975) for Tarbert Landing, Miss.; corrected to Head of Passes using appropriate time lag and converted to metric units.

Table 3-4
Mississippi River and Nearshore Gulf of Mexico Suspended Matter Metal Concentrations
(Mean and Standard Deviation for Number of Samples Analyzed)

Dates of Collection	Sample Location	No. of Samples	Fe (%)	Al (%)	Mn (μg/g)	Zn (μg/g)	Pb (μg/g)	Cu (μg/g)	Co (μg/g)	Cr (μg/g)	Ni (μg/g)	Cd (μg/g)
29 May–2 June 1974	River (2,3,5,6)	7	5.00 (±0.07)	9.11 (±0.34)	1400 (±62)	209 (±5)	51.6 (±3.4)	42.2 (±1.9)	21.7 (±1.5)	81.5 (±18.7)	59.9 (±3.6)	1.5 (±0.2)
	Gulf (8,9)	3	4.80 (±0.17)	8.50 (±0.76)	1346 (±138)	231 (±9)	74.5 (±8.7)	48.9 (±5.7)	18.2 (±3.3)	—	67.8 (±1.4)	2.4 (±0.6)
26 Feb–1 Mar 1975	River (2,3,5,6)	12	4.61 (±0.34)	8.59 (±0.34)	1190 (±54)	171 (±16)	42.5 (±4.3)	40.0 (±3.4)	19.6 (±1.4)	68.1 (±17.4)	56.6 (±2.9)	1.3 (±0.3)
	Gulf (7,8,9)	6	5.21 (±0.62)	9.42 (±0.64)	1000 (±170)	211 (±29)	77.2 (±22.3)	69.2 (±11.9)	19.8 (±1.7)	62.5 (±9.8)	62.7 (±6.2)	1.5 (±0.4)
10–14 July 1975	River (1,2,3)	6	4.61 (±0.07)	8.71 (±0.32)	1332 (±65)	171 (±9)	42.2 (±5.8)	44.8 (±3.3)	22.3 (±1.3)	68.0 (±12.5)	50.3 (±1.9)	—
	Gulf (7,8,9)	8	4.83 (±0.29)	9.08 (±0.92)	1227 (±141)	236 (±22)	47.4 (±4.3)	47.9 (±4.3)	23.0 (±1.6)	83.0 (±13.4)	54.3 (±3.5)	—
7–9 Sept 1975	River (1,3)	4	3.53 (±0.17)	6.35 (±0.33)	1700 (±188)	259 (±12)	42.9 (±2.2)	58.2 (±3.5)	18.1 (±1.4)	69.9 (±3.3)	49.4 (±4.6)	—
	Gulf (5,6,8)	4	4.46 (±0.28)	7.92 (±0.86)	1308 (±440)	233 (±64)	51.3 (±3.8)	51.4 (±5.5)	22.1 (±1.7)	74.2 (±18.7)	49.8 (±5.1)	—

of metal content for the first three data sets is also reflected in the chlorinity, pH, organic carbon, and silica data, but not so well in the TSM and clay/silt values (Table 3-3). In fact, the TSM concentrations for the initial three cruises ranged from 108-500 mg/l, yet all the metal concentrations were very similar in samples with the lowest TSM to those with the highest. This uniformity was for samples varying from 48-65% clay, thus implying that within these ranges the metal concentrations are independent of clay content. Similarly, Turekian and Scott (1967) noted no significant differences in trace element content for river sediment in the 20-2 μm, 2-0.2 μm, and <0.2 μm ranges.

Metal concentrations for river suspended matter collected at low flow showed several differences relative to the first three sample sets. TSM (11 mg/l) and dissolved silica (19 μM) were markedly lower at this time whereas the POC/TSM ratio was nine times higher (8.3×10^{-2} vs. 0.9×10^{-2} in previous samples). Iron and Al concentrations were about 25% lower, whereas Mn, Zn and Cu were 30-40% higher. There was no significant difference in the Pb, Ni, Co and Cr concentrations. The decreased Fe and Al content in the low flow samples may well be due to a simple dilution by organic matter, which keeps the Al/Fe ratio for these samples (1.80) statistically the same as that for the previous three sets (1.82, 1.86, 1.89). However, the ratios of Mn, Zn, Pb, Cu, Cr and Ni to Fe are higher This altered pattern of heavy metal distribution probably indicates some association of these metals with the increased organic matter.

Data from Carpenter (in Troup and Bricker, 1975) show greatly elevated suspended metal concentration in the Susquehanna River during periods of low flow, which were related to increased organic matter content during these periods. We do not observe, however, the extreme range in trace metal content of the suspended matter reported by Carpenter, who found order of magnitude changes. This is probably due to the much higher suspended matter concentrations and much greater flow of the Mississippi, which makes it less sensitive to man-introduced metals or other perturbances. Most other investigations of heavy metals in river suspended matter (Turekian and Scott, 1967; Windom *et al.*, 1971; Gibbs, 1973) do not permit identification of seasonal variations. Furthermore, since the reported metal concentrations are so variable, because of differences in mineralogy, grain size, organic content, and total suspended and dissolved loads, any comparisons with our data are difficult.

The annual addition of particulate heavy metals to the Gulf of Mexico by the Mississippi River is given in Table 3-2. These were calculated by using an average sediment transport of 2.7×10^{14} g/yr for the period 1950-1974 and averages of the river suspended matter metal concentrations shown in Table 3-5, excluding those for the 6.8×10^6 l/sec period. This exclusion is reasonable because less than 5% of the annual load is carried at these low flow periods and the chemistry is atypical then. The average suspended matter load for the last 10 years is 13% lower than the 25 year average used in these calculations,

Table 3-5
Comparison of Metal Concentrations in Suspended Matter and Bottom Sediments

	Fe (%)	Al (%)	Mn (μg/g)	Zn (μg/g)	Pb (μg/g)	Cu (μg/g)	Co (μg/g)	Cr (μg/g)	Ni (μg/g)	Cd (μg/g)	As (μg/g)
River susp. matter (3 seasons, $n = 25$)	4.74	8.80	1307	184	45.4	42.3	21.2	72.5	55.6	1.4	14.6
$\dfrac{\text{Metal}}{\text{Fe}}$ (× 10³)		1,860	27.6	3.9	0.96	0.89	0.45	1.3	1.2	0.03	—
Nearshore Gulf susp. matter (3 seasons, $n = 17$)	5.02	9.20	1191	226	66.4	55.6	20.1	72.8	58.5	2.0	—
$\dfrac{\text{Metal}}{\text{Fe}}$ (× 10³)		1,830	23.7	4.5	1.3	1.1	0.40	1.4	1.2	0.04	—
Gulf deltaic sediments ($n = 30$)	4.33	7.83	675	160	35.1	29.2	18.9	79.6	39.3	1.2	12.3
$\dfrac{\text{Metal}}{\text{Fe}}$ (× 10³)		1,810	15.6	3.7	0.81	0.67	0.44	1.8	0.91	0.03	—

and would, of course result in a corresponding reduction in the calculated flux.

The particulate flux in Table 3-2 represents suspended matter only and does not include bed load transport. However, bed loads have been estimated to be only 10 - 20% of the suspended load (Holle, 1950; Fisk *et al.*, 1954) and because they have much lower trace metal content (U.S.G.S., 1975) would probably contribute less than 10% to the totals given in the table.

Physico-chemical interactions involving heavy metals reported to occur across the freshwater/seawater interface may affect both the ultimate area of deposition of the metals and their availability to nearshore marine organisms. Desorptive processes would make metals more available to organisms and delay their removal to the sediments while adsorptive processes would have an opposite effect. Lowman *et al.* (1966) observed that river dissolved metals became predominantly "particulate" upon admixture with seawater, while Kharkar *et al.* (1968) found significant desorption of several heavy metal radioisotopes from clays exposed to seawater. Evans and Cutshall (1973) observed desorption of ^{54}Mn and ^{65}Zn from river suspended matter which was mixed with seawater but no loss of ^{51}Cr, ^{124}Sb or ^{46}Sc. Windom (1975) and others have noted an exponential decrease in dissolved iron with increase in salinity at river mouths. Such variables as the amount and nature of the suspended matter, pH, dissolved metal concentrations, and biological activity have been shown to complicate sorptive reactions at river mouths and have been discussed by several authors (Murray and Murray, 1973; O'Connor and Kester, 1975).

A comparison of our suspended matter data for the Mississippi River and areas immediately outside the river mouth is given in Tables 3-4 and 3-5. The river suspended matter metal concentrations were very uniform while the seawater samples showed much greater variability. This variability is certainly related to the wide range of chlorinity (0.25-20.0 mg/l), pH (8.15-8.55), TSM (1.6-75 mg/l), DOC (0.79-4.94 mg/l) and plankton concentrations observed at the various locations and depths sampled right outside the river.

In addition to the observed variability among samples, individual river particulate metals also appeared to behave differently upon mixing with Gulf of Mexico water. Fe, Al, Co, Ni and Cr concentrations were very similar in river and Gulf suspended matter, while Mn content generally decreased seaward and Zn, Pb, Cu and Cd concentrations were either similar or higher in Gulf samples. These observations argue against extensive desorption of any of these metals except Mn and agree with the laboratory study of O'Connor and Kester (1975). However, the decreased Mn concentrations in Gulf suspended matter, which were up to 40% lower than those in the river, suggest desorption of Mn similar to that observed by Evans and Cutshall (1973). In a number of instances, the Zn, Pb, Cu and Cd concentrations were higher in Gulf suspended matter, suggesting that under certain conditions (e.g., pH changes from less than 8.0 to 8.5 or plankton blooms) uptake of these metals may occur.

The above observation may have been complicated by the variable exposure times of the suspended matter to seawater, some as short as one or two hours. The deltaic sediments collected, however, have certainly reached equilibrium and may give a more representative picture of sorptive processes. Zn, Co, Cd, Pb and Cr are statistically the same in river suspended material and surface delta sediments from near the river mouth; however, Mn, Cu and Ni are significantly lower, suggesting that desorption and or postdepositional migration of these metals occurs.

Plankton

It has long been known that marine plankton greatly concentrate certain trace elements from seawater (Vinogradov, 1953; Goldberg, 1957). There has also been considerable speculation on the importance of plankton in transporting trace elements from surface to deep water or from water to sediments (Goldberg, 1965; Brewer, 1975). Martin and Knauer (1973) list, with appropriate references to other workers, the following transport mechanisms: (1) the sinking of skeletal structures and organic detritus, (2) the moulting of crustacean exoskeletons, (3) vertical migrations of zooplankton, (4) incorporation of elements into fast sinking fecal pellets and (5) passage of elements to a higher trophic level. Thus, plankton are potentially a key factor in removing pollutant metals from seawater. To assess the influence of Mississippi River derived material on trace metal concentrations in plankton and the importance of plankton in removing and transporting metals, phytoplankton and zooplankton from the Mississippi Delta and the northwest Gulf of Mexico (Figure 3-6) were analyzed as part of the present study.

Phytoplankton trace metal concentrations were quite variable (Table 3-6) as has been reported in other studies (I.D.O.E., 1972; Martin and Knauer, 1973). This is probably not a function of the species composition of the samples because Martin and Knauer quoting Riley and Roth (1971) note that phytoplankton trace metal content does not depend on taxonomy. Nevertheless, the samples used here were identified, and are discussed below.

While we cannot presently make any definitive statements about metal variations with species or location, there does seem to be some dependence on the latter. Samples 8, 9 and 11, for example, were 90% *Skeletonema costatum* and were all collected in February 1975. Sample 11, from very near the river mouth had a somewhat higher Pb content; and the Pb to Al ratio, which makes an approximate accounting for clay derived Pb in the samples, progressively drops by a factor of three in moving away from the river. Other elements do not show this pattern. The phytoplankton from Corpus Christi Bay had Pb values similar to the delta samples, but since they contained much more clay their Pb to Al ratios are lower.

Sample 10, also from very near the river mouth, but collected during a

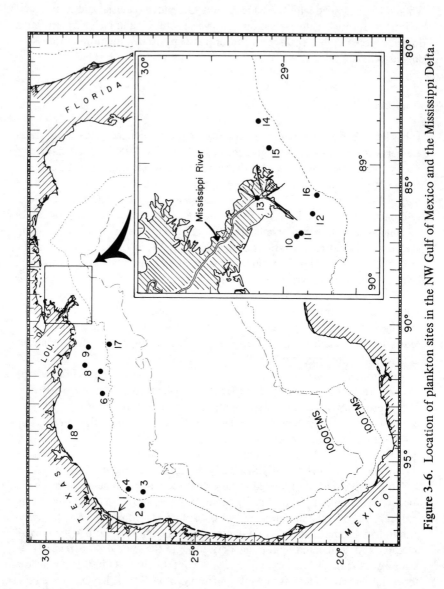

Figure 3–6. Location of plankton sites in the NW Gulf of Mexico and the Mississippi Delta.

Table 3-6
Heavy Metal Distribution in *Sargassum* and Mixed Phytoplankton from the Northwest Gulf of Mexico and the Mississippi Delta
(*Concentrations in μg/g Dry Weight*)

Location	Al	As	Cd	Co	Cu	Fe	Pb	Mn	Ni	Zn
1[a]	13,450	5.1	1.1	4.0	6.2	7,550	20.1	181	7.9	40
8	6,431	47.0	1.8	4.0	11.0	3,514	12.9	77.2	4.8	74
9	6,396	52.0	1.5	6.6	25.2	5,886	21.3	135	1.1	129
10	1,072	5.5	4.3	<0.5	5.1	1,094	5.8	19.4	0.9	55
11	3,036	—	0.2	1.0	6.6	2,887	29.0	115	4.8	22
12	3,894	—	1.4	2.5	6.0	3,115	13.7	80.9	11.0	52
14	1,364	2.9	<0.05	1.8	1.2	1,277	2.5	21.2	4.5	13
17[b]	33	82.0	1.7	0.9	5.1	61	9.1	4.5	15.6	34
18[b]	903	40.0	4.6	4.1	10.6	685	39.2	21.4	2.6	87

Source: Sims, 1975.

[a]Median of 10 samples from Corpus Christi Bay
[b]Sargassum

plankton "bloom" in May 1974, was 99% *Rhizosolenia delicatula*. It was much lower in Pb content than the previous samples, but was enriched in Cd. Sample 14, collected at the same time, should be much less influenced by the river. It was more mixed in species composition, but was more than 50% *Rhizosolenia* and it was low in Pb content and had no detectable Cd. It was also lower than any of the other samples in Cu and Zn, but not in Co and Ni. Sample 12, which was also collected in May 1974, but from an area influenced by the river, had high Pb and Cd values and a different species composition being 90% *Nitzschia pungens*. Thus, based on the limited number of samples analyzed, there is some indication of an enrichment in the elements Pb and Cd near, and to the west of, the river mouth. These samples were higher in Pb content than the median of the phytoplankton samples from Monterey Bay analyzed by Martin and Knauer (1973). They were also much higher in Mn, but this may be due to the greater clay content of these samples as the Mn is in approximately the ratio to Fe expected for clay.

While zooplankton trace metal concentrations seem to be more dependent on the taxonomy of the sample than is the case for phytoplankton (Martin and Knauer, 1973), zooplankton are equally susceptible to clay incorporation and contamination during sampling and handling. The trace metal concentrations in both groups will vary with productivity—that is, the concentration will be less when the organisms are multiplying very rapidly, thereby exposing any sample to seawater for a shorter time and perhaps depleting the water. All these factors make interpretation of the data difficult, and call for a large sampling effort to insure statistical validity. Despite these complications some generalizations can be made on the zooplankton data presented in Table 3-7.

The zooplankton trace metal concentrations showed no pattern that would indicate an adverse effect from the Mississippi River, but as is discussed above it is necessary to consider some of the factors which can complicate the gross distribution pattern. The three samples collected offshore from Corpus Christi were enriched in Pb, Cd and Cu compared to samples from the immediate Mississippi River delta area and those from offshore Louisiana. The samples from Corpus Christi were predominantly copepods (Table 3-8), unlike most of the other samples, but this cannot explain all of their metal enrichment because one sample from near the Mississippi River had a high copepod component, but low trace metals. Likewise, clay contamination can explain some, but not all of the enrichment, because one of the three Corpus Christi samples was very low in Al (clay). Data on 74 additional zooplankton samples from South Texas analyzed in our laboratory (Horowitz and Presley, 1976 and unpublished data) show large variations in most trace metals, but generally higher Pb, Cd and Cu concentrations than the Mississippi River delta samples. Furthermore, the values are not greatly different from values given by Martin and Knauer (1973). Thus, the river does not appear to grossly contaminate zooplankton. Nevertheless, samples near the river mouth showed some interesting features.

Table 3-7
Heavy Metal Distribution in Mixed Zooplankton from the Northwest Gulf of Mexico and the Mississippi Delta
(Concentrations in µg/g Dry Weight)

Location	Al	As	Cd	Co	Cu	Fe	Pb	Mn	Ni	Zn
2	1,252	7.6	2.4	0.9	74.0	799	15.3	12.6	2.0	155
3	4,266	6.9	4.4	1.5	23.1	3,663	62.5	105	6.1	200
4	500	7.3	4.4	2.0	25.6	977	16.5	21.8	2.9	133
6	75	–	2.9	2.1	8.9	77	8.5	7.5	7.8	135
7	103	5.8	1.9	<0.5	7.5	122	1.2	13.7	2.8	86
8	340	12.0	1.5	<0.5	8.2	305	2.3	9.2	2.7	75
9	225	4.9	1.0	<0.5	4.3	270	6.2	10.3	1.0	41
10	314	1.9	2.9	<0.5	6.1	397	2.5	16.3	1.4	139
11	266	9.0	0.9	<0.5	8.6	300	< 0.5	8.4	8.2	107
12	426	3.9	2.4	1.1	6.5	532	< 0.5	28.0	3.5	116
14	6,000	6.5	1.2	<0.5	3.5	4,035	5.1	114	7.4	52
15	4,620	6.4	2.6	0.8	6.6	4,760	8.3	70.4	6.6	68
16	51	23.1	0.4	<0.5	35.3	62	3.0	4.7	<0.5	49
17	44	29.5	2.5	0.7	9.2	237	7.4	7.7	1.4	57

Source: Sims, 1975.

Table 3-8
Taxonomic Distribution of Dominant Groups in Mixed Zooplankton from the Northwest Gulf of Mexico and the Mississippi Delta
(Percent Composition by Dry Weight)

								Location							
Location	2	3	4	6	7	8	9	10	11	12	14	15	16	17
Date	12-74	12-74	12-74	5-74	5-74	2-75	2-75	5-74	2-75	5-74	5-74	5-74	2-75	2-75
Amphipoda			31	6										
Chaetognatha		19	15	44	2		13		16					
Coelenterata	1									4	95	53		
Copepoda	94	78	54	50	2	18	31	2	56	8		22		40
Ctenophora								98		3				
Decopoda					1		33		27			28		
Fish Larvae						82				85				
Ostracoda		3					21							
Salpida					95									
Penaeid Decopod													99	
Euphausia														26

Source: Sim, 1975.

Samples 10, 11 and 12 (Figure 3-6), which were collected in relatively muddy water very near the river mouth, were either about equal to, or lower than, samples 14 and 15 in metal content. These latter samples were collected to the east of the delta, outside of the normal influence of the river, and in clear water. What is even more surprising is that these two clear water samples were greatly enriched in Al, Fe and Mn, in about the ratios to be expected from clay incorporation. Examination of these samples showed them to be quite distinct taxonomically (Table 3-8), being predominantly jellyfish-like organisms. Apparently detrital particles readily adsorb on their surfaces, and they might become even more enriched in trace metals if they drifted into more turbid water. On the other hand, particles must stick to fish larva and crustaceans to a much lesser extent, because these organisms were only enriched in Fe, Mn and Al by a factor of two or so over concentrations in similar organisms from the deep clear water of Monterey Bay (Martin and Knauer, 1973). It is thus necessary to consider both taxonomy and location before judging the influence of the river on trace metal concentrations in zooplankton.

The overall productivity in the delta area is high (450 gC/m^2/yr) (El-Sayed, personal communication), but the distribution of species, grazing rate and fate of the produced carbon is not well known. In the open sea most organic carbon is destroyed before it is buried in the sediments (Menzel, 1974 and references therein). Such destruction also seems to be true here, despite the shallow water, because our unpublished data from more than 20 cores in the area show the organic carbon content of the sediments to be similar to that of the river suspended matter. Both are commonly only about one-half percent organic carbon. This would be expected near and to the west of the delta where detrital sediment is accumulating at rates of 5 g/cm^2/yr or faster, because this rapid detrital sedimentation would effectively dilute the 0.045 gC/cm^2/yr being produced, even if all of it survived to be buried. However, the sedimentation rate drops very rapidly to 0.5 g/cm^2/yr or less outside the 100 m isobath, and to the east of the delta, yet the organic carbon percentage in the sediments goes up only slightly. Obviously, little of the carbon is being buried. By similar reasoning, plankton are responsible for little trace metal enrichment of the sediments, which are relatively constant in composition over an area of measured or implied great variation in detrital sedimentation rate. If plankton, which must be produced fairly uniformly over this small area near the delta, were making a significant contribution to the sediments then the sediment trace metal content would vary with sedimentation rate.

Sediments

Coastal sediments provide the major sink for continentally derived heavy metals (Goldschmidt, 1937; Goldberg and Arrhenius, 1958). When they accumulate in

an undisturbed fashion, sediments also provide a history of the flux of materials
to the marine environment. Thus the finding of higher concentrations of heavy
metals in surface sediments than in deeper (older) sediments might indicate recent inputs of anthropogenic origin. Such an approach was used by Bruland *et
al.* (1974) for four Southern California basins. The areal distribution of heavy
metal concentrations around a point source such as the Mississippi River mouth
also gives information on possible influences by man and on the physical and
chemical behavior of the added metals. For these reasons we have studied sediment cores from the Mississippi River delta and the adjacent northwest Gulf of
Mexico.

Previous work on trace metal concentrations in sediments from this area has
been discussed by Young (1954), Potter *et al.* (1963), Tieh *et al.* (1973), Holmes
(1973) and Trefry and Presley (1976). Potter *et al.* collected eight samples from
directly west of Southwest Pass. The total and the less than 2 μm fraction was
analyzed separately with the following results:

	Total Sediment (μg/g)	<2μm Fraction (μg/g)
Cr	72	65
Cu	33	33
Ga	20	22
Ni	42	42
Pb	24	19
V	134	157

We find sediments in this area to be essentially free of sand size particles and 60
to 70% clay (< 2 μm). Nevertheless, Potter's data suggest that differences in the
metal content of the less than 2 μm fraction and the coarser fraction in these
fine grained sediments are small. Total concentrations reported by Young and
Davis are similar to Potter's values for the elements determined in common.
Holmes surveyed an extensive area of the delta and found a wide range of concentrations, with higher values for most elements near the mouth of SW Pass.

Trefry and Presley (1976) applied a HNO_3-HCl leaching technique to 72
sediment samples from the NW Gulf of Mexico, 50 of them from the Mississippi
River delta. They found values similar to those of previous investigators and also
wide variation in concentration of all metals, which seemed to be a function of
factors such as grain size, organic carbon and calcium carbonate content and
mineralogy. To partially compensate for these variables in a simple way, metal
concentrations were normalized to Fe. The resulting Zn to Fe and Ni to Fe
scatter plots with 95% prediction intervals are shown in Figures 3-7 and 3-8.
Cu and Mn showed similar patterns. These elements correlate well with Fe for
the entire NW Gulf of Mexico and for all depths in the sediment column. This
relationship is assumed to represent the natural conditions in the continental
shelf sediments of the northwest Gulf. Any anthropogenic input of these metals

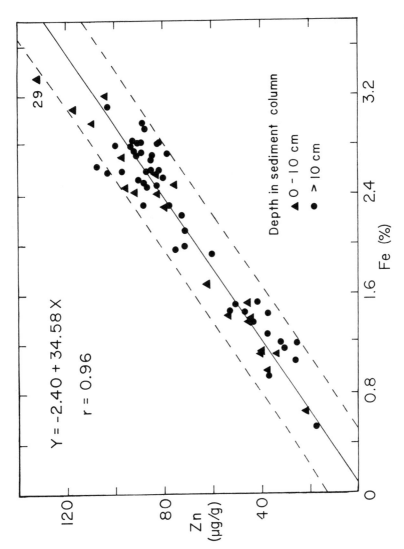

Figure 3-7. Zn/Fe scatter plot for NW Gulf of Mexico sediments (from Trefry and Presley, 1976).

MARINE POLLUTANT TRANSFER

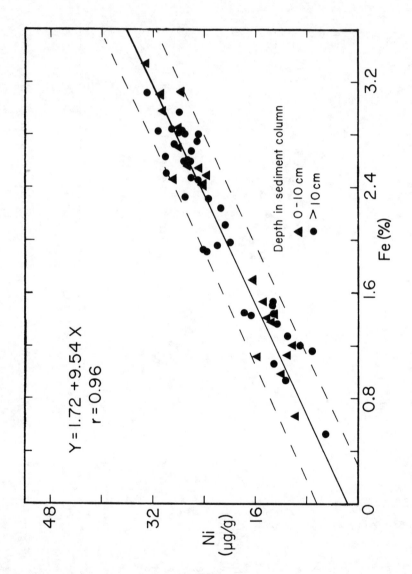

Figure 3–8. Ni/Fe scatter plot for NW Gulf of Mexico sediments (from Trefry and Presley, 1976).

would cause points to deviate from the regression line in a positive y-direction, because it is unlikely that enough Fe would be added to cause a corresponding increase in its concentration. An example of such a deviation can be seen in the Pb vs. Fe scatter plot (Figure 3-9) where 8 of the 72 samples analyzed deviated from the overall linear trend by more than twice the analytical precision. All of these anomalies are from samples taken in the immediate sediment transport pathway of the Mississippi River, and are believed to be due to anthropogenic lead input. Of the other metals determined only Cd behaved similarly to Pb. The depths of the Pb and Cd anomalies were dependent on the distance from the mouth of the river. Station 11, closest to the Pass, had deviations to a depth of 60 cm while at station 8 and 8A, at respectively greater distances, the excesses went to 50 and 30 cm. Sedimentation rates of several cm/yr for this area (Coleman, 1975) suggest a 10-20 year history for this Pb excess. In deeper water, at station 33, only the 0-5 cm samples contained excess Pb. This is consistent with reported sedimentation rates for the area of a mm or so per year (Ludwig, 1971).

We have recently analyzed a new set of samples from near South and Southwest Passes and from deeper water (Figure 3-3). Total Cu and Pb concentrations and the ratios to Fe for station 19 from this set are plotted versus depth in the sediment column in Figures 3-10 and 3-11. Iron and Al concentrations were constant for the entire core, thus changes in the Cu/Fe or Pb/Fe values are due to changes in Cu or Pb at a time of constant Fe. It can be seen that Cu concentration is constant with depth, but Pb decreases to about 30 cm where it levels off at 60% of the surface value. Cadmium concentrations in the bottom 40 cm of this core (0.4 μg/g) were half that in the top 20 cm, while Co, Cr, and Ni concentrations were uniform over the entire sample length. Sedimentation rates of about 1 cm/yr have been estimated for this area (Coleman, 1975) and thus the Pb and Cd excesses appear to have been accumulating for at least 25 years.

Lead concentrations in sediments very near Southwest and South Passes (Δ Stations 8 and 9 in Figure 3-3) are uniform over the length of the core (Figure 3-12), but are at levels (40 μg/g) similar to those at the top of the core just discussed. Cadmium concentration is also uniform, and is 50% higher than the previous high values. Sedimentation rates for this area are estimated to be 20 cm/yr or higher (Coleman, 1975) and thus only the past three to five years were sampled.

In summary, sediment that is less than 25 to 50 years old appears to be enriched in Pb and Cd near the Mississippi River mouth, compared to older sediment or that farther away from the river. This observation strongly suggests that the distribution of these two elements is being influenced by man, but none of the other elements analyzed showed this pattern. The suspended matter data are consistent with this interpretation since Pb and Cd concentrations in it are similar to or higher than the elevated sediment values found in recent sediments.

Sediment from the Mississippi River, and presumably any associated pollutant

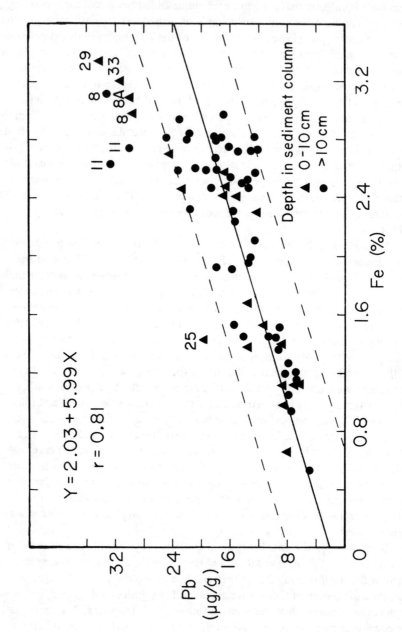

Figure 3–9. Pb/Fe scatter plot for NW Gulf of Mexico sediments (from Trefry and Presley, 1976).

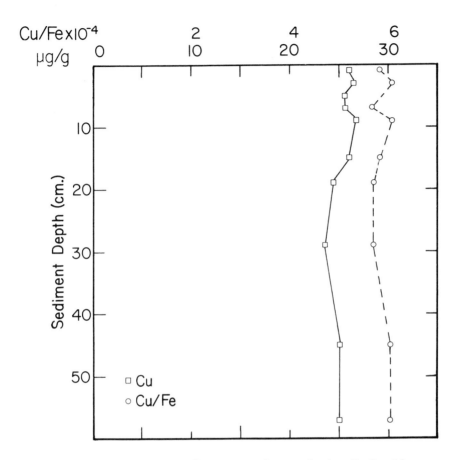

Figure 3-10. Sediment Cu concentrations vs. depth at Station 10.

metals, are spread over at least 35–40% of the Gulf of Mexico floor (Davies, 1972). However, a very high percentage of the river-derived carried sediment settles quickly near the river mouth. Fisk *et al.* (1954), for example, gives figures suggesting that more than 95% of the sediment is deposited within his 1800 km² outline of the modern delta. This area is only 0.1% of the total Gulf of Mexico. We have arrived at a similar, but less extreme, result by considering the few available sediment accumulation rates for Gulf of Mexico cores. Our analysis shows that more than 90% of the sediment can be accounted for in an area around the delta that is less than 1% of the area of the Gulf.

Conclusions

The Mississippi River is the major U.S. outlet to the ocean and man is using it to

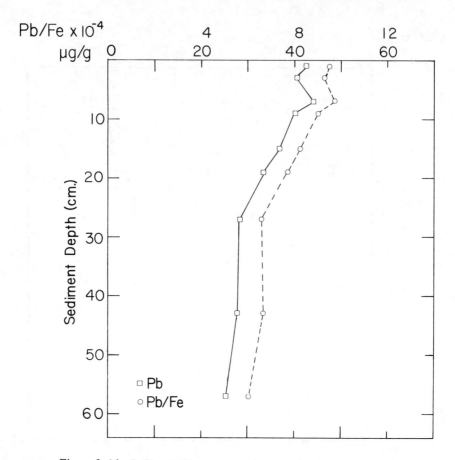

Figure 3-11. Sediment Pb concentrations vs. depth at Station 10.

carry enormous loads of industrial and municipal wastes. Although fish kills and drinking water contamination have been related to organic wastes in the river, there is little evidence of excessive metal levels in the lower Mississippi River. Dissolved metal concentrations determined by this and other studies are quite low and account for only a small percentage (< 1-10%) of the total heavy metal load of the river. This is most likely due to specific adsorption of the metals to the abundant river suspended matter, a process known to be effective at the pH levels involved (7.5-8.0).

Analysis of 29 river suspended matter samples taken at four different times show that heavy metal concentrations are very uniform, despite differences in sample location and depth and variations in total suspended matter and clay content. The only exception to this was found at very low river flow and total suspended matter concentration. At this time, a 25% decrease in Fe and Al content

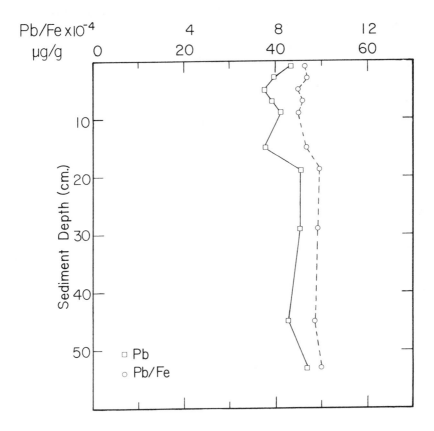

Figure 3-12. Sediment Pb concentrations vs. depth at Station 8.

seems to be directly related to a comparable increase in organic matter. Higher than average concentrations of Mn, Cu and Zn at low flow suggest a significant association of these metals with the organic fraction.

Heavy metal concentrations of nearshore Gulf suspended matter are more variable than those found in the river, yet overall there is little difference in the concentrations of most metals between the two. Changes that have been observed suggest that there is some desorption of Mn across the freshwater/salt-water interface and perhaps uptake of Zn, Cd, Pb and Cu under certain conditions.

Although some plankton samples from near the mouth of the Mississippi River appear to have elevated Pb and Cd content, the overall metal concentrations of the plankton are quite similar to those reported from areas thought to be unaffected by man. Even though productivity is high, it is overshadowed by the large suspended sediment load of the Mississippi River, and the plankton play an insignificant role in transporting heavy metals to the deltaic sediments.

The sedimentary record, which provides one of the best means for assessing anthropogenic metal input, shows that there has been a 60% increase in Pb and a 100% increase in Cd flux to the sediments over the past 25-30 years. The Cd concentration of present-day suspended material is three times that of mid-1900s sediment. On the other hand, there seems to be no significant recent increase of Mn, Ni, Co, Zn, Cr or Cu from the Mississippi.

Acknowledgments

We especially thank R. R. Sims, Jr. for making his plankton data available to us. We are also grateful to R. F. Shokes, E. M. Marek, S. L. Parker, C. W. Lindau, T. R. McKee, R. A. Feely, M. C. Dobson, and A. H. Tsai for their assistance with some of the sampling, analytical and technical aspects of this study. Our work was supported by the National Science Foundation, Office for the International Decade of Ocean Exploration (Grant No. GX-42576) and the Office of Naval Research (Grant No. N00014-75-C-0537).

References

Andren, A. W. and R. C. Harriss (1975). Observations on the association between mercury and organic matter dissolved in natural waters. *Geochim. Cosmochim. Acta, 39,* 1253-1258.

Barnard, W. M. and M. J. Fishman (1973). Evaluation of the use of the heated graphite atomizer for the routine determination of trace metals in water. *At. Absorption Newslett., 12,* 118-124.

Brewer, P. G. (1975). Minor elements in sea water. In *Chemical Oceanography,* Vol. 1, (J. P. Riley and G. Skirrow, eds.). Academic Press, New York, pp. 415-496.

Brooks, R. R., B. J. Presley and I. R. Kaplan (1967). The APDC-MIBK extraction system for the determination of trace elements in saline water by atomic absorption spectrometry. *Talanta, 14,* 809-816.

Bruland, K. W., K. Bertine, M. Koide and E. D. Goldberg (1974). History of metal pollution in Southern California coastal zone. *Env. Sci. Technol., 8,* 425-432.

Coleman, J. M. (1975). Louisiana State Univ., personal communication.

Curtis, W. F., J. K. Culbertson and E. B. Chase (1973). Fluvial sediment discharge to the ocean from the conterminous United States. *U.S. Geol. Surv. Circ. 670,* 17 pp.

Davies, D. K. (1972). Deep sea sediments and their sedimentation, Gulf of Mexico. *Am. Assoc. Petrol. Geol. Bull., 56,* 2212-2239.

Davis, D. R. (1968). *The measurement and evaluation of certain trace metal concentrations in the nearshore environment of the northwest Gulf of Mexico and Galveston Bay.* Ph.D. Dissertation, Texas A&M University, College Station, 70 pp.

Durum, W. H. and J. Haffty (1961). Occurrence of minor elements in water. *U.S. Geol. Surv. Circ. 445,* 11 pp.

Durum, W. H., J. D. Hem, and S. G. Heidel (1971). Reconnaissance of selected minor elements in surface waters of the United States, October 1970. *U.S. Geol. Surv. Circ. 643,* 49 pp.

Dyrssen, D., C. Patterson, J. Ui and G. F. Weichart (1972). Inorganic chemicals. In *A Guide to Marine Pollution,* (E. D. Goldberg, ed.), Gordon and Breach Sci. Pub., New York, pp. 41-58.

Edmunds. W. M., D. R. Giddings and M. Morgan-Jones (1973). The application of flameless atomic absorption in hydrogeochemical analysis. *At. Absorption Newslett., 12,* 45-49.

El-Sayed, S. Z. (1975). Texas A&M Univ., personal communication.

Evans, D. W. and N. H. Cutshall (1973). Effects of ocean water on the soluble-suspended distribution of Columbia River radionuclides. In *Radioactive Contamination of the Marine Environment,* IAEA, Vienna, pp. 125-140.

Everett, D. E. (1971). Hydrologic and Quality Characteristics of the Lower Mississippi River. *Louisiana Dept. Public Works–U.S. Geol. Survey,* 48 pp.

Feely, R. A. (1974). *Chemical characterization of the particulate matter in the near bottom nepheloid layer of the Gulf of Mexico.* Ph.D. Dissertation, Texas A&M University, College Station, 145 pp.

Fisk, H. N., E. McFarlan, C. R. Kolb and L. J. Wilbert (1954). Sedimentary framework of the modern Mississippi delta. *J. Sediment. Petrol. 24,* 76-99.

Folk, R. L. (1968). *Petrology of Sedimentary Rocks,* Hemphills, Austin, Tex. 170 pp.

Fredericks, A. D. and W. M. Sackett (1970). Organic carbon in the Gulf of Mexico. *J. Geophys. Res., 75,* 2199-2206.

Garrels, R. M. and F. T. Mackenzie (1971). *Evolution of Sedimentary Rocks,* W. W. Norton, New York, 397 pp.

Gibbs, R. J. (1973). Mechanisms of trace metal transport in rivers. *Science, 180,* 71-73.

Goldberg, E. D. (1957). Biogeochemistry of trace metals. In *Treatise on Marinecology and Paleoecology, Vol. I,* (J.W. Hedgpeth, ed.), pp. 345-358. Geol. Soc. Am. Mem. 67, Washington, D. C.

Goldberg, E. D. (1965). Minor elements in sea water. In *Chemical Oceanography, Vol. 1* (J. P. Riley and G. Skirrow, eds.), Academic Press, New York, pp. 163–196.

Goldberg, E. D. and G. O. S. Arrhenius (1958). Chemistry of Pacific pelagic sediments. *Geochim. Cosmochim., Acta 13*, 153–212.

Goldschmidt, V. M. (1937). The principles of distribution of chemical elements in minerals and rocks. *J. Chem. Soc., 1937*, 655–673.

Goldschmidt, V. M. (1958). *Geochemistry*, Oxford Univ. Press, 730 pp.

Holle, C. G. (1952). Sedimentation at the mouth of the Mississippi River. In *Proc. Second Conf. on Coastal Eng.*, Univ. of Calif., Berkeley, pp. 111–129.

Holmes, C. W. (1973). Distribution of selected elements in surficial marine sediments of the northern Gulf of Mexico continental shelf and slope. *U.S. Geol. Surv. Prof. Paper 814*, 7 pp.

Horowitz, A. and B. J. Presley (1976). Trace metal concentrations and partitioning in zooplankton, neuston, and benthos from the South Texas Outer Continental Shelf. *Arch. Environ. Pollut. Toxicol.* (in press).

I.D.O.E. (1972). *Baseline Studies of Pollutants in the Marine Environment.* Natl. Sci. Found., Washington, D. C., 799 pp.

Iseri, K. T. and W. B. Langbein (1974). Larger rivers of the United States. *U.S. Geol. Surv. Circ. 686*, 10 pp.

Kharkar, D. P., K. K. Turekian and K. K. Bertine (1968). Stream supply of dissolved silver, molybdenum, antimony, selenium, chromium, cobalt, rubidium and cesium to the oceans. *Geochim. Cosmochim. Acta, 32*, 285–298.

Kopp, J. E. and R. C. Kroner (1967). *Trace Metals in Water of the United States.* Fed. Water Poll. Control Admin., Div. Poll. Surveillance.

Leifeste, D. K. (1974). Dissolved-solids discharge to the oceans from the conterminuous United States. *U. S. Geol. Surv. Circ. 685*, 8 pp.

Lowman, F. G., D. K. Phelps, R. McClin, V. Roman de Vega, I. Oliver de Padovani and R. J. Garcia (1966). Interactions of the environmental and biological factors on the distribution of trace elements in the marine environment. In *Disposal of Radioactive Wastes into Seas, Oceans and Surface Waters*, IAEA, Vienna, pp. 249–265.

Ludwig, C. P. (1971). *The micropaleontological boundary between Holocene and Pleistocene sediments in the Gulf of Mexico.* M.S. Thesis, Texas A&M Univ., College Station, 143 pp.

Martin, J. H. and G. A. Knauer (1973). The elemental composition of plankton. *Geochim. Cosmochim., Acta 37*, 1639–1653.

Menzel, D. W. (1974). Primary productivity, dissolved and particulate organic matter, and the sites of oxidation of organic matter. In *The Sea, Vol. 5* (E. D. Goldberg, ed.), Wiley-Interscience, New York, pp. 659–678.

Murray, C. N. and L. Murray (1973). Adsorption-desorption equilibria of some radionuclides in sediment-fresh-water and sediment-seawater systems. In *Radioactive Contamination of the Marine Environment,* IAEA, Vienna, pp. 105-124.

NAS (National Academy of Science) (1975). *Assessing Potential Ocean Pollutants,* NAS, Washington, D.C., 438 pp.

O'Connor, T. P. and D. R. Kester (1975). Adsorption of copper and cobalt from fresh and marine systems. *Geochim. Cosmochim. Acta, 39,* 1531-1544.

Potter, P. E., N. F. Shimp and J. Witters (1963). Trace elements in marine and fresh-water argillaceous sediments. *Geochim. Cosmochim. Acta, 27,* 669-694.

Riley, J. P. and I. Roth (1971). The distribution of trace elements in some species of phytoplankton grown in culture. *J. Mar. Biol. Ass. U. K., 51,* 63-72.

Sims, R. R., Jr. (1975). *Selected chemistry of primary producers, primary consumers and suspended matter from Corpus Christi Bay and the northwest Gulf of Mexico.* M.S. Thesis, Texas A&M Univ., College Station, 65 pp.

Slowey, J. F. and D. W. Hood (1971). Copper, manganese and zinc concentrations in Gulf of Mexico waters. *Geochim. Cosmochim. Acta, 35,* 121-138.

Strickland, J. D. H. and T. R. Parsons (1972). *A Practical Handbook of Seawater Analysis.* Fisheries Research Board of Canada, 310 pp.

Tieh, T. T., T. E. Pyle, D. H. Eggler and R. A. Nelson (1973). Chemical variations in sedimentary facies of an inner continental shelf environment, northern Gulf of Mexico. *Sediment. Geol., 9,* 101-115.

Trefry, J.H. and B.J. Presley (1976). Heavy metals in sediments from San Antonio Bay and the northwest Gulf of Mexico. *Environ. Geol., 1, no. 5* (in press).

Troup, B. N. and O. P. Bricker (1975). Processes affecting the transport of materials from continents to oceans. In *Marine Chemistry in the Coastal Environment,* (T. M. Church, ed.), ACS Symposium Series 18, Washington, D.C., pp. 133-151.

Turekian, K. K. (1969). The oceans, streams, and atmosphere. In *Handbook of Geochemistry Vol. I* (K. H. Wedepohl, ed.), Springer-Verlag, Berlin, pp. 297-323.

Turekian, K. K. and M. R. Scott (1967). Concentrations of Cr, Ag, Mo, Ni, Co and Mn in suspended material in streams. *Environ. Sci. Technol., 1,* 940-942.

U.S. Army Corps of Engineers (1939). Study of materials in transport, passes of the Mississippi River. Tech. Memo 158-1, Waterways Exp. Station, Vicksburg, Miss., 36 pp.

U.S. Army Corps of Engineers (1950-1975). Stages and Discharges of the

Mississippi River and Tributaries and Other Watersheds in the New Orleans District, U.S. Army Corps of Engineers, New Orleans.

U.S. Geol. Survey (1972, 1973, 1974, 1975). *Water Resources Data for Louisiana.* U.S. Geol. Survey, Water Resources Division, Baton Rouge.

Vinogradov, A. P. (1953). *The Elementary Chemical Composition of Marine Organisms,* Sears Found. for Mar. Res., Yale Univ., New Haven, 647 pp.

Vogel, A. I. (1971). *A Textbook of Quantitative Inorganic Analysis,* Longman Ltd, London, 1216 pp.

Windom, H. L. (1975). Heavy metal fluxes through salt marsh estuaries. In *Estuarine Research* (L. E. Cronin, ed.), vol. 1, Academic Press, New York, pp. 137–152.

Windom, H. L., K. C. Beck and R. Smith (1971). Transport of trace metals to the Atlantic Ocean by three Southeastern rivers. *Southeastern Geol., 12,* 169–181.

Young, E. J. (1954). *Studies of trace elements in sediments.* Ph.D. Thesis, MIT, Cambridge, Mass.

4

Trace Metals in the Marine
Atmosphere: Sources
and Fluxes

R. A. Duce, G. L. Hoffman,
B. J. Ray, I. S. Fletcher,
G. T. Wallace, J. L. Fasching,
S. R. Piotrowicz, P. R. Walsh,
E. J. Hoffman, J. M. Miller
and *J. L. Heffter*

Introduction

There has been increasing concern that considerable quantities of pollutants
reaching the marine environment are transported via the atmosphere. In coastal areas, primary attention has been given to direct water discharge of urban and
industrial wastes, but atmospheric input to these regions may also be large and in
some cases dominant for certain substances. For example, the input of several
trace metals to Lake Michigan appears to be primarily from air pollution fallout
from Chicago (Winchester and Nifong, 1972; Winchester, 1972; Skibin *et al.*,
1973). High concentrations of lead in surface seawater of the Los Angeles basin,
compared to deep water, are attributed to automotive aerosol fallout (Tatsumoto
and Patterson, 1963) and the record of the pollution history appears to exist in
the underlying marine sediments (Bruland *et al.*, 1974; Chow *et al.*, 1973). Patterson and Settle (1974) found that the atmospheric deposition of Pb into a
12000 Km^2 area of the Southern California Bight accounted for about 45% of
the pollutant Pb input, the remaining 55% being from wastewater, storm runoff,
and river input. In a study of trace metals over the New York Bight, Duce *et al.*
(1975a) estimated that approximately 13% of the input of pollutant Pb to a
10000 Km^2 area of the Bight was from atmospheric fallout, the rest being due to
barge dumping, runoff, sewage, and river input. It was estimated that atmospheric sources accounted for approximately 8% of the Zn, approximately 5%
of the Fe, and approximately 2% of the Cd input into this area of the New York
Bight.
   There is also evidence that considerable quantities of lead and perhaps other
trace metals, DDT, PCB, low molecular weight petroleum hydrocarbons and
other organic substances are transported to the open ocean by the atmosphere,
either as particles or in the gas phase (SCEP, 1970; FAO, 1971). The high lead
content of Greenland ice has been attributed to the burning of tetraethyllead
in automobile fuel in populated areas of the northern hemisphere, as has the
atmospheric lead concentrations over the central Pacific Ocean (Murozumi *et al.*,
1969; Chow *et al.*, 1969; Hoffman *et al.*, 1972). It has been suggested that sulfur
found in recent Greenland ice cores is the result of atmospheric transport of
sulfur produced from fossil fuel burning (Weiss *et al.*, 1971, 1975). High concentrations of atmospheric vanadium in particles in the tropospheric westerlies over
the North Atlantic are believed to result from the burning of heavy fuel oils rich
in vanadium porphyrin complexes (Duce and Hoffman, 1976). DDT and PCB

are known to be present in the atmosphere over the North Atlantic in both the vapor phase and on particles (Risebrough *et al.* 1968; Seba and Prospero, 1971; Bidleman and Olney, 1974; Harvey and Steinhauer, 1974). It is clear that the long-range atmospheric transport of many pollutants, both gaseous and particulate, has been well documented.

While this long-range transport of atmospheric pollutants is a generally accepted fact, there have been few concerted quantitative efforts to relate atmospherically transported pollutants to the general problem of marine pollution, especially of the open ocean. There is virtually no information on the mean concentration of these substances in the atmosphere in the westerlies over the North Atlantic, where pollutant transport from the heavily industrialized North American continent would be a maximum. Previous work in this area has been limited to a small number of samples collected from ships without systematic control of geographic location and without consideration of rapidly changing meteorological conditions. The flux of most of these atmospheric pollutants into the ocean is completely unknown as is the rate of their reintroduction into the atmosphere by spray and bubble breaking. Very little is known concerning the mechanisms of deposition of these pollutants into the ocean from the atmosphere. For example, the relative efficiencies and the parameters controlling the removal of atmospheric particulate matter to the ocean surface via rainfall and dry fallout remain a major unanswered question.

The primary objectives of the atmospheric and sea surface chemistry program described in this paper are (1) to determine the sources for and evaluate the significance of the atmospheric transport of heavy metals to mid-ocean regions over the North Atlantic, and (2) to assess the rate of their removal to the ocean surface and the possible significance of their recycling back into the atmosphere on sea salt particles produced by bubbles bursting at the air/sea interface. The remainder of this paper will consist of three sections. The first will describe the distribution of atmospheric trace metals found in Bermuda during 1973 and 1974. The second section will describe several studies underway to attempt to ascertain the sources for the atmospheric trace metals observed in Bermuda, and the third section will consider several models of atmospheric trace metal input to the ocean near Bermuda and the relationship between these model calculations and possible sources for the metals.

Atmospheric Trace Metal Distribution Studies in Bermuda

Experimental

Sample Collection. The primary atmospheric sampling site for this study is on the island of Bermuda. Additional atmospheric samples have been and will be collected from the University of Rhode Island's research vessel *R/V Trident* over wide areas of the North Atlantic. While atmospheric sampling from a ship has some advantages, particularly for short-term preliminary studies and quick investigations of a particular geographical area, the problems of shipboard contamination (Moyers *et al.*, 1972) and the difficulty of collecting sufficient air samples to obtain any meaningful statistics at a given location precludes ship sampling for any detailed study of atmospheric transport. The most practical way to obtain reasonable values for atmospheric concentrations and fluxes of pollutants out over the ocean is to have a stationary sampling site in operation over a relatively extended period of time. Bermuda is an ideal location for such a site.

In 1972, a 20 meter high aluminum walk-up tower was constructed on the southwest coast of Bermuda at 32°14'52"N, 64°51'58"W. The base of the tower is located approximately 40 meters above the sea on a rather steep bluff. Tests made with smoke flares at the base and intermediate levels of the tower under moderate (15-20 Kts) onshore winds indicate that air blowing up the face of the bluff passes well beneath the top of the tower. Thus, under normal sampling conditions, local particles from surf breaking against the bluff and soil particles from the bluff itself will definitely not be collected at the top of the tower. The location of the tower site in Bermuda is shown in Figure 4-1. This site is ideal for atmospheric sampling when the surface winds are from the southeast through southwest to northwest, as local winds from these directions should cause no contamination from the island itself. The wind is from these directions over 70% of the time in all months except September–November. From December through March the wind is primarily westerly and from April to September it is primarily southerly.

Atmospheric collection equipment is located on the top level of the tower. Particulate samples for trace metal analysis were collected during 1973 and 1974 on double Whatman 41 filters (20 × 25 cm) mounted in all plastic filter holders in polyvinyl chloride rain shelters. During the summer of 1975, several cascade impactor samples were collected in Bermuda using a Sierra Hi Volume cascade impactor. The particles were deposited on Whatman 41 filter paper in each stage, and the fifth stage was followed by a final filtration through Whatman 41 filters. Air sampling pumps are separated from the filters and impactors by 8 meters of 7.6 cm diameter fiberglass tubing, the pumps being located at a lower level of the tower. This avoids contamination from the pumps themselves (Hoffman and Duce, 1971).

Extreme care must be taken during the collection and handling of atmospheric particulate samples in remote areas. The atmospheric concentrations of all the substances of interest are so low that even the slightest contamination can render a sample useless. Aside from the instrumental safeguards to minimize contamination, described below, personnel collecting the samples must be thoroughly familiar with precautions to be taken during sample collection and handling and must be aware of the many sources of contamination that can arise. The hands must be covered during any manipulation of these samples and

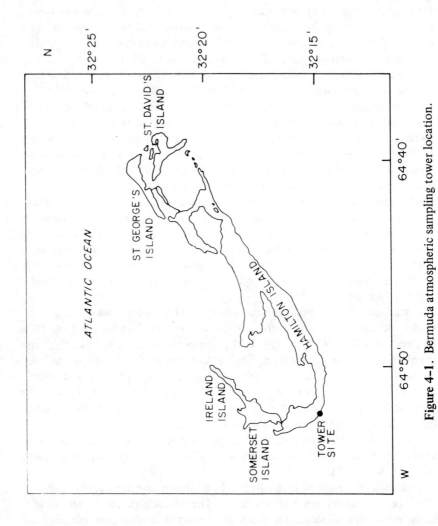

Figure 4-1. Bermuda atmospheric sampling tower location.

maximum use is made of a laminar flow clean station when changing filters, etc.

In any atmospheric collection program of this type it is critical that as much as possible be known about the past history of the air being sampled. The pumps used to collect the samples on the Bermuda tower are controlled by wind direction through a directional air sampler control using high amperage relays for use with standard high volume pumps. This directional control allows the pumps to be operated automatically when the wind is blowing from some pre-determined sector of the compass, thus avoiding the local contamination which can occur when the wind is off the land. All filter holders have a door which automatically closes over the filter when the pump is shut down due to out-of-sector winds. This prevents filter contamination from nonmarine air when the pumps are not operating.

An Environment/One Model Rich 100 recording condensation nucleus counter (Environment/One Corp., Schenectady, New York) is also located near the top of the tower. Since local pollution particles are excellent condensation nuclei (CN), short-term rapid variations in CN counts can indicate air mass changes and/or possible local contamination of samples. The CN counter is integrated into the directional air sampling control system so that collection oc-curs only when the wind is off the ocean *and* CN counts are near background for marine air, 300–400 per cm^3. This is an added safeguard against collection of local pollution. Recorders for wind speed and direction, CN counts, rainfall, and pump operating times are housed in a small air-conditioned building near the base of the tower. Filters are changed at intervals ranging from one to three days in a laminar flow clean station in the building.

Analytical Techniques. Details of the analytical techniques vary somewhat for the various sample matrices and experimental configurations described in this paper. The individual publications cited for the various studies can be consulted for these details. The basic procedure utilized for the analysis of atmospheric particulate matter for trace metals by atomic absorption spectrophotometry and neutron activation analysis will be presented here as an illustration of the tech-niques used.

Atomic Absorption Analysis. Whatman 41 filter samples are cut into quarters in a laminar flow clean station and one quarter is folded and inserted into a tef-lon beaker (25 ml capacity). The filter matrix is destroyed in a low temperature asher (LFE Model LTA-505). The ashing conditions are: O_2 flow 50 cc/min, RF power 50 watts. After ashing the residue is transferred to a precleaned 7 ml polyvial. This transfer is facilitated by adding 0.5 ml of ultra-pure concentrated HF to the teflon beaker and slowly picking up the residue into the 0.5 ml drop of HF as it is rolled around the inside of the teflon beaker. The HF will not wet the inside of the teflon beaker and can be quantitatively transferred into the polyvial. Then 0.5 ml of ultra-pure concentrated HNO_3 is added to the teflon beaker and the acid drop is again rolled around in the beaker and quantitatively transferred to the polyvial. The polyvial is capped and allowed to stand at room temperature for at least one week to insure dissolution of the particulate matter. The vial is opened after the dissolution period and 5 ml of distilled demineralized H_2O is added. The vial is recapped and allowed to stand for several days at room temperature. The final acid concentration is approximately 1.6 N in HNO_3 and 2.0 N in HF. The samples can then be analyzed directly for the elements of interest (Na, K, Ca, Mg, Al, Fe, Pb, Zn, Mn, Cu, Cr, and Cd) by flameless and flame atomic absorption using a Perkin-Elmer Model 503 unit with HGA-2100 flameless attachment.

Neutron Activation Analysis. One quarter of a double 20 X 25 cm Whatman 41 filter is pelletized in a stainless steel press with a nylon liner and titanium (99.5%) end caps at 4000 psi. The resulting pellet (3/4" diameter X 1/4" thick) is wrapped in aluminum foil and irradiated for seven hours per day for five days in the Rhode Island Nuclear Science Center Reactor at a thermal neutron flux of $4 \times 10^{12} \, n/cm^2/sec$. After a decay period of four weeks the aluminum foil is removed and the pellet is dissolved in 5 ml of water and counted without chemical separation two inches above a 23% Ge(Li) detector (FWHM resolution of 2.3 Kev for the 1332 Kev gamma ray of ^{60}Co) for a period of 40,000 seconds. The nuclides used for element identification were 5.24 year ^{60}Co, 27.0 day $^{233}Pa(Th)$, 71.3 day $^{58}Co(Ni)$, 45 day ^{59}Fe, 27.8 day ^{51}Cr, 60.9 day ^{124}Sb, 12.2 yr ^{152}Eu, 253 day ^{110m}Ag, 245 day ^{65}Zn, 83.9 day ^{46}Sc, 32.5 day ^{141}Ce, 121 day ^{75}Se, and 46.9 day ^{203}Hg. Occasionally it is possible to determine 18.66 day ^{86}Rb, 6.7 day ^{177}Lu, 44.6 day ^{181}Hf, and 2.07 yr ^{134}Cs. Standard solutions were spotted on Whatman 41 filter paper to create a matrix identical to and processed in the same manner as the samples.

For As analysis one quarter of the 20 X 25 cm Whatman 41 filter is pelletized, wrapped in dilute acid washed aluminum foil and irradiated with standards under Cd for sixteen hours at an epithermal flux of 1×10^{11} neutrons/cm^2/sec. Twenty-four hours after irradiation, the samples are unwrapped, placed in a beaker with 15 ml of distilled demineralized water and 2 ml of 10% H_2SO_4 in concentrated HNO_3, and the solution is slowly heated to about 45°C until the sample is completely dissolved. One milliliter of 5 mg/ml As carrier is added to the solution, along with 35 milliliters of concentrated HCl. Then 100 milligrams of thioacetamide are added and the solution is heated at 45°C for 20–25 minutes. After cooling, the precipitated As_2S_3 and sample particulates are filtered through a 0.4 μm pore size, 47 mm Nuclepore filter and washed thoroughly with 6 N HCl and acetone. After counting for thirty minutes to obtain the ^{76}As ($t_{1/2}$ = 26.4 hr) activity, the precipitate is stored for reirradiation to obtain an experimental yield. These yields are generally better than 85%.

Trace Metal Concentrations

Atmospheric particulate samples were collected on double Whatman 41 filters during April to December 1973 and April to October 1974 at the Bermuda tower site. On the basis of local meteorological conditions, CN counts, and inside vs. outside sector sampling times, 60 samples in 1973 and 74 samples in 1974 were selected for chemical analysis as being most representative of locally uncontaminated marine air at Bermuda. All of these samples were analyzed for Na, Ca, Mg, K, Fe, Al, Pb, Zn, Mn, Cu, Cr, and Cd by atomic absorption. For the 1973 samples a subset of these 60 have been analyzed by neutron activation analysis for Hg, Ce, Se, Ni, As, Co, Sb, Sc, Th, Ag, and Eu. Neutron activation analysis of a subset of the 74 samples from 1974 is still underway. NAA for As in 40 samples from 1974 and 20 samples from 1973 has been completed. Because of possible volatility losses of Hg during neutron activation, the Hg concentrations should be considered lower limits.

   The geometric mean atmospheric trace metal concentrations and the concentration ranges obtained to date for the 1973 and 1974 sampling periods in Bermuda are presented in Table 4-1. The agreement of the 1973 and 1974 mean concentration is generally quite good. Note that the concentration ranges for each year span two to three orders of magnitude. This illustrates quite dramatically the necessity of collecting a large number of samples over an extended period of time to obtain any meaningful information on the "climatological mean" concentration of any of these metals.

Enrichment Factors

There are a number of natural as well as anthropogenic sources for trace metals in the marine atmosphere. As described previously, much of the work in this investigation is directed toward determining which sources are important for which metals.

   Obvious natural sources for atmospheric trace metals include the ocean and weathering of the earth's crust. Sodium is used as a reference element to estimate the contribution of the sea to any metal component of the marine aerosol. An enrichment factor, EF_{sea} for an element X in the atmospheric particles relative to bulk seawater, can be calculated as follows:

$$EF_{sea} = \frac{(X/Na)_{air}}{(X/Na)_{sea}} \tag{4.1}$$

Table 4–1

Geometric Mean Atmospheric Trace Metal Concentrations for Bermuda

Element	1973			1974		
	No. of Samples	Mean[a] (ng/SCM[b])	Range (ng/SCM[b])	No. of Samples	Mean[a] (ng/SCMb)	Range (ng/SCM[b])
Na	60	1500	210-7800	74	2200	440-13000
Mg	60	220	30-850	74	270	60-1500
Al	60	140	3-2700	74	120	7-2200
Ca	59	140	6-1100	74	160	2-740
K	60	130	17-970	74	140	26-540
Fe	60	94	4-1900	74	100	16-1100
Pb	60	3.5	0.10-32	74	3.3	0.20-71
Zn	60	3.2	0.2-32	74	1.9	0.1-32
Mn	60	1.2	0.03-27	74	1.8	0.17-20
Cu	60	0.90	≤0.08-24	74	1.2	0.07-15
Hg	29	>0.49	≤0.02-5.0	–	–	–
Cr	60	0.28	≤0.04-3.3	–	–	–
Ce	29	0.21	0.005-3.4	–	–	–
Cd	60	0.19	≤0.01-2.5	73	0.13	0.005-3.2
Se	29	0.13	≤0.02-0.62	–	–	–
Ni	29	0.080	≤0.02-1.5	–	–	–
As	20	0.072	0.012-0.46	40	0.093	0.010-0.96
Co	29	0.042	≤0.005-0.50	–	–	–
Sb	29	0.030	≤0.001-0.30	–	–	–
Sc	29	0.022	0.002-0.40	–	–	–
Th	29	0.020	0.002-0.21	–	–	–
Ag	29	0.0034	≤0.002-0.08	–	–	–
Eu	29	0.0031	≤0.0002-0.05	–	–	–

[a]Means calculated using ≤ values as actual concentrations.

[b]SCM = Standard Cubic Meter.

where $(X/Na)_{air}$ and $(X/Na)_{sea}$ refer, respectively, to the ratio of the concentration of X to that of Na in the atmosphere and in bulk seawater; EF_{sea} values near unity for any element X would suggest the sea as a likely source for that element in the atmosphere. Calculation of EF_{sea} values for samples collected from ships from the North Atlantic indicate that atmospheric trace metals, aside from Na, which clearly originate predominantly from bulk seawater are K, Mg, and Ca (E. Hoffman *et al.,* 1974).

In an attempt to determine the contribution of crustal weathering to any metal component of marine aerosols, Al, which comprises over 8% of the average crustal material, is used as a reference element. An enrichment factor, EF_{crust}, for any element X in the atmospheric particles relative to the crust, can be calculated as follows:

$$EF_{crust} = \frac{(X/Al)_{air}}{(X/Al)_{crust}} \qquad (4.2)$$

where $(X/Al)_{air}$ and $(X/Al)_{crust}$ refer, respectively, to the ratio of the concentration of X to that of Al in the atmosphere and in the average crustal material; EF_{crust} values near unity for any element X suggest that crustal material is the probable source for that element in the atmosphere in remote areas. However, this approach will not distinguish between crustal material injected into the atmosphere by natural processes and that injected as a result of man's activities, for example, increased exposure of soil surfaces due to agriculture and other land-clearing operations; crushed stone, sand, and gravel operations; and soil particles suspended by vehicle traffic. Coal and even fly ash also have a relative concentration of many trace metals similar to that of average crustal material (von Lehmden et al., 1974; Bertine and Goldberg, 1971). In addition, mean crustal ratios can only be used as a crude approximation of the relative composition of crustal material areosols, owing to differing types of crustal material and soils in various source areas and uncertainties concerning fractionation during weathering processes (Hoffman et al., 1969). Thus EF_{crust} variations from unity up to an order of magnitude may still indicate a crustal material source for the elements. It is clear that enrichment factors must be used with caution in attempting to ascertain atmospheric trace metal sources.

Geometric mean values for EF_{crust}, calculated from the individual EF_{crust} values for each sample, for the 1973 and 1974 Bermuda samples are presented in Table 4-2. It is clear that the first ten elements (Mn, Fe, Al, Sc, Ni, Cr, Co, Th, Eu, and Ce) all have EF_{crust} values not far from unity, suggesting strongly that they originate from simple crustal weathering. The other elements (Cu, Zn, As, Ag, Hg, Pb, Sb, Cd, and Se), however, show varying degrees of enrichment relative to the crust, suggesting that some other source or sources may be primarily responsible for their presence in the marine atmosphere. It is interesting to note that the EF_{crust} values for Pb in Bermuda are quite similar to those found by Murozumi et al. (1969) in some polar snows which fell in the nineteenth and twentieth centuries in Greenland.

It should also be noted that the range of EF_{crust} values for any particular element was much less than the range of concentration values observed for that element over the extended sampling periods reported here.

In summary, the ocean is apparently the major source for Na, Mg, Ca, and K in marine aerosols, and crustal weathering is apparently the major source for Al, Mn, Fe, Sc, Ni, Cr, Co, Th, Eu, and Ce. The elements Cu, Zn, As, Ag, Hg,

Table 4-2
Geometric Mean Atmospheric Trace Metal Enrichment Factors for Bermuda

Element	1973 EF_{crust}	1974 EF_{crust}
Mn	0.73	2.3
Fe	0.97	1.2
Al	1.0	1.0
Sc	1.1	–
Ni	1.1	–
Cr	1.7	–
Co	1.8	–
Th	2.1	–
Eu	2.5	–
Ce	3.6	–
Cu	9.6	12
Zn	26	18
As	22	36
Ag	52	–
Hg	>65	–
Pb	170	180
Sb	180	–
Cd	570	400
Se	2600	–

Pb, Sb, Cd, and Se, however, are anomalously enriched in the marine atmosphere relative to either bulk seawater or average crustal material. It is thus these anomalously enriched trace metals which are of most interest relative to possible anthropogenic effects and geochemical processes.

Atmospheric Trace Metal Source Identification Studies

In what ways can we attempt to identify possible sources other than bulk seawater and crustal weathering for these trace metals? Several approaches have been followed, and they are described below.

Northern-Southern Hemisphere Comparison

Duce *et al.* (1975b) compared EF_{crust} values for atmospheric particulate samples collected in the North Atlantic westerlies (from Bermuda and *R/V Trident*) with samples collected at the geographic South Pole (Zoller *et al.*, 1974).

A graphical representation of this comparison is presented in Figure 4-2. Over the North Atlantic the elements Zn, Cu, Cd, Sb, Pb, and Se, as well as V, are apparently present in the atmospheric particulate matter in concentrations too high to be explained in terms of normal crustal weathering processes. It is apparent, within the statistical variations presented, that the enrichment factors at the South Pole are in good agreement with those from the North Atlantic with the exception of V. The significantly higher mean V enrichment factor found for samples collected in the North Atlantic westerlies (which is not apparent in samples collected in the North Atlantic northeasterly trades, where EF_{crust} is 1.4 (Duce et al., 1974)) is almost certainly due to the transport of pollution aerosols from the east coast of North America across the Atlantic on the prevailing westerlies. The source for V in these pollution aerosols is very likely the combustion of heavy fuel oils containing V-porphyrin complexes (Zoller et al., 1973). Details of V over the North Atlantic and its flux to the ocean are presented elsewhere (Duce and Hoffman, 1976).

The general agreement between the EF_{crust} values for samples collected at the South Pole and in the North Atlantic westerlies is potentially significant relative to the source of these trace metals. With respect to the highly enriched elements at the South Pole, Zoller et al. (1974) have pointed out that these are also the more volatile elements and have suggested that the relatively high volatility of these elements and many of their compounds might well be associated with their high enrichment factors. Zoller et al. (1974) have also suggested that high-temperature processes, either natural or anthropogenic, could be responsible for the enrichment of these metals observed in Antarctica. Comparison of the enrichment factors in Figure 4-2 suggests that long-range transport of non-crustal anthropogenic pollution may not be the cause of the high enrichment factors observed in the two locations. It has been estimated that approximately 90% of the particulate pollutants in the global troposphere are injected in the northern hemisphere (Robinson and Robbins 1971). Interhemispheric mixing times in the troposphere are relatively long, on the order of six to twelve months (Newell et al., 1974) and the atmospheric residence time for particles in the troposphere is generally less than a week but certainly no longer than three to four weeks (Poet et al., 1972). Thus, there is not sufficient time for pollution source trace metals to be transported from their primary source area in the northern hemisphere to the southern hemisphere to give similar EF_{crust} values in both areas before the particles are removed from the atmosphere. This suggests that the anomalously high atmospheric concentrations of some of these trace metals may originate primarily from some natural source. Of course, alternative explanations are also possible. For example, the smaller land area in the southern hemisphere probably results in a lower natural production of atmospheric crustal material. This, combined with the lower pollution production in the southern hemisphere, could possibly account for the similarity in the ratios of pollution elements to Al in both hemispheres.

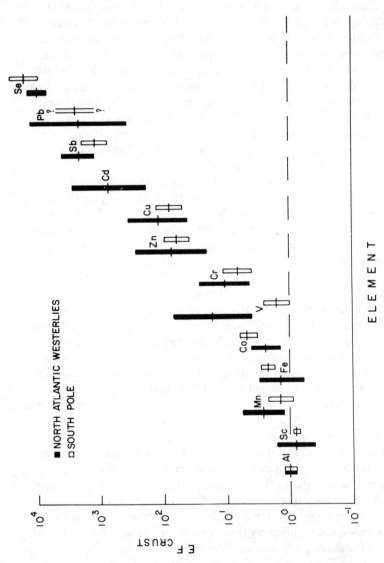

Figure 4-2. EF crust values for atmospheric trace metals collected in the North Atlantic westerlies from *R/V Trident* and Bermuda, and at the South Pole. The horizontal bars represent geometric mean enrichment factors, and vertical bars represent the geometric standard deviation (after Duce *et al.*, 1975).

The suggestions above may not apply to Pb, and the close agreement of the Pb enrichment values at the northern and southern hemisphere locations may be coincidental. In the calculation of the mean EF_{crust} value for Pb at the South Pole the lowest atmospheric Pb concentration observed was used because of concern that other samples may have been contaminated with this element. The more recent 1973-1974 measurements, made with the improved collection techniques at the Bermuda tower described above, suggest that background atmospheric Pb concentrations over North Atlantic areas may be significantly lower than those previously measured from *Trident* by our group and by others, possibly as a result of contamination during sample collection. Indeed, the Pb EF_{crust} values for the 1973 and 1974 Bermuda samples appear significantly lower than measured previously, either from *Trident* or at the South Pole. Improved collection techniques similar to those in use at the Bermuda tower have recently been instituted at the South Pole and a new comparison should be possible soon. The most valuable information relative to northern-southern hemisphere distributions, however, will come from a comparison of the extended series of marine samples from Bermuda and our recently completed tower station in American Samoa.

Air Parcel Trajectory Analysis

During the 1974 sampling program at Bermuda, the NOAA Geophysical Monitoring for Climatic Change (GMCC) group at the Air Resources Laboratories (ARL), Silver Spring, Maryland, provided us with six hourly near-surface (300 to 1200 meter) air parcel trajectory analyses. The trajectories were computed using the Regional-Continental Scale Trajectory Program developed at ARL. Five to ten day backward trajectories were computed. Details of this program are given by Heffter *et al.* (1975).

The air parcel trajectories obtained from this program were compared to the sampling intervals for the 74 Bermuda samples in 1974. The samples were separated into three groups according to the direction from which the air mass came to Bermuda. The three directional sectors used to define the sample groups are presented in Figure 4-3. New geometric mean concentrations and crustal enrichment factors were calculated for each of the groups to determine whether there was any significant difference in the distribution of these samples as a function of air parcel trajectory. The results are presented in Figure 4-4. The metals Pb and Zn appear to have significantly higher concentrations and definitely have higher enrichment factors when the air parcel sampled had recently passed over the North American continent. This suggests that anthropogenic sources are potentially significant for those metals. The obviously crustal elements, (e.g., Al, Fe, and Mn) do not show any significant variation, suggesting that the North American continent is not a particularly strong source for direct transport of normal weathering products to this region of the Atlantic.

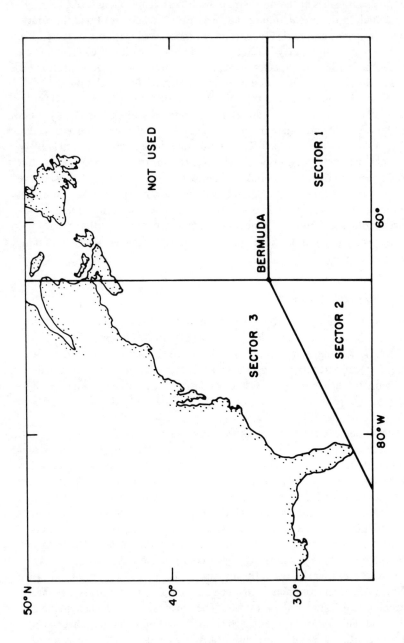

Figure 4-3.   Air parcel trajectory sectors utilized with the 1974 Bermuda trace metal data.

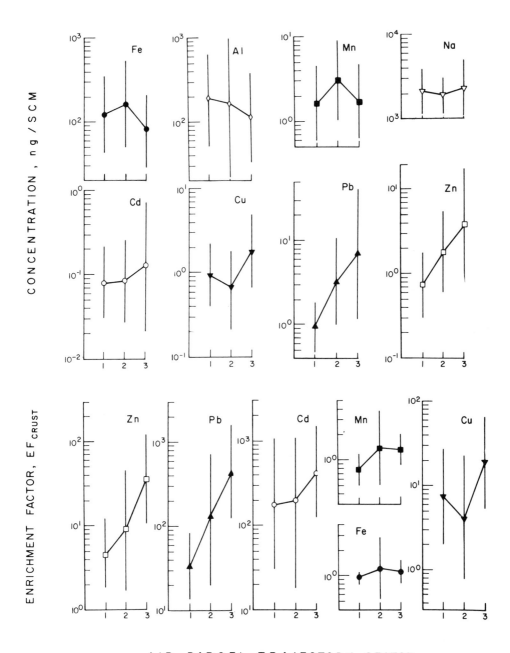

AIR PARCEL TRAJECTORY SECTOR

Figure 4-4. Trace metal concentrations and enrichment factors as a function of air parcel trajectory sector.

Particle Size Distribution and Volatility

Knowledge of the distributions of trace metals as a function of particle size can often give valuable information on the source of the various trace metals (Rahn, 1971). For example, particles produced from crustal weathering are generally larger than 1 μm in radius, and the major mass of atmospheric sea salt particles is also generally in particles with radii greater than 1 μm. However, natural or anthropogenic particles produced from gas phase reactions or gas-phase condensation and coagulation, ultimately attain radii of only a few tenths of a micrometer in a few days (Junge, 1963).

Results of the analysis of two high-volume cascade impactor samples from Bermuda are presented in Figure 4-5 for the metals Al, Fe, Na, Mn, Cr, Cd, Zn, Cu, Sb, and Pb. The cascade impactor separates the atmospheric particle into six size fractions which are deposited on Whatman 41 filters. The equivalent aerodynamic radius cutoffs at 50% collection efficiency for particles with a density of 1 g/cc at the flow rate used are as follows: Stage 1 = 3.6 μm; Stage 2 = 1.5 μm; Stage 3 = 0.75 μm; Stage 4 = 0.48 μm; Stage 5 = 0.25 μm; final filter = <0.25 μm. It is apparent that the major mass of Fe, Al, Na, Mn, and Cr is found on particles of radius greater than 1 μm, consistent with a crustal weathering or seawater source for these elements. The elements Cd, Zn, Pb, Sb, and Cu have an apparent secondary concentration maximum in the same size as the other group. However, since the cascade impactor was located inside a shelter on the top of the tower, it is very likely that the collection system discriminated against the largest particles. It does appear, however, that the major mass of Cd, Zn, Pb, Sb, and Cu is on particles with radii less than 1.0 μm.

Figure 4-6 presents the EF$_{crust}$ values as a function of particle size for these two impactor samples. Again, the separation of elements into two groups is clearly evident — those with EF$_{crust}$ near unity show little variation with particle size, while the anomalously enriched elements have much higher values for EF$_{crust}$ on the small particles. These results strongly support the concept of a vapor phase or at least initially very small particles playing a significant role in the atmospheric cycle of these anomalous metals. Measurements of these metals in urban air (Gladney *et al.*, 1974; Lee *et al.,* 1972) show similar size distributions. Analyses of these impactor samples for additional elements is presently underway.

It is interesting to note the order of volatility of the trace metals measured in Bermuda air. The volatility, as measured by the elemental boiling point, decreases in the order Hg > As > Se > Cd > Zn > Eu > Pb > Sb > Ag > Mn > Al > Cu > Cr > Fe > Sc > Co > Ni > Ce > Th. A similar order of volatility is found for many inorganic compounds of these elements (Bertine and Goldberg, 1971). Note that eight of the nine anomalously enriched elements in Bermuda (Table 4-2) are included in the nine most volatile elements. Both anthropogenic sources (such as fossil fuel combustion and other high temperature industrial

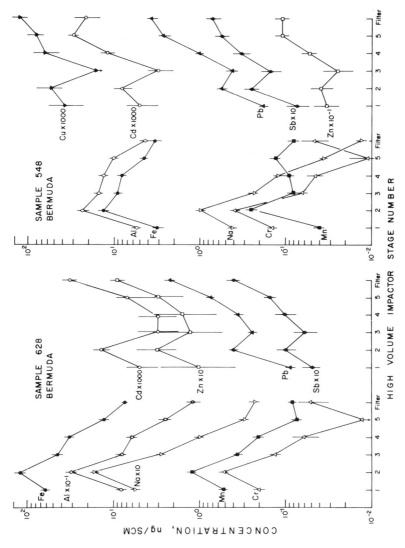

Figure 4-5. Trace metal concentration as a function of particle size in Bermuda marine air.

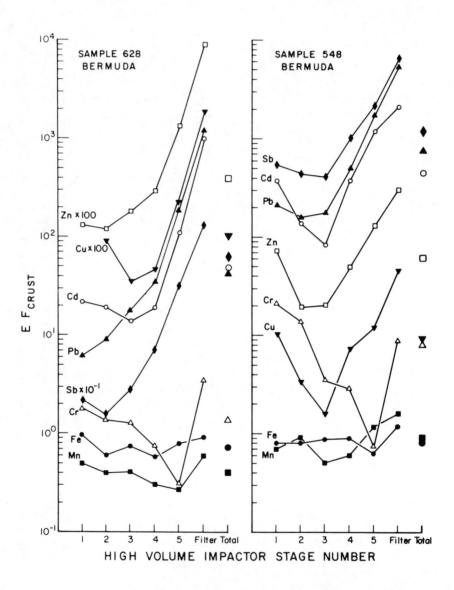

Figure 4-6. EF$_{crust}$ values as a function of particle size in Bermuda marine air.

processes) and natural processes (such as volcanism and biological methylation can result in at least transient volatile species for many of these trace
metals.

Volcanoes

As mentioned previously, one possible natural high-temperature source for these
enriched trace metals is volcanism. Mroz and Zoller (1975) have discussed this
possible source in some detail and have measured EF_{crust} values in atmospheric
particles collected a few meters from active volcanic activity during an eruption
on Heimaey, Iceland. They find significant enrichment of the elements Se, Sb,
and Zn. We have analyzed an atmospheric sample collected by Dr. Bruce Finlayson and Dr. John Naughton of the Department of Chemistry at the University of
Hawaii only 1-2 meters from an active fumarole on Kilauea Volcano, Hawaii.
The values for EF_{crust} observed in this sample are presented in Table 4-3. If
this table is compared with Table 4-2, it is apparent that the same general order
of enrichment factors is observed. An exception is Sb, which we did not find enriched at Kilauea. Mroz and Zoller (1975) also found lower EF_{crust} values than
expected for this element in the Iceland volcano samples. The general similarity
in enrichment factors does not, of course, prove that volcanism is the source for
any of the anomalously enriched atmospheric trace metals, as the source
strength for the volcanic metals remains unknown. It does, however, suggest
that this source may be potentially significant. The question of the relative importance of volcanism to the global atmospheric chemistry of these metals is
being pursued in our laboratory.

Sea to Air Transport

Overview. The ocean is probably the major natural source of atmospheric particulate matter. However, we have seen that the trace metal composition of atmospheric particles over the North Atlantic is generally not similar to sea water and
we have evaluated other possible sources for these trace metals. It has been suggested that part of the relative difference in chemical composition of the atmospheric particles and seawater may be due to chemical and physical processes
occurring during the passage of bubbles through the water and the production of
atmospheric sea salt particles when these bubbles burst at the air/sea interface
(Duce et al., 1972; Piotrowicz et al., 1972; Van Grieken et al., 1974; Peirson et
al., 1974; Wallace and Duce, 1975).

It is generally believed that most of the sea salt particles with atmospheric
residence times longer than a few minutes are produced by bubbles breaking at
the sea surface. Blanchard and Woodcock (1957) investigated the various

Table 4-3

EF_{crust} **Values for Kilauea Fume Sample**

Element	EF_{crust}
Sb	<0.4 (?)
Th	0.7
Al	1.0
Sc	1.5
Mn	2.5
Fe	2.9
Co	4.0
Eu	3.8
Ni	23
Cr	27
Pb	100
Cu	100
Zn	140
Cd	320
Ag	360
Se	56,000
Hg	310,000

mechanisms for bubble production in the ocean, e.g., breaking waves, and rain drops and snowflakes striking the water surface. They suggested that, except under local conditions, breaking waves, or whitecaps, are by far the most import-ant source for these bubbles. Boyce (1951) showed that relatively few salt particles were produced by the mechanical disintegration of the water in a break-ing wave, but that a considerably greater number of particles were produced a few seconds later when the air bubbles resulting from the wave action burst at the sea surface. Using high-speed photography, Kientzler *et al.* (1954) showed that a bubble at a seawater surface forms a jet which ejects two to five droplets into the air. Blanchard (1963) found that the diameter of these jet droplets was approximately 10% of the diameter of the bubbles from which they were formed. Mason (1957) and Blanchard (1963) found that a significant but vari-able number of smaller droplets, called film droplets, were also produced by the shattering bubble film cap.

When bubbles break at the surface of the ocean they skim off a very thin layer of the air/sea interface to produce the atmospheric film and jet droplets. MacIntyre (1972) calculated that the material present in the first jet droplet produced by a bursting bubble was originally spread over the interior of the bubble surface at a thickness equal to approximately 0.05% of the bubble diameter. Thus, the top or first jet droplet from a 100 μm diameter bubble is composed of material from a surface layer only 0.05 μm thick. The second jet droplet is produced from the next 0.05 μm layer, etc. With a bubble size distribution in the sea ranging from approximately 50 μm to perhaps 1500 μm diameter, the top jet droplets produced from these bubbles strip off approximately the top 0.025 to 0.75 μm of the air/water interface. The chemical composition of this extremely thin stripped surface layer is determined both by the composition of the ocean surface layer before the bubble arrives at the surface (the surface microlayer) and the composition of the bubble itself.

It is interesting to note that when the bubble cap or film shatters to produce film droplets, it is probably no thinner than 2 μm, since these bubble caps show no visible interference patterns which become apparent for thinner films (MacIntyre, 1972). As MacIntyre (1974) points out, this suggests that the jet droplets, which are generally larger than the film droplets, may be "sampling," or may be composed of, a much thinner layer of the water surface than the smaller film droplets. It should also be noted that the smaller jet droplets are apparently composed of material from a thinner layer of the surface than the larger jet droplets. Unfortunately, while we now have a rather good qualitative picture of these processes occurring in the upper few hundred micrometers of the ocean surface, the detailed hydrodynamics remain largely unknown. MacIntyre (1974) presents an excellent review of our present understanding of these processes. We have undertaken several types of studies to ascertain the potential importance of sea surface phenomena to the atmospheric distribution of trace metals.

Surface Microlayer Composition. Since the atmospheric particles are produced from such a thin layer of the ocean surface, determination of the chemical composition of this surface microlayer is critical. There is already considerable evidence that the trace metal content of the sea surface microlayer is considerably different from that of subsurface water at 20–40 cm depths (Duce *et al.*, 1972; Piotrowicz *et al.*, 1972; Barker and Zeitlin, 1972; Szekielda *et al.*, 1972; and G. Hoffman *et al.*, 1974).

In our studies open ocean surface microlayer samples were collected by the screen technique of Garrett (1965) using a 75 × 75 cm polyethylene screen (20 mesh) mounted in a Plexiglas frame. After the screen had been rinsed in seawater several times, it was submerged in the water and then passed back up through the water surface with the screen approximately parallel to the water surface. The frame was allowed to drain for 10 seconds, and then the screen was drained for 60 seconds into a polyethylene container through a polyethylene funnel. This collector samples the top 150–300 μm of the water surface. Polyethylene gloves were worn throughout the collection, and great care was taken to ensure that the sample was not contaminated by the collection boat (a rubber raft) or the individuals doing the collecting.

Subsurface samples were collected from a depth of 20-40 μm by submerging a polyethylene bottle below the surface and then removing the cap. Polyethylene gloves were worn throughout the collection. After collection, the samples were filtered in a laminar flow clean bench in the ship's laboratory using 0.4 μm, 47 mm diameter Nuclepore filters.

All particulate heavy metals investigated in the surface microlayer to date (lead, aluminum, iron, vanadium, copper, cadmium, chromium, and manganese) are significantly enriched compared to water approximately 20-40 cm below the surface. Typical surface microlayer and subsurface concentrations of these particulate trace metals found on a cruise of *R/V Trident* from Dakar to Bermuda are shown in Figure 4-7. It appears that only in regions with extremely high atmospheric particle loadings, e.g., the Sahara dust plume in the·northeast trades off West Africa, does the direct deposition of atmospheric particles affect the surface microlayer particulate enrichment for most trace metals (G. Hoffman *et al.,* 1974). Other areas, e.g., between Bermuda and the Azores, have rather similar enrichments in the surface microlayer, but atmospheric particle loadings are several orders of magnitude less. It appears that the enrichment of these particulate trace metals found in the surface microlayer are probably derived from transport to the surface from subsurface water, most likely by bubble flotation (see next section), at least in areas of the ocean which are not dominated by large fluxes of atmospheric continental material. G. Hoffman *et al.* (1974) have calculated that the lifetime of directly deposited atmospheric dust in the oceanic surface microlayer is relatively short (i.e., a few seconds at most) under winds of five to ten meters per second. Not only are these higher concentrations of trace metals potentially important to marine neuston communities, but it would be expected that enrichment of trace metals in the surface microlayer should result in their enrichment in atmospheric sea salt particles generated when bubbles break through this microlayer (Piotrowicz *et al.,* 1972; E. Hoffman and Duce, 1974).

Laboratory Bubble Flotation Studies.   In a second attempt to ascertain the importance of the transport of trace metals from the ocean to the atmosphere, laboratory studies were undertaken to investigate the potential significance of bibble transport of particulate trace metals to the air/sea interface. Surface samples of seawater were collected over a six-month period in the West Passage of Narragansett Bay, Rhode Island, a relatively unpolluted estuary. Collection was made using a plastic bucket and line from a drifting boat such that contamination from the boat was highly improbable. The samples were passed through a 300 μm nylon mesh net to remove the larger zooplankton and organic debris and collected in thoroughly cleaned acid-washed polyethylene carboys. Salinities at the stations sampled have been observed to range from 25 to 30 $^\text{O}$/oo (Hicks, 1959; Smayda, 1973).

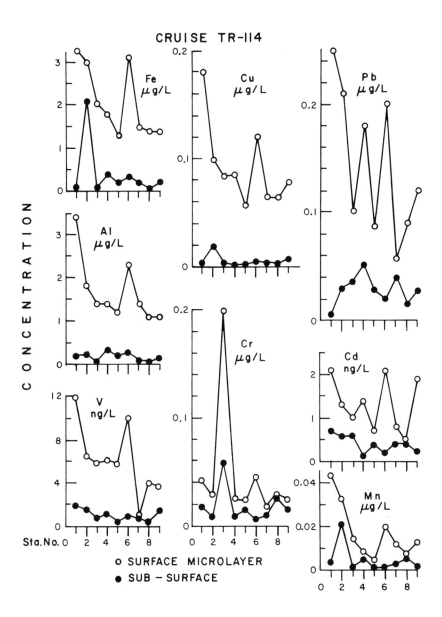

Figure 4-7. Oceanic particulate trace metal concentration between Dakar (left) and Bermuda (right). Data obtained from *R/V Trident* Cruise TR-114. Open circles represent surface microlayer concentrations; filled circles represent subsurface (20 to 40 cm) concentrations.

The flotation procedure has been described elsewhere in detail (Wallace and Wilson, 1969; Wallace et al., 1972). Briefly, N_2 bubbles of approximately 1 mm diameter were allowed to rise through the sample contained in an all-glass column 70 mm outside diameter and 2.4 m high. Any froth or foam which accumulated at the surface was drawn off and its contents analyzed for the substance of interest. Typically, 30-70 ml of collapsed froth were collected from the column which contained approximately 7 liters of the original sample. Collection of the froth was accomplished within 15 minutes of the initiation of bubbling. The original sample and the bubble stripped residue were also analysed, thus allowing a budget to be calculated for the flotation process.

Samples for particulate organic carbon (POC) analysis were collected in duplicate and analyzed as described elsewhere (Wallace et al., 1972). Particulate trace metal (PTM) samples were collected in duplicate on 47 mm 0.4 μm Nuclepore filters. After filtration each filter was thoroughly washed with three 1-3 ml portions of deionized H_2O to rid the filter of residual sea salt. After being washed and dried at 60°C for 15 minutes, the filters were stored frozen in clean plastic petri dishes until analyzed. Additional details of the experimental and analytical techniques are presented by Wallace and Duce (1975).

Recoveries of POC in the foam ranged from 30-59% while those of Al, Mn, Fe, V, Cu, Zn, Ni, Pb, Cr, and Cd were generally greater than 50%. Wallace and Duce (1975) extrapolated these laboratory results to obtain a crude order of magnitude estimate of the bubble transport of PTM to the ocean surface under open ocean conditions. Details of the assumptions made in calculating these estimated fluxes are presented by Wallace and Duce (1975), and include consideration of bubble surface area available for adsorption of particulates, bubble size, and open ocean POC and PTM concentrations. Results of this flux calculation are presented in Table 4-4. Wallace and Duce (1975) suggested that association of the POC and PTM may be the primary mechanism by which the sea surface microlayer is enriched in particulate trace metals. Thus bubble flotation in seawater appears to be an effective means of transporting and concentrating particulate trace metals at the air/sea interface. Enrichment of these substances on the atmospheric particles produced when these bubbles burst would seem likely.

Other laboratory studies have investigated the transfer of trace metals and organic carbon into the atmosphere by bursting bubbles. Van Grieken et al. (1974) added carrier free quantities of inorganic [65] Zn, [75] Se, and [22] Na to unfiltered coastal seawater and collected the particles produced by bursting bubbles of air generated below the water surface. Both [75] Se and [65] Zn showed enrichment in the bulk aerosol relative to the seawater, with [65] Zn generally having the greater enrichment.

E. Hoffman and Duce (1976) investigated factors influencing the organic carbon content of atmospheric sea salt particles generated in the laboratory. They found that the organic matter on the atmospheric particles was highly

Table 4-4

Transport of Particulate Trace Metals to the Sea Surface Via Bubbles

Particulate Trace Metal	Calculated Bubble Transport to the Air/Sea Interface (10^{-15} g/cm^2 sec)
Al	1.3
Mn	0.21
Fe	26
V	0.26
Cu	0.88
Zn	12
Ni	0.16
Pb	0.44
Cr	0.57
Cd	0.023

enriched relative to the bulk seawater from which it was generated. The enrichment increased with increasing bubble path length before bursting, pointing out the importance of bubble scavenging and transport of this material. (Similar results were observed by Blanchard and Syzdek (1974) for fresh water bacteria.) The organic carbon content of the aerosols in the model study of Hoffman and Duce (1976) apparently originated from the dissolved or colloidal fraction of the organic matter in seawater.

Bubble Interfacial Microlayer Sampler (BIMS). A third line of research directed toward increasing our understanding of the transport of trace substances from the ocean to the atmosphere involves the Bubble Interfacial Microlayer Sampler (BIMS). With this instrument we hope to obtain direct evidence of the importance or nonimportance of enrichment of trace metals on atmospheric particles generated at the air/sea interface. The BIMS, shown in Figure 4-8, is suspended between the twin hulls of a 4 meter long catamaran. It produces bubbles approximately 1000 μm in diameter by forcing compressed nitrogen at a flow rate of approximately 7 1/min through seven glass frits, 120 mm in diameter, at adjustable depths down to 50 cm beneath the sea surface. With modification, this depth can be extended to one meter. These bubbles rise and burst at the sea surface creating jet and film drops in the atmosphere enclosed within the truncated pyramid of the BIMS. The ambient marine atmosphere is excluded from the interior of the BIMS by clean air curtains in the front and back, wind screens on either side, and a slight positive pressure inside. The artificially produced sea salt particles are collected from the enclosed atmosphere at the top of the truncated pyramid on 20 × 25 cm Whatman 41 filters, using a commercial high volume air sampling pump. The exhaust from the pump is filtered through a

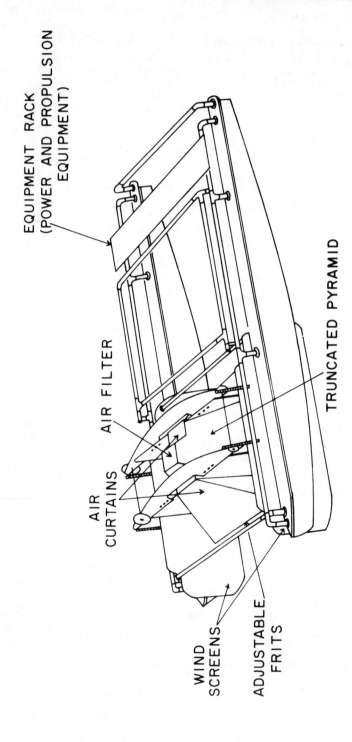

Figure 4–8. The Bubble Interfacial Microlayer Sampler (BIMS) suspended between the twin hulls of a 4-meter long catamaran.

Delbag Microsorban® filter, which offers little resistance to the air flow but maintains a high particle removal efficiency. This air is then recycled back into the BIMS as the air supply for the air curtains (Fasching *et al.,* 1974).

A series of BIMS samples has been collected near the mouth of Narragansett Bay. The BIMS generated atmospheric sea salt particles and samples of Narragansett Bay seawater were analyzed for Cu, Zn, and Fe. A summary of the results in presented in Table 4-5. All three elements were significantly enriched on the BIMS generated sea salt particles. We can use this data to make a crude estimate of the annual flux of these metals from the ocean to the atmosphere. An example of the calculation for Fe is given below.

From Table 4-5 we will assume the mean value for EF_{sea} for Fe of 150 applies to open ocean as well as to coastal locations. We will assume an average total Fe concentration of 2.0 μg/l (Brewer, 1975) and a Na concentration of 10.6 g/l in these open ocean regions. Thus, the calculated Fe/Na ratio on atmospheric sea salt particles generated on the open sea is

$$\frac{2.0 \times 10^{-6} \text{g Fe/l}}{10.6 \text{ g Na/l}} \times 150 = 2.8 \times 10^{-5} \text{g Fe/g Na} \tag{4.3}$$

Eriksson (1959) has calculated that the global flux of sea salt particles with radii less than 20 μm from the sea to the atmosphere is 1×10^{15} g/yr, resulting in a Na flux of 3.6×10^{14} g/yr. Thus the total Fe flux from sea to air would be 1×10^{10} g/yr. Similar calculations for Cu and Zn using seawater concentrations of 0.5 and 5 μg/l respectively (Brewer, 1975) yielded the fluxes reported in Table 4-5. Also presented in Table 4-5 are the estimated crustal weathering and anthropogenic fluxes for these elements to the atmosphere. For the weathering calculation we assume a total production of atmospheric particles due to weathering of 2×10^{14} g/yr (Robinson and Robbins, 1971) and use the crustal

Table 4-5
Preliminary BIMS Enrichment Factors and Fluxes

Element	BIMS EF_{sea}	Estimated Fluxes to the Atmosphere		
		Marine (10^{10} g/yr)	Crustal Weathering[a] (10^{10} g/yr)	Anthropogenic (10^{10} g/yr)
Cu	200	0.33	1.1	1.6[b]
Fe	150	1.0	1100	800[b]
Zn	200	3.3	1.4	4.6[c]

[a]Using crustal abundances of Taylor (1964).

[b]NAS. (1975).

[c]See text.

abundances of Taylor (1964). The global atmospheric anthropogenic Fe and Cu production were taken from NAS (1975). The total anthropogenic input of CU and Fe to the atmosphere was 7.6 and 5.7 times, respectively, that due to fossil fuel (NAS, 1975; Bertine and Goldberg, 1971). The fossil fuel mobilization of Zn to the atmosphere reported by Bertine and Goldberg (0.7 × 10^{10} g/yr) was then multiplied by 6.5, assuming that the ratio of fossil fuel Zn to total anthropogenic Zn was approximately the same as this ratio for Cu and Fe

The fluxes to the atmosphere from the three sources, crustal weathering, man, and fractionated sea salt, are of the same order of magnitude for Cu and Zn, suggesting that the sea cannot be ruled out as a potentially significant source for these elements in the marine atmosphere. Iron, which is apparently significantly enriched on sea salt particles, is still largely from weathering and man's activities. Thus, in most areas the Fe enrichment in marine atmospheric particles produced at the air/sea interface is swamped by atmospheric Fe from other sources. The observed EF_{crust} value of approximately 1 for Fe supports this concept. The potential importance of the sea as a source for other trace metals in the atmosphere is presently being investigated with the BIMS. It is apparent that our understanding of the distribution of the anomalously enriched trace metals in the marine atmosphere will be incomplete until we can determine the significance of chemical processes occurring at the air/sea interface.

Pattern Recognition and Factor Analysis

Statistical techniques such as pattern recognition and factor analysis can give us additional information which may aid in understanding possible sources of these trace metals. Factor analysis is a method of reducing the number of measured variables within a data set using linear combinations based on a product-moment correlation matrix. For example, consider 12 elements from the 1973 Bermuda samples: Na, Mg, Ca, K, Al, Fe, Cr, Mn, Cu, Zn, Cd, and Pb. Factor analysis (FA) diagonalizes the correlation matrix for these 12 elements and solves for eigenvalues and vectors which are linear combinations of the original 12 variables such that the variance of each variable is successively maximized in each vector. Basically, the technique is grouping linear combinations of similar variances from variables that are highly correlated. The results of FA of the 1973 Bermuda samples is shown in Table 4-6. Four vectors, or processes, are contained in the original 12 elements measured. One possible interpretation of these four vectors is that they represent different source functions for the particles in the atmosphere. Vector 1, for example, appears generally to represent crustal material, as it is primarily associated with Fe, Al, Cr, and Mn, with some indication of Zn, Cd, Mg, Ca, and K. Vector 2 is primarily associated with Cu, Zn, and Cd with some association with Na, Mg, and lesser with other elements. It is tempting to associate this with a possible fractionated, perhaps small

Table 4-6
Factor Analysis of 1973 Bermuda Atmospheric Samples

	Vector 1	Vector 2	Vector 3	Vector 4
Fe	+0.97	+0.15	+0.11	
Al	+0.96			
Cu		+0.85		+0.26
Cr	+0.87	+0.15	+0.14	
Zn	+0.14	+0.61	+0.37	+0.48
Mn	+0.83	+0.13	+0.27	
Pb		+0.19		+0.95
Cd	+0.22	+0.86	+0.23	
Na		+0.21	+0.96	
Mg	+0.43	+0.28	+0.84	
Ca	+0.66		+0.63	
K	+0.76	+0.16	+0.60	
Eigenvalues	4.589	2.081	2.682	1.243
Relative Percent of Eigenvalues	43.3	19.6	25.3	11.7

Cumulative Percent for the four vectors = 88.3%

particle, sea salt aerosol. Vector 3 is largely associated with Na, Mg, Ca, and K, and is likely the large particle sea salt aerosol. Vector 4 is associated primarily with Pb, with some Zn and Cu, and may be associated with an anthropogenic aerosol, as suggested by the trajectory analyses. As indicated in Table 4–6 these four vectors can account for all but 12% of the sample set variance, and this is approximately equal to the analytical uncertainty of the data.

We can obtain a three-dimensional plot of these interrelationships by using the correlation of the 12 elements in the first three vectors as the X, Y, and Z axes, respectively. This is shown in Figure 4-9. The height of the corner posts represent zero on the Z axis. The crustal elements Al, Fe, Cr, and Mn are along one edge and vary back to Na through Ca, K, and Mg. Both continental dust and sea salt are known to be significant sources for Ca, K, and Mg (E. Hoffman et al., 1974) and the relative importance of each source is shown by its nearness to Fe or Na. The four elements Cd, Zn, Pb, and Cu are grouped together. Recall that vector 4, which is associated with most of the Pb and much of the Zn and Cu, is not represented in this plot.

As another example of the use of these statistical methods, pattern recognition techniques such as K-nearest neighbor (KNN) and hierarchial clustering (Kowalski, 1975) were used to analyze the 1974 samples relative to the NOAA generated air parcel trajectories. Of the original 74 samples, trajectories were available for 60, and of these, only 54 fell clearly into one of the source areas

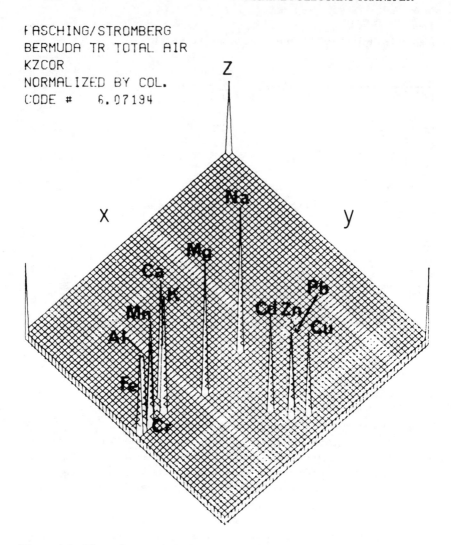

Figure 4-9. **Three-dimen**sional plot of the 1973 Bermuda atmospheric trace
metal data using the correlations in vectors 1, 2, and 3.

shown in Figure 4-3. Thus, the initial data base was these 54 samples. KNN and
clustering techniques showed that 16 of the 54 samples are apparently not
related to air parcel trajectory, i.e., these samples, if they were in group 1, were
more similar to samples in group 2 or 3. These samples were excluded and the
data set reduced to 38 samples. The separability or difference of the three
groups was tested using hyperplane separator and nonlinear mapping techniques,
which showed clearly that the three groups were different and separable from

each other. It was then assumed that the air parcel trajectory value (1, 2, or 3) was an independent variable and its correlation with the various elements measured in the 38 samples was determined. The results, shown in Figure 4-10, indicate that a significant correlation at $P < 0.01$ exists for Pb, Zn, and $P < 0.05$ for Cu, i.e., there is approximately a 99% and 95% probability, respectively, that a real correlation exists for these elements to air parcel trajectory. Note that these are the same elements in the 1973 samples associated with vector 4, the vector tentatively assigned to an anthropogenic source.

Atmospheric Fluxes of Trace Metals to the Ocean

With the data available from Bermuda it is possible to make some crude estimates of the potential significance of atmospheric input of trace metals to the ocean in the Bermuda area. The mean atmospheric concentrations observed in Bermuda are not necessarily representative of the global marine atmosphere. Some of the obviously crustal elements, such as Al, Fe, and Mn, present primarily on the larger particles, will have lower concentrations in more remote mid-ocean locations. The mean Fe concentration in Hawaii, for example, was observed to be 12 ng/SCM over a nine-month period (Hoffman, *et al.* 1972) compared to an average of 130 ng/SCM in Bermuda. The concentration of small particle elements such as Pb, however, were rather similar in the two locations – a mean of 3 ng/SCM in Hawaii and 3.5 ng/SCM in Bermuda. Since no information is yet available for marine areas in the southern hemisphere and relatively little from the North Pacific, we are limited to the Bermuda area. However, calculations of atmospheric fluxes to the ocean in this area can still give us an insight into the potential significance of general atmospheric transport to the ocean. Duce (1975), in a background paper prepared for the NAS/SCOR Workshop on Tropospheric Transport of Pollutants to the Ocean, Miami, Florida, 8-12 December 1975, and Duce and Hoffman (1976) utilized three different models, all very simple and thus very crude, to estimate the flux of vanadium from the atmosphere to the ocean. These models are used below and are applied to the Bermuda atmospheric data:

Model #1: In this model we assume that the atmospheric particles are distributed uniformly to a height of 5000 meters and there are 40 rainfalls/year which remove these particles (Bruland *et al.*, 1974). The fluxes thus calculated are presented in Table 4-7.

Model #2: In this model we use the data of Cambray *et al.* (1975), who investigated the total removal of atmospheric trace metals to the North Sea in rain plus dry fallout using a continuously open rain collector. A washout factor, W, which is the ratio of the rain concentrations in $\mu g/l$ to the air concentrations in ng/SCM can be calculated from their data. This washout factor can then be applied to the observed atmospheric concentrations of trace metals in other

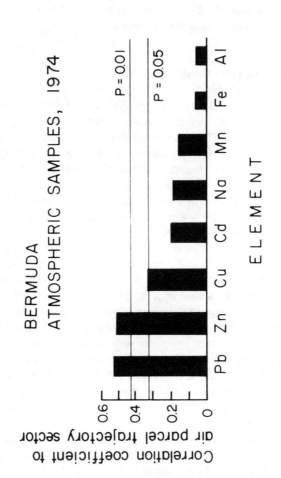

Figure 4–10. Correlation coefficients of the 1974 Bermuda atmospheric trace metal concentration vs. modified air parcel trajectory sector.

Table 4-7
Estimated Fluxes of Trace Metals in the Vicinity of Bermuda

Element	Mean Bermuda Atmospheric Concentration (ng/SCM)	Atmosphere to the Sea			Removal from the Sea by Sedimentation (10^{-15} g/cm² sec)[a]	Model 1 Sedimentation
		Model 1 (10^{-15} g/cm² sec)	Model 2 (10^{-15} g/cm² sec)	Model 3 (10^{-15} g/cm² sec)		
Al	130	75	1300	590	130	0.6
Fe	100	58	530	250	90	0.6
Pb	3.3	1.9	3.9	2.6	0.019	100*
Zn	2.5	1.5	16	4.2	0.07	21*
Mn	1.5	0.87	5.5	1.9	1.8	0.5
Cu	1.0	0.58	13	–	0.07	8.3*
Hg	≥ 0.5	≥ 0.29	–	–	0.0009	>300*
Cr	0.3	0.17	1.2	0.57	0.18	0.9
Cd	0.15	0.09	–	–	0.0005	150*
Se	0.1	0.06	0.2	0.03	0.0006[b]	100*
As	0.1	0.06	0.5	0.08	0.06[b]	1.0
Ni	0.08	0.05	0.3	–	0.14	0.4
Co	0.04	0.02	0.2	0.18	0.007	2.9
Sb	0.03	0.02	0.05	0.01	–	–
Sc	0.02	0.01	0.1	0.07	0.009	1.1

[a]From Bertine and Goldberg (1971).

[b]Determined from river input data (Bertine and Goldberg, 1971).

* Anomalous input/output ratios.

regions, e.g. Bermuda, to calculate an approximate total removal of these metals by rain-dry fallout. At the land stations around the North Sea, the average annual rainfall is approximately 80 cm/yr (Cambray *et al.*, 1975) while that at Bermuda is approximately 140 cm/yr. The different quantity, frequency, and intensity of rainfall at the two locations will affect the relative importance of rain vs. dry removal, but it is not clear to what extent the total removal would be significantly altered. Possible trace metal particle size distribution differences in the two areas also argue against a simple extrapolation to open ocean areas. However, this approach will still give us an order of magnitude estimate of the flux, and these values are reported in Table 4-7.

Model #3: In this model mean dry deposition velocities for most of the trace metals are calculated from the mean atmospheric concentrations and dry deposition values attained at four stations around the periphery of the North Sea (Cambray *et al.*, 1975). The calculated dry deposition velocities are presented in Table 4-8. These dry deposition velocities are multiplied by the atmospheric concentrations at Bermuda to obtain a dry deposition to the ocean surface. We then assume that rainfall removal of particles is approximately twice that of dry fallout (estimates range from approximately equal contributions to rainfall accounting for some 90% of the total removal) so the total flux is three times the dry fallout flux. These fluxes are presented in Table 4-7.

The estimates from the three models agree within an order of magnitude, with Model 2 generally giving the highest fluxes and Model 1 the lowest. Considering the simplicity of the models and the many and varying assumptions made in the calculations, the agreement among the results for any one metal is rather remarkable.

Table 4-8
Dry Deposition Velocities Calculated from North Sea Data

Metal	Deposition Velocity (cm/sec)
Al	1.5
Fe	0.87
Pb	0.25
Zn	0.56
Mn	0.42
Cr	0.64
Se	0.08
As	0.26
Co	1.5
Sb	0.10
Sc	1.2

Source: Cambray *et al.*, 1975.

It is interesting to compare the estimated atmospheric flux of these metals to the ocean near Bermuda with the total input of these metals to the ocean in that area from all sources. Perhaps the best way to make a first-order comparison is to utilize information on total sedimentation of these elements to the sea floor. This approach was utilized by Duce (1975) and Duce and Hoffman (1976) for vanadium and during discussions of the input fluxes of trace metals to the global ocean from the atmosphere at the NAS/SCOR Workshop on Tropospheric Transport of Pollutants to the Ocean. Bertine and Goldberg (1971) have calculated the total input of each element to the ocean using sedimentation data. Assuming the sedimentation is uniform over the ocean floor (which obviously is not the case, but it will satisfy our requirements here) we calculate the total fluxes from the sea, which are then equivalent to the input into the sea, given in Table 4-7. Assuming these figures are applicable in the Bermuda area, and using the calculations from Model 1, in the final column of Table 4-7 we find the ratio of the atmospheric input of trace metals to the sea near Bermuda to the total input of trace metals as indicated by sedimentation. The ratios for individual metals suggest that the atmosphere may be a significant source for all trace metals, although the uncertainties in the calculations make it impossible to assign any specific percentage input to the ocean near Bermuda from the atmosphere. Most important are deviations from realistic values of this "input/ output" ratio which can point out elements which do not conform to these simple models. This in turn can suggest possible processes or sources not considered by the models which may be critical in understanding the exchange of these metals between the atmosphere and the ocean. The "input/output" ratio for Zn, Pb, Cu, Hg, Cd, and Se are anomalously high. This is also the case if we calculate "input/output" ratios using the fluxes calculated from Models 2 or 3. These metals also have high values for EF_{crust}. Several possible explanations exist for the high "input/output" ratios for these elements. Three possibilities considered during discussions at the NAS/SCOR Workshop in Miami are as follows:

Anthropogenic Sources. The removal rate of these metals from the ocean is determined from sedimentation data on a geological time scale. Thus, the total ocean removal would not reflect the removal of increased amounts of pollution source metals in recent times. The atmospheric input would thus be greater than the apparent removal. There is evidence, presented earlier, that Pb and perhaps Cu, Cd and Zn may have a significant anthropogenic contribution in Bermuda.

Chemical Enrichment on Atmospheric Particles Produced at the Air/Sea Interface. If there is a significant quantity of trace metals being reinjected into the atmosphere from the ocean surface during bubble breaking, much of the calculated input of these metals to the ocean may in fact be simply recycled salt, and thus the input flux may be greatly inflated relative to a true net flux to the ocean. This would be particularly important for metals which are highly enriched on the sea salt particles produced by bursting bubbles. Evidence presented earlier suggests that bubbles can effectively transport particulate trace metals in sea water to the air/sea interface and that enrichment does occur for Zn, Cu, and Fe on the atmospheric sea salt particles produced when the bubbles burst, although the Fe is masked by the large weathering and anthropogenic flux. Thus the results using the bubble flotation column and BIMS and these calculations are not inconsistent.

Different Particle Size Distributions. Different particle size distributions for the various trace metals will result in different atmospheric residence times. The metals found primarily on the smaller particles should have longer atmospheric residence times since rain and dry fallout remove the larger particles more efficiently. Models 2 and 3 account for this to some extent, as measured deposition velocities and washout efficiencies are used. Model 1 does not take this factor into account. Thus the prediction of fluxes for the small particle elements would be relatively too high compared to the large particle elements. As described previously, Cu, Zn, Pb, Sb, and Cd in Bermuda are present primarily on the small particles while Fe, Al, Mn, and Cr are large particle elements.

These differences may be quite significant for some of the small particle elements. For example, Patterson *et al.* (1976) in their discussion of Pb input into the sea in this volume utilized a dry deposition velocity to the ocean surface of 0.025 cm/sec for Pb. This value was obtained from remote field measurement of a deposition velocity of 0.10 cm/sec to a horizontal surface for atmospheric Pb and a reduction of this deposition velocity by a factor of four to account for the lower probability of capture of the particles by water compared to a dry surface. This lower deposition velocity, if valid, would result in a Pb flux to the ocean in Bermuda of about an order of magnitude less than indicated for Model 3 in Table 4-7. Rain removal would also be less efficient for the small particle elements. Thus assumptions that all elements are removed from the atmosphere with the same efficiency are obviously gross oversimplifications.

We cannot distinguish among these possible explanations for the high "input/output" ratios at present, and in fact other unknown factors undoubtedly play an important role. It is significant, however, that the deviations observed from expected values for this ratio can be at least partially explained by experimental evidence presented earlier in this paper on the properties and distribution of these elements at the sea surface and in the atmosphere.

Conclusions

While our understanding of the sources of trace metals present in the marine atmosphere and the fluxes of these metals in both directions across the air/sea interface is far from complete, a somewhat consistent picture is beginning to emerge from the various studies underway. We now have extensive atmospheric data from marine air in Bermuda during 1973 and 1974 for 23 metals (Na, Mg, Ca, K, Fe, Al, Pb, Zn, Mn, Cu, Hg, Cr, Ce, Cd, Se, Ni, As, Co, Sb, Sc, Th, Ag, and Eu). The atmospheric concentrations of most of these trace metals fluctuate over two to three orders of magnitude from day to day. It is also apparent that such atmospheric elements as Na, Mg, Ca, and K are largely from the ocean and Fe, Al, Mn, Ce, Cr, Ni, Co, Sc, Th, and Eu are primarily due to transport of continental weathering products to marine areas. The sources for Pb, Zn, Cu, Hg, Cd, Se, As, Sb, and Ag, however, are more difficult to determine as they are anomalously enriched, i.e. they are present in concentrations higher than expected from normal crustal weathering or transport of bulk seawater into the atmosphere. Air parcel trajectory analysis and statistical techniques suggest strongly that the Pb, probably a significant fraction of the Zn and possibly of the Cu and Cd may be from anthropogenic sources on the North American continent. Comparison of EF_{crust} values for samples collected over the North Atlantic and at the South Pole, however, suggest that natural processes may be important for some of these anomalously enriched metals as well. Particle size distribution studies in Bermuda indicate the anomalous elements analyzed to date are generally found on particles with radii less than 1 μm, suggesting that vapor phase processes may be important for these metals. This could include both anthropogenic and natural processes. One high temperature natural process which may be important is volcanism, and EF_{crust} values similar to those found in ambient atmospheric particles were found in a sample collected in the fume from Kilauea volcano, Hawaii. Low temperature volatilization may also be important, for example biological methylation of some metals (Wood, 1974) or direct volatilization from the crust (Goldberg, personal communication).

There is now good experimental evidence that Fe, Zn, and Cu are fractionated, or enriched, in atmospheric sea salt particles produced by bubbles bursting at the air/sea interface. Calculations suggest that on a global scale the ocean contributes quantities of Zn and Cu to the atmosphere via this fractionation process which may be of the same order as the atmospheric Zn and Cu from crustal weathering and anthropogenic sources. While Fe is enriched on atmospheric sea salt particles, the contribution of Fe from the sea is still apparently small compared with other sources. Further work on the potential significance of chemical fractionation at the air/sea interface for other anomalously enriched metals is underway.

Using three crude models, we have attempted to calculate the flux of trace metals from the atmosphere to the ocean in the Bermuda area. The agreement of the fluxes calculated by these three models was generally within an order of magnitude. By comparing the calculated atmospheric input to the ocean with removal rates from the ocean by sedimentation, it was apparent that transfer via the atmosphere could be a significant route for trace metal input to the ocean.

Acknowledgments

We thank the director and staff of the Bermuda Biological Station and the Rhode Island Nuclear Science Center for their aid in the collection and analysis of these samples. We also thank the officers and men of the U.S. Navy Tudor Hill Laboratory for use of their facilities and space in Bermuda and we thank the crew of *R/V Trident* for their help in sample collection. We are grateful to Paul Deslaurier, Randolph Borys, Kenneth Rahn, William Fitzgerald, William Graham, Clifford Weisel, Dana Kester and Janice Stearns for their aid in the collection and analysis of the samples and interpretation of the data. And finally we thank the National Science Foundation, Office for the International Decade of Ocean Exploration for their financial support under NSF/IDOE Grant GX-33777.

References

Barker, D. R. and H. Zeitlin (1972). Metal-ion concentrations in sea-surface microlayer and size-separated atmospheric aerosol samples in Hawaii. *J. Geophys. Res., 77,* 5076-86.

Bertine, K. K. and E. D. Goldberg (1971). Fossil fuel combustion and the major sedimentary cycle. *Science, 173,* 233-5.

Bidleman, T. F. and C. E. Olney (1974). Chlorinated hydrocarbons in the Sargasso Sea atmosphere and surface water. *Science, 183,* 516-18.

Blanchard, D. C. (1963). Electrification of the atmosphere by particles from bubbles in the sea. *Progress in Oceanography, 1,* Oxford, England, Pergamon, 71-202.

Blanchard, D. C. and L. D. Syzdek (1974). Bubble tube: apparatus for determining rate of collection of bacteria by an air bubble rising in water. *Limnol. Oceanog., 19,* 133-8.

Blanchard, D. C. and A. H. Woodcock (1957). Bubble formation and modification in the sea and its meteorological significance. *Tellus, 9,* 145-58.

Boyce, S. G. (1951). Source of atmospheric salts. *Science, 113,* 620-21.

Brewer, P. B. (1975). Minor elements in seawater. *Chemical Oceanography,* 2nd ed., Vol. 1, 415-96.

Bruland, K. W., K. Bertine, M. Koide, and E. D. Goldberg (1974). History of metal pollution in Southern California coastal zone. *Environ. Sci. and Technol., 8,* 425-31.

Cambray, R. S., D. F. Jefferies and G. Topping (1975). An estimate of the input of atmospheric trace elements into the North Sea and the Clyde Sea (1972-73). United Kingdom Atomic Energy Authority Harwell Report AERE-R7733, 30 pp.

Chow, T. J., J. L. Earl and C. F. Bennett (1969). Lead aerosols in marine atmosphere. *Environ. Sci. and Technol., 3,* 737–40.

Chow, T. J., K. W. Bruland, K. Bertine, A. Soutar, M. Koide and E. D. Goldberg (1973). Lead pollution: Records in Southern California coastal sediments. *Science, 181,* 551–2.

Duce, R. A. (1975). Atmospheric vanadium transport to the ocean. Background paper prepared for the NAS/SCOR Workshop on Tropospheric Transport of Pollutants to the Ocean, Miami, Florida, 12 pp., 8–12 December.

Duce, R. A., J. G. Quinn, C. E. Olney, S. R. Piotrowicz, B. J. Ray and T. L. Wade (1972). Enrichment of heavy metals and organic compounds in the surface microlayer of Narragansett Bay, Rhode Island. *Science, 176,* 161–3.

Duce, R. A., G. L. Hoffman, J. L. Fasching and J. L. Moyers (1974). The collection and analysis of trace elements in atmospheric particulate matter over the North Atlantic Ocean. *WMO Special Environmental Report No. 3, Observation and Measurement of Atmospheric Pollution,* WMO-No. 368, Geneva, 370–79.

Duce, R. A., G. W. Wallace and B. J. Ray (1975a). Atmospheric trace metals over the New York Bight. Final Report, NOAA-MESA Grant No. 01-4-022-36, 42 pp.

Duce, R. A., G. L. Hoffman and W. H. Zoller (1975b). Atmospheric trace metals at remote northern and southern hemisphere sites: pollution or natural? *Science, 187,* 59–61.

Duce, R. A., and G. L. Hoffman (1976). Atmospheric vanadium transport to the ocean. *Atmos. Environ.* (in press).

Eriksson, E. (1959). The yearly circulation of chloride and sulfur in nature: meteorological, geochemical, and pedological implications. Part 1. *Tellus, 11,* 375–403.

FAO Fisheries Reports, No. 99, Supplement 1 (1971). Report of the seminar on methods of detection, measurement, and monitoring of pollutants in the marine environment. Food and Agricultural Organization of the United Nations, Rome, 123 pp.

Fasching, J. L., R. A. Courant, R. A. Duce and S. R. Piotrowicz (1974). A new surface microlayer sampler utilizing the bubble microtome. *J. Rech. Atmos., 8,* 649–52.

Garrett, W. D. (1965). Collection of slick-forming materials from the sea surface. *Limnol. Oceanog., 10,* 602–5.

Gladney, E. S., W. H. Zoller, A. G. Jones and G. E. Gordon (1974). Composition and size distributions of atmospheric particulate matter in the Boston area. *Environ. Sci. and Technol., 8,* 551–7.

Harvey, G. R. and W. G. Steinhauer (1974). Atmospheric transport of polychlorobiphenyls to the North Atlantic. *Atmos. Environ., 8,* 777-89.

Heffter, J. L., A. D. Taylor and G. J. Ferber (1975). A regional-continental scale
transport, diffusion, and deposition model. *NOAA Technical Memorandum
ERL-ARL 50.*

Hicks, S. D. (1959). Physical oceanography of Narragansett Bay. *Limnol.
Oceanog., 4,* 316-27.

Hoffman, E. J. and R. A. Duce (1974). The organic carbon content of marine
aerosols collected on Bermuda. *J. Geophys. Res., 79,* 4474-7.

Hoffman, E. J., G. L. Hoffman and R. A. Duce (1974). Chemical fractionation
of alkali and alkaline earth metals in atmospheric particulate matter over the
North Atlantic. *J. Rech. Atmos., 8,* 675-88.

Hoffman, E. J. and R. A. Duce (1976). Factors influencing the organic carbon
content of atmospheric sea salt particles: a laboratory study. *J. Geophys.
Res., 81,* 3667-3670.

Hoffman, G. L., R. A. Duce and W. H. Zoller (1969). Vanadium, copper, and
aluminum in the lower atmosphere between California and Hawaii.
Environ. Sci. and Technol., 3, 1207-10.

Hoffman, G. L. and R. A. Duce (1971). Copper contamination of atmospheric
particulate samples collected with Gelman Hurricane air samplers. *Environ.
Sci. and Technol., 5,* 1134-6.

Hoffman, G. L., R. A. Duce and E. J. Hoffman (1972). Trace metals in the
Hawaiian marine atmosphere. *J. Geophys. Res., 77,* 5322-9.

Hoffman, G. L., R. A. Duce, P. R. Walsh, E. J. Hoffman, J. L. Fasching and B. J.
Ray (1974). Residence time of some particulate trace metals in the oceanic
surface microlayer: significance of atmospheric deposition. *J. Rech.
Atmos., 8,* 745-59.

Junge, C. E. (1963). *Air Chemistry and Radioactivity,* New York: Academic
Press, 382 pp.

Kientzler, C. F., A. B. Arons, D. C. Blanchard and A. H. Woodcock (1954).
Photographic investigation of the projection of droplets by bubbles bursting
at a water surface. *Tellus, 6,* 1-7.

Kowalski, B. R. (1975). Measurement analysis by pattern recognition. *Anal.
Chem., 47,* 1152A-1162A.

Lee, R. E. Jr., S. S. Goranson, R. E. Enrione and G. B. Morgan (1972). National
air surveillance cascade impactor network. II. Size distribution measurements of trace metal components. *Environ. Sci. and Technol., 6,* 1025-30.

MacIntyre, F. (1972). Flow patterns in breaking bubbles. *J. Geophys. Res., 77,*
5211-28.

MacIntyre, F. (1974). Chemical fractionation and sea-surface microlayer processes. In *The Sea, 5, Marine Chemistry,* New York: Wiley, 245-99.

Mason, B. J. (1957). The oceans as a source of cloud-forming nuclei. *Geofisica Pura e Applicata, 36,* 148-55.

Moyers, J. L., R. A. Duce and G. L. Hoffman (1972). A note on the contamination of atmospheric particulate samples collected from ships. *Atmos. Environ., 6,* 551-6.

Mroz, E. J. and W. H. Zoller (1975). Composition of atmospheric particulate matter from the eruption of Heimaey, Iceland. *Science, 190,* 461-4.

Murozumi, N., T. J. Chow and C. Patterson (1969). Chemical concentrations of pollutant lead aerosols, terrestrial dusts, and sea salts in Greenland and Antarctic snow strata. *Geochim. Cosmochim. Acta, 33,* 1247-94.

National Academy of Sciences (1975). *Assessing Potential Ocean Pollutants.* NAS, Washington, D. C., 438 pp.

Newell, R. E., B. J. Boer, Jr., and J. W. Kidson (1974). An estimate of the inter-hemispheric transfer of carbon monoxide from tropical general circulation data. *Tellus, 26,* 103-7.

Patterson, C. C. and D. Settle (1974). Contribution of lead via aerosol deposition to the Southern California Bight. *J. Rech. Atmos., 8,* 957-60.

Patterson, C., D. Settle, B. Schaule and M. Burnette (1976). Transport of pollutant lead to the oceans and within ocean ecosystems. This volume.

Peirson, D. H., P. A. Cawse and R. S. Cambray (1974). Chemical uniformity of airborne particulate material, and a maritime effect. *Nature, 251,* 675-9.

Piotrowicz, S. R., B. J. Ray, G. L. Hoffman and R. A. Duce (1972). Trace metal enrichment in the sea-surface microlayer. *J. Geophys. Res., 77,* 5243-54.

Poet, S. E., H. E. Moore and E. A. Martell (1972). Lead 210, Bismuth 210, and Polonium 210 in the atmosphere: Accurate ratio measurement and application to aerosol residence time determination. *J. Geophys. Res., 77,* 6515-27.

Rahn, K. A. (1971). Sources of trace elements in aerosols – an approach to clean air. Ph.D. Thesis, University of Michigan, 309 pp.

Risebrough, R. W., R. J. Huggett, J. J. Griffin and E. D. Goldberg (1968). Pesticides: transatlantic movements in the northeast trades. *Science, 159,* 1233-5.

Robinson, E., and R. C. Robbins (1971). *Emissions, Concentrations, and Fate of Particulate Atmospheric Pollutants,* Pub. No. 4076, American Petroleum Institute, Washington, D. C., 108 pp.

SCEP, *Man's Impact on the Global Environment,* MIT Press, 319 pp. (1970).

Seba, D. B. and J. M. Prospero (1971). Pesticides in the lower atmosphere of the Northern Equatorial Atlantic Ocean. *Atmos. Environ., 5,* 1043-50.

Skibin, D., R. C. Staley and J. W. Winchester (1973). Comments on water pollution in Lake Michigan from pollution aerosol fallout. *Water, Air, and Soil Pollution, 2,* 405-7.

Smayda, T. J. (1973). The growth of *Skeletonema costatum* during a winter-spring bloom in Narragansett Bay, Rhode Island. *Norw. J. Bot., 20,* 219-47.

Szekielda, K. H., S. L. Kupferman, V. Klemas and D. F. Polis (1972). Element enrichment in organic films and foam associated with aquatic frontal systems. *J. Geophys. Res., 77,* 5278-82.

Tatsumoto, M. and C. C. Patterson (1963). The concentration of common lead in seawater. *Earth Science and Meteoritics* (J. Geiss and E. D. Goldberg, eds.), Amsterdam: North-Holland, 74-89.

Taylor, S. R. (1964). Abundance of chemical elements in the continental crust: a new table. *Geochim. Cosmochim. Acta, 28,* 1273-86.

Van Grieken, R. E., T. B. Johansson and J. W. Winchester (1974). Trace metal fractionation effects between sea water and aerosols from bubble bursting. *J. Rech. Atmos., 8,* 611-21.

von Lehmden, D. J., R. H. Jungers and R. E. Lee, Jr. (1974). Determination of trace elements in coal, fly ash, fuel oil, and gasoline — a preliminary comparison of selected analytical techniques. *Anal. Chem., 46,* 239-45.

Wallace, G. T., Jr. and D. F. Wilson (1969). Foam separation as a tool in chemical oceanography. Naval Research Lab, Washington, D. C., Report No. 6958, 19 pp.

Wallace, G. T., Jr., G. I. Loeb and D. F. Wilson (1972). On the flotation of particulates in sea water by rising bubbles. *J. Geophys. Res., 77,* 5293-301.

Wallace, G. T., Jr., and R. A. Duce (1975). Concentration of particulate trace metals and particulate organic carbon in marine surface waters by a bubble flotation mechanism. *Marine Chemistry, 3,* 157-81.

Weiss, H. V., M. Koide and E. D. Goldberg (1971). Selenium and sulfur in a Greenland ice sheet: relation to fossil fuel combustion. *Science, 172,* 261-3.

Weiss, H. V., K. Bertine, M. Koide and E. D. Goldberg (1975). The chemical composition of a Greenland glacier. *Geochim. Cosmochim. Acta, 39,* 1-10.

Winchester, J. W. (1972). A chemical model for Lake Michigan pollution: considerations of atmospheric and surface water trace metal inputs. In *Nutrients in Natural Waters* (H. E. Allen and J. R. Kramer, eds.), New York: Wiley.

Winchester, J. W. and G. D. Nifong (1972). Watter pollution in Lake Michigan by trace elements from pollution aerosol fallout. *Water, Air, and Soil Pollution, 1,* 50-64.

Wood, J. M. (1974). Biological cycles for toxic elements in the environment. *Science, 183,* 1049-52.

Zoller, W. H., G. E. Gordon, E. S. Gladney and A. G. Jones (1973). The sources and distribution of vanadium in the atmosphere. *Trace Elements in the Environment,* Advances in Chemistry Series, No. 123, American Chemical Society, Washington, D. C., 31-47.

Zoller, W. H., E. S. Gladney and R. A. Duce (1974). Atmospheric concentrations and sources of trace metals at the South Pole. *Science, 183,* 198-200.

5

Mercury Studies of Seawater and Rain: Geochemical Flux Implications

W. F. Fitzgerald

Introduction

Our investigations associated with the NSF/IDOE Pollutant Transfer Program were concerned initially with the distribution and speciation of Hg in open ocean and coastal surface waters. These efforts have been gradually expanding to include other aspects of the global Hg cycle. In particular, we are attempting to determine the fluxes of Hg associated with marine atmospheres, rain water, and submarine volcanism.

In this present paper, I will summarize the results of our seawater studies within the context of present knowledge about the sources, sinks, and interactions of Hg in the oceans. This comparative exercise for Hg transfer in the marine environment will indicate where additional investigations would be most fruitful, and how our present studies are designed to correct certain deficiencies in geochemical calculations and interpretations with regard to Hg in the environment.

Seawater Investigations

In our preliminary studies concerned with sources, transport and chemical forms of Hg in the marine environment, we examined the amounts and distribution of total Hg, reactive Hg, and Hg strongly associated with organic material in the surface microlayer and subsurface waters of the northwest Atlantic Ocean. The mercury analyses in seawater were carried out using a cold-trap preconcentration modification of the closed system reduction-aeration flameless atomic absorption procedure described by Hatch and Ott (1968). The cold-trap is created by the immersion in liquid N_2 of a glass U-tube packed with glass beads. After reduction with $SnCl_2$, purging, and trapping, the Hg is removed from the glass column by controlled heating, and the gas phase absorption of the eluting Hg is measured. This procedure has been applied to both shipboard and laboratory analyses of Hg in seawater. The precision of analysis reported as a coefficient of variation is 20% at 10 ng Hg/l and 15% at 25 ng Hg/l; the detection limit is 2 ng Hg/l for a 100 ml seawater sample. Analytical details associated with this method can be found in Fitzgerald, Lyons, and Hunt (1974), and in Fitzgerald (1975).

Experimentally, the mercury measurements in seawater have been divided into two fractions — reactive and total Hg. The reactive fraction represents the amount of mercury measured in preacidified raw unfiltered seawater at approximately pH 1. The total mercury measurement is carried out on aliquots of the preacidified seawater samples in which the organic matter has been destroyed by ultraviolet photo-oxidation (Armstrong, Williams and Strickland, 1966). The amount of Hg determined as the difference between the "reactive" and "total Hg determinations" represents a very stable organo-mercury association.

The results of our first investigations are summarized in Table 5-1 (Fitzgerald and Hunt, 1974). In the open ocean surface waters for a Bermuda to Narragansett, R. I., transect, we found a mean total Hg concentration of 7 ng/l and a range between 6 to 11 ng/l. Interestingly, no Hg enrichments were apparent in the surface microlayer when compared to subsurface water. Also, no significant differences were found for Hg determinations in preacidified open-ocean seawater ("reactive Hg") and the total Hg measurements in photo-oxidized ("organic free") aliquots. Thus, chemical species of Hg characterized by a strong association with organic matter were not observed in these open ocean surface waters.

Observations in coastal surface water showed a substantial increase in both reactive and total Hg (Table 5-1). Thus, the Hg concentrations appear to decrease seaward with increasing distance from terrestrial sources. This apparent Hg gradient in the northwest Atlantic Ocean suggested that much of the Hg present in shelf and slope surface waters is locally derived. The concentration pattern for organically associated Hg also indicates either a local source or the production of such species in the coastal zone. However, the relative contribution of the principle sources of Hg, continental runoff and atmospheric inputs, remains to be ascertained.

These initial and limited observations have been followed by further studies of Hg in the same region, and in the coastal waters of the New York Bight and the Georges Bank area. The results of these investigations have been tabulated in Tables 5-2 and 5-3. Twenty-four analyses of open ocean surface waters at four stations between 40° and 32°N at ~66°W yielded a mean concentration of 8 ng/l with a standard deviation of 3 ng Hg/l (Table 5-2). This mean concentration agrees favorably with the average concentrations of 7 ng Hg/l found in the earlier open ocean investigation. Johnson and Braman (1975) report Hg concentrations of 10 ng/l for surface Sargasso Sea water and <10 ng/l for 20 "deep" samples collected 200 miles south of Bermuda (~29°N, 65°W) and analyzed within an hour. Observations of Hg in concentrations >100 ng/l in this region of the northwest Atlantic were not confirmed (e.g., Fitzgerald, Gordon and Cranston, 1974).

Our open-ocean surface water measurements were made on samples obtained by hand from a small rubber work boat away from the influence of the oceanographic research vessel. A sampling procedure was developed and tested for seawater collections at depth for Hg determinations (Fitzgerald and Lyons,

Table 5-1
Mercury Concentrations in Surface Seawater Versus Geographical Region of the Northwest Atlantic Ocean

Ocean Region	Hg Concentration (ng/1)					
	Surface Microlayer			Subsurface		
	Reactive	Total	Organically Associated[a]	Reactive	Total	Organically Associated[a]
Open Ocean IDOE Stations 1-6, Trident 137	2-10	5-10		2-11	6-11	
Continental Slope IDOE Station 7, Trident 137	8	42	34	12	41	29
Continental Shelf IDOE Station 8, Trident 137	34	89	55	41	122	81
Long Island Sound[b] (August–October, 1972)				21-33	45-78	14-50

Source: Fitzgerald and Hunt, 1974.

[a]Mercury strongly associated with organic material is determined as the difference between the total Hg measurement in photo-oxidized sea water and the amount of Hg determined in raw acidified seawater.

[b]Fitzgerald and Lyons, 1973.

Table 5-2

Mercury Concentrations Versus Depth for Four Stations in the Northwest Atlantic Ocean: Trident Cruise 152, May 1974

	Hg *Concentration* (ng/l)			
Depth (meters)	*Station 1* 39°56.6'N 66°18.2'W	*Station 2* 36°35.3'N 66°03.2'W	*Station 3* 32°54.4'N 66°07.0'W	*Station 4* 32°03.0'N 64°54.0'W
0	7	6	10	9[11]
25	5[6]			
100		9[12]	8	4
250	7	10	8[5]	10
500	6	12	12	12[9]
750	8	10[10]	10	9

NOTE: Square brackets indicate total Hg measurements in photo-oxidized preacidified seawater samples.

1975). The Hg concentration obtained at various depths to 750 meters at the four open ocean stations on the Bermuda–Rhode Island transect are shown in Table 5-2. The concentrations of Hg at depth show little variation from the surface values. The observed variability is principally a reflection of the experimental precision limits. In confirmation of the initial study, no significant differences were observed between the reactive Hg determination and the total Hg measurements carried out with about a third of the samples.

The increases in Hg concentrations in coastal waters observed previously (Table 5-1) are also evident at stations in the New York Bight and Block Island Sound (Table 5-3). These concentrations (27–45 ng Hg/l) are slightly higher than British coastal waters (Burton and Leatherland, 1971), but agree with the low end of the range for Hg concentrations found by Windom, Taylor and Waiters (1975) in southeastern Atlantic continental shelf surface waters. Moreover, greater than 50% of the Hg in coastal waters can be associated strongly with organic material. At present, the understanding of the role of organo-Hg species in the marine geochemistry of Hg is very limited and speculative. For example, the Georges Bank stations did not reveal this organo-Hg fraction. Nevertheless, it appears probable that the isolation and indentification of organic Hg chemical species may provide a very useful means of tracing certain parts of the Hg cycle in the oceans.

In summary, the results from these studies examining the amounts and distribution of Hg in coastal and open ocean waters in the northwest Atlantic Ocean are the following:

1. Concentrations of Hg in these open ocean surface waters and at various depths to 750 meters are small (average between 7–8 ng Hg/l) and rather

Table 5-3
Mercury Concentrations in Coastal Waters off the Northeastern United States

Oceanic Region	Time	Location	Depth	Hg Concentration (ng/l)	
				Reactive	Total
N.Y. Bight	Oct 1973				
Acid Dumping Grounds		40°22.0'N 73°34.7'W	13 m	11	37
			23	10	37
		40°19.2'N 73°32.8'W	8	15	27
			18	9	28
Control		40°21.0'N 73°09.8'W	13	12	30
Block Island Sound		41°15.0'N 71°30.0'W	10	33	45
Georges Bank	May 1974	40°52.0'N 70°18.1'W	~15 cm	8	10
		42°20.5'N 67°13.8'W	~15	5	11
		41°12.4'N 66°30.7'W	~15	5	9

uniformly distributed; no enrichment was apparent in the surface microlayer nor has the presence of Hg strongly associated with organic matter been observed.

2. Concentrations of total Hg generally increase in coastal waters, often approaching 50 ng/l, and a substantial fraction of the total Hg is characterized by a very stable association with organic material; relatively small concentrations of Hg may also be found in coastal waters (e.g., Georges Bank) suggesting either variable inputs (continental runoff or atmospheric) or seasonal changes in biological uptake.

Mercury Distribution – Coastal Zone

The presence in the coastal region of both variable amounts of Hg and strong organo-Hg associations (Table 5-3) is a reflection of the dynamic character of this regime. In addition to physical mixing processes, mercury in coastal waters will be significantly affected by variable inputs associated with the major sources (continental runoff and atmospheric deposition) and by variations in biological and geochemical removal routes. These processes vary on a relatively short time scale; consequently, measured changes in Hg concentrations are thus far difficult to decipher. It is clear, however, that studies designed to isolate and identify organo-Hg species in the marine environment should confine initial efforts to the coastal zone.

In addition, we may anticipate that the atmospheric contribution of Hg to coastal waters may be considerable. Young et al. (1973), for example, suggest that a doubling of Hg inputs has occurred since 1950 to the varved sediments of the Santa Barbara Basin, while Windom, Taylor and Waiters (1975) conclude that the atmospheric deposition of Hg from continental sources is sufficient to alter significantly the Hg concentrations in southeastern U.S. coastal waters. This estimate is based in part on the concept that the atmospheric Hg is effectively washed out of the atmosphere in rainstorms (Weiss, Koide and Goldberg, 1971). At present, we are collecting and analyzing local coastal rainfall for Hg. We wish to expand this work to include measurements of Hg in coastal and open ocean atmospheres, and rains. These studies are discussed in the section entitled *Rain Water and Dry Fallout*.

Open Ocean Hg Distribution

A summary of recent observations for the amounts and distributions of Hg in various open ocean regions is presented in Table 5-4. Although the average concentrations of Hg found in many of these investigations are similar, there exists a significant spatial and vertical variability for the quantities of Hg associated

Table 5-4
Recent Determinations of Mercury in the Oceans

Source	Region	Number of Observations	Hg Concentration (ng/l)	
			Range	Mean
Leatherland et al. (1971)	NE Atlantic Ocean (0–4660 m)	9	< 3–20	13
Leatherland et al. (1973)	NE Atlantic Ocean (0–4030 m)	11	17–142	54
Gardner (1975)	Subtropical N. Atlantic Ocean (surface)	10	10–54	35
	SE Atlantic Ocean (surface-upwelling)	5	6–25	14
	NE Atlantic Ocean (surface)	7	tr –34	15
	Icelandic coastal waters (0–800 m)	36	4–142	26
Chester et al. (1973)	World Ocean (surface)	28	0.5–127	47
Olafsson (1974)	Icelandic coastal waters (0–1000 m)	5	13–18	14
Fitzgerald and Hunt (1974)	NW Atlantic Ocean (surface)	8	2–11	7
Fitzgerald (1975)	NW Atlantic Ocean (0–750 m)	20	4–12	8
Burton and Jones (1974)[a]	NW Atlantic Ocean and Caribbean			10
Williams et al. (1974)	NE Pacific Ocean (0–5000 m)	28	12–37	24
	New Zealand to Ross Sea (0–5000 m)	48	50–150	98
Sugawara, K. (1974)	NW Pacific Ocean (surface, 10 m)	13	0–23	11
Matsunaga et al. (1975)	Kuroshio, Oyashio regions and (0–1200 m) Japan Sea	52	3.6–5.6	5.0
Carr et al. (1972)	Greenland Sea	21	16–364	125
Gardner and Riley (1974)	Off Iceland	30	12–225	71
Robertson (1974)[b]	W. Atlantic	>1000	2–400	
Carr et al. (1974)	Bottom Mid-Atlantic Ridge	15	870–1420	1090

[a]Personal communication
[b]Taken from Carr et al. (1974)

with open-ocean seawater. A concentration of 30 ng Hg/l has been commonly
used in global Hg models to establish the ocean reservoir for Hg, to estimate
residence times for Hg transfer in the marine environment and to assess impact
of anthropogenic introductions of Hg (see, for example, Garrels, MacKenzie and
Hunt, 1975; Weiss, Koide and Goldberg, 1971). The variability noted in Table
5-4 appears sufficient to render inappropriate the use of a concentration of 30
ng Hg/l as representative of bulk ocean water.

The difficulties associated with oceanic trace metal sampling and analysis
are well known. Thus, some of the variations in oceanic Hg concentrations may
be experimental artifacts. A discussion of various analytical problems that may
accompany the sampling, preservation and analysis of Hg in seawater can be
found in Bothner and Robertson (1975), Fitzgerald and Lyons (1975), Matsunaga, Nishimura, and Konishi (1975), Litman, Finston and Williams (1975)
and Chilov (1975).

Some differences in the surface water concentrations reflect localized
oceanic phenomena such as upwelling while other variations may reflect variations in atmospheric inputs. For example, Gardner (1975) suggests that the
enhanced atmospheric flux of Hg associated with the westerly airflow from
the industrialized United States, may yield the observed Hg increases in North
Atlantic surface water relative to the South Atlantic. An evaluation of the latter process must await additional data for Hg fluxes associated with dry fallout,
rainwater and air/sea gas exchange in the open ocean.

Volcanic Sources of Hg

The largest reported concentrations of Hg in seawater are at depth in the North
Atlantic and South Polar seas. It has been suggested that these elevated levels
of Hg are produced either by natural inputs of Hg associated with submarine
vulcanism (Carr, Jones and Russ, 1974; Williams et al., 1974), or through leaching and weathering of basalts on the sea floor (Gardner, 1975). Quite strikingly,
Olafsson (1975) found concentrations of Hg to be as large as 478 ng/l in Icelandic coastal seawater (for sixteen sites) near the lava front associated with the
Heimaey eruption. This is about 35 times greater than Icelandic coastal waters
not influenced by active vulcanism (Olafsson, 1974).

It does appear that volatile Hg emission associated with sea floor vulcanism
is sufficient to elevate the Hg levels normally present in ambient seawater. Indeed, anomalies in Hg concentrations in seawater from depth may be a sensitive
indicator of such tectonic activity. The question of whether Hg content of deep
ocean water is influenced as significantly as the studies noted in Table 5-4 suggest, must be verified. Substantial quantities of Hg are observed in the atmosphere near active volcanoes in Hawaii and Iceland (Siegel and Siegel, 1975).
Olafsson (1975) estimated a Hg flux of 7×10^5 g Hg associated with 6×10^{14} g

of ejecta at Heimaey. This quantity of Hg is considerably below estimates of Hg mobilized through weathering (2.1×10^9 g/yr), soil degassing ($< 8.9 \times 10^9$ g/yr) and man's activities (8×10^9 g/yr, McNeal and Rose, 1974).

Bostrom and Fisher (1969) suggested submarine volcanic Hg introduction and adsorption as mechanisms that could explain the high values for Hg in sediments on the crest of the East Pacific Rise. Aston *et al.* (1972), however, found "no apparent geographic trends in the distribution of Hg in North Atlantic deep-sea sediments". The question of the flux and fate of Hg associated with seafloor magmatic activities remains quite intriguing.

Although the major source of Hg in the atmosphere and to the oceans would appear to be the natural degassing from the crust and upper mantle (Weiss, Koide and Goldberg, 1971), the mercury measurements in the Greenland ice sheet on which the outgassing hypothesis is based have been questioned (Dickson, 1972; Patterson *et al.*, 1972; Carr and Wilkniss, 1973). It appears probable that the Greenland ice sheet is affected by enhanced deposition of Hg due to volcanic activity on nearby Iceland (Siegel, Siegel and Thorarinsson, 1973). Additional and independent studies of natural volcanic gas Hg contributions to the atmosphere will test the significance of the potential under-sea volcanic Hg input as well as the contribution of Hg from these potentially large natural sources to the global atmospheric cycle of Hg.

Siegel and Siegel (1975) have suggested that volatile metal emissions (e.g., Hg, Se, Tl) associated with volcanic eruptions and exhalations may be a potentially toxic and unwelcome by-product that must be considered in this era of increased geothermal power source exploration. Geochemical studies of Hg and other elements in volcanic regions may provide useful data for evaluating the hazards associated with volatile metal emissions. It is necessary, however, to examine volcanoes such as Mauna Loa in systematic detail to better estimate the magnitude and the extent to which such inputs enter and affect the natural geochemistry of Hg and other metals (Mroz and Zoller, 1975).

Rain Water and Dry Fallout

There is a need to provide additional and independent estimates of Hg inputs to the ocean based on annual oceanic rainwater and dry fallout calculations. For example, the annual rainfall Hg flux to the earth's surface has been estimated both on the basis of total washout of atmospheric Hg approximately 35 times a year and by using average precipitation concentrations of Hg and annual global rainfall (Weiss, Koide and Goldberg, 1971). This calculation, which yields a substantial rainfall flux of Hg (10^{10}-10^{11} g/yr) to the oceans, is based on very limited data. Measurements of Hg concentrations in marine atmospheres are scarce, and the only values for Hg concentrations in rain that McNeal and Rose

(1974) could find were determined by Stock and Cucuel in 1934.[a] Moreover, the assumption of total washout for atmospheric Hg with rains has not been verified. Williston, as noted by Kothny (1973), did not observe the gaseous mercury levels to be significantly reduced during rainstorms. The decreases in Hg concentrations observed after rainstorms were attributed to increased ventilation (Kothny, 1973).

Recently, we have been collecting rainfall from Cape Cod and outside our laboratory. The results of this preliminary study are presented in Table 5-5.

Two important features of these rainwater measurements are evident: (1) the concentrations of Hg in the rainwater are small, averaging 11±6 ng/l, and (2) there does not appear to be total rain washout of Hg from the atmosphere during a prolonged rainstorm. The latter observation agrees with Williston's observations and supports the recent studies of Johnson and Braman (1974) who found the total Hg present in the atmosphere near the surface to be composed principally of several volatile chemical Hg species. An average fractionation based on 54 studies of atmospheric Hg speciation yielded particulate Hg at 4% and volatile species at 96%.

If the Hg concentrations in oceanic rains are near the 10 ng/l level, then the rainfall flux of mercury to the marine environment is closer to 30×10^8 g/yr, rather than 180×10^8 g/yr obtained using 60 ng Hg/l as average rainfall Hg levels (Garrels, MacKenzie and Hunt, 1975). Also, the apparent gaseous nature of Hg species in the atmosphere would suggest a longer residence time for Hg in the atmospheric reservoir. We are planning to continue this study with accompanying measurements of atmospheric Hg concentrations and a more precise identification of air mass trajectories. Preliminary analysis suggests that the variations in Hg concentrations for these rainwater collections can be correlated with wind direction.

In summary, although the natural geochemical cycle of Hg and man's augmentation of natural Hg fluxes in the environment have received increasing attention, only limited information is available for certain important aspects of the global Hg cycle. For example, volatile emissions of Hg from natural sources such as volcanoes and seafloor tectonic activity should be more precisely quantified. Atmospheric fluxes of Hg via rainfall, dry fallout, and sea surface exchange in the coastal zone and in the open ocean require further study. Oceanic regions such as the North Atlantic and South Polar waters with anomalous Hg concentrations should be reinvestigated. Finally, one wonders about the geochemical role and biological interactions of organo-Hg species in the atmosphere and in coastal waters.

---

[a]Additional determination of mercury concentrations in rain water are now available; see, for example, Schlesinger, Reiners and Knopman (1974).

Table 5-5
Rainwater Concentrations of Hg from Cape Cod and Groton, Connecticut, 1975

Location	Date of Collection[a]	Rainfall Collected	Collection Period	Hg Concentration[b] (ng/1)
Hyannis, Mass.	Sept. 12	0.57 inches	2000-2100 hrs (unfiltered)[d]	4; 10; 3; 6
Hyannis, Mass.	Oct. 17-18	0.17	1930-0830 (filtered)	17; 19
Hyannis, Mass.	Oct. 18	0.97	1300-1800	13; 6
	Oct. 18-19	0.85	1800-0200	13; 4
Avery Pt., Groton	Nov. 12-13	1.10	1400-1000	5; 2
	Nov. 13	0.93	1000-1700	17; 12; 14
	Nov. 13-14	0.21	1700-1000	12; 21
Avery Pt., Groton	Nov. 23-24	0.29[c]	1400-1000	13; 12
			All rains: mean and standard deviation 19 analyses	11±6

[a] Analyses conducted December 4-12, 1975.

[b] Blanks containing acidified distilled deionized water are below detection limit for Hg (<2 ng/1).

[c] Collection started three hours after rain began.

[d] Two simultaneous collections; first hour of storm.

Acknowledgment

Our seawater sampling was conducted with the assistance of the captain and crew of the *R/V Trident* and our colleagues at the Graduate School of Oceanography, University of Rhode Island. Thomas Fogg of the Marine Sciences Department/Institute of the University of Connecticut has been making the recent Hg measurements in seawater, rainwater and basalts. This work was supported in part by the IDOE/NSF Grant GX-33777.

References

Armstrong, F. A. J., P. M. Williams and J. D. H. Strickland (1966). Photo-oxidation of organic matter in seawater by ultraviolet radiation, analytical and other applications, *Nature, 211,* 481–483.

Aston, S. R., D. Bruty, R. Chester and J. P. Riley (1972). The distribution of mercury in North Atlantic deep sea sediments, *Nature Physical Science, 237,* 125.

Boström, K. and D. E. Fisher (1969). Distribution of mercury in East Pacific sediments. *Geochim. et Cosmochim. Acta, 33,* 743–745.

Bothner, M. H. and D. E. Robertson (1975). Mercury contamination of sea water samples stored in polyethylene containers. *Anal. Chem., 47,* 592–595.

Burton, J. D., and T. M. Leatherland (1971). Mercury in a coastal marine environment, *Nature, 231,* 440–442.

Carr, R. A., J. B. Hoover, and P. E. Wilkniss (1972). Cold-vapor atomic absorption for mercury in the Greenland Sea. *Deep Sea Res., 19,* 747–752.

Carr, R. A. and P. E. Wilkniss (1973). Mercury in the Greenland ice sheet: Further data. *Science, 181,* 843–844.

Carr, R. A., M. M. Jones and E. R. Russ (1974). Anomalous mercury in near-bottom water of a mid-Atlantic Rift Valley. *Nature, 251,* 89-90.

Chester, R., D. Gardner, J. P. Riley and J. Stoner (1973). Mercury in some surface waters of the world ocean. *Marine Pollut. Bull., 4,* 28–29.

Chilov, S. (1975). Determination of small amounts of mercury. *Talanta, 22,* 205–232.

Dickson, E. M. (1972). Mercury and lead in the Greenland ice sheet: A reexamination of the data. *Science, 177,* 536-538.

Fitzgerald, R. A., D. C. Gordon, Jr., and R. E. Cranston (1974). Total mercury in sea water in the northwest Atlantic Ocean. *Deep-Sea Res., 21,* 139-144.

Fitzgerald, W. F. and W. B. Lyons (1973). Organic mercury compounds in coastal waters. *Nature, 242,* 452-453.

Fitzgerald, W. F. and C. D. Hunt (1974). Distribution of mercury in surface micro-layer and in subsurface waters of the northwest Atlantic. *J. Rech. Atmos., 8,* 629-637.

Fitzgerald, W. F., W. B. Lyons and C. D. Hunt (1974). Cold-trap preconcentration method for the determination of mercury in sea water and in other natural materials. *Anal. Chem., 46,* 1882-1885.

Fitzgerald, W. F. and W. B. Lyons (1975). Mercury concentrations in open-ocean waters: Sampling procedure. *Limnol. Oceanogr. 20,* 468-471.

Fitzgerald, W. F. (1975). Mercury analyses in seawater using cold-trap preconcentration and gas phase detection, in *Analytical Methods of Oceanography.* (T.R.P. Gibb, ed.), Advances in Chemistry Series, American Chemical Society, pp. 99-109.

Gardner, D. and J. P. Riley (1974). Mercury in the Atlantic around Iceland. *J. Cons. Int. Explor. Mer., 35,* 202-204.

Gardner, D. (1975). Observations on the distribution of dissolved mercury in the ocean. *Mar. Pollut. Bull., 6,* 43-46.

Garrels, R. M., F. T. MacKenzie and C. Hunt (1975). Chemical cycles and the global environment: assessing human influences. Los Altos, California: W. Kaufman, Inc., 206 pp.

Hatch, W. R. and W. L. Ott (1968). Determination of sub-microgram quantities of mercury by atomic absorption spectrophotometry. *Anal. Chem., 40,* 2085-2087.

Johnson, D. L. and R. S. Braman (1974). Distribution of atmospheric mercury species near ground. *Environ. Sci. Technol., 8,* 1003-1009.

Johnson, D. L. and R. S. Braman (1975). The speciation of arsenic and the content of germanium and mercury in members of the pelagic Sargassum community. *Deep-Sea Res., 22,* 503-507.

Kothny, E. L. (1973). The three-phase equilibrium of mercury in nature. In *Trace Metals in the Environment,* (E.L. Kothny, ed.), Advances in Chemistry Series, no. 123, American Chemical Society, pp. 48-80.

Leatherland, T. M., J. D. Burton, M.J. McCartney, and F. Culkin (1971). Mercury in north-eastern Atlantic Ocean water. *Nature, 232,* 112.

Leatherland, T. M., J. D. Burton, F. Culkin, M. J. McCartney and R. J. Morris (1973). Concentrations of some trace metals in pelagic organisms and of mercury in Northeast Atlantic Ocean water. *Deep Sea Res., 20,* 679-685.

Litman, R., H. L. Finston and E. T. Williams (1975). Evaluation of sample pretreatments for mercury determinations. *Anal. Chem. 47,* 2364-2369.

McNeal, J. M. and A. W. Rose (1974). The geochemistry of mercury in sedimentary rocks and soils in Pennsylvania. *Geochim. et Cosmochim. Acta, 38,* 1759-1784.

Matsunaga, K., M. Nishimura and S. Konishi (1975). Mercury in the Kuroshio and Oyashio regions and the Japan Sea. *Nature, 258*, 224-225.

Mroz, E. J. and W. H. Zoller (1975). Composition of atmospheric particulate matter from the eruption of Heimaey, Iceland. *Science, 190*, 461-464.

Olafsson, J. (1974). Determination of nannogram quantities of mercury in seawater. *Anal. Chim. Acta, 68*, 207-211.

Olafsson, J. (1975). Volcanic influence on seawater at Heimaey. *Nature, 255*, 138-141.

Patterson, C. C., H. V. Weiss, M. Koide and E. D. Goldberg (1972). Mercury and lead in the Greenland ice sheet: A reexamination of the data. *Science, 177*, 538.

Schlesinger, W. H., W. A. Reiners and D. S. Knopman (1974). Heavy metal concentrations and depositions in bulk precipitation in montane ecosystems of New Hampshire, U.S.A. *Environ. Pollut., 6*, 39-47.

Siegel, B. Z., S. M. Siegel and F. Thorarinsson (1973). Icelandic geothermal activity and mercury of the Greenland ice cap. *Nature, 241*, 526.

Siegel, S. M. and B. Z. Siegel (1975). Geothermal hazards, mercury emission. *Environ. Sci. Technol., 9*, 473-474.

Stock, A. and F. Cucuel (1934). Die Verbreitung des Quechsilbers. *Naturwiss. 22*, 390-393.

Sugawara, K. (1974). Ocean as a reservoir of elements. *J. Oceanog. Soc. Japan, 30*, 243-250.

Weiss, H. V., M. Koide and E. D. Goldberg (1971). Mercury in a Greenland ice sheet: Evidence of recent input by man. *Science, 174*, 692-694.

Williams, P. M., K. J. Robertson, K. Chew and H. V. Weiss (1974). Mercury in the South Polar Seas and in the Northeast Pacific Ocean. *Marine Chem., 2*, 287-299.

Windom, H. L., F. E. Taylor and E. M. Waiters (1975). Possible influence of atmospheric transport on the total mercury content of southeastern Atlantic Continental Shelf surface waters. *Deep-Sea Res., 22*, 629-633.

Young, D. R., J. N. Johnson, A. Sontar and J. D. Issacs (1973). Mercury concentrations in dated varved marine sediments collected off Southern California. *Nature, 244*, 273-275.

6

Cadmium and Mercury Transfer in a Coastal Marine Ecosystem

H. L. Windom, W. S. Gardner, W. M. Dunstan and *G. A. Paffenhofer*

Introduction

This paper summarizes the results of studies of cadmium and mercury transfer through the coastal ecosystem of the southeastern Atlantic coast (Georgia Embayment, Figure 6-1). Involved was a determination of input rates of these metals from the continents and their distribution within important components of the ecosystem. This information along with the results of laboratory studies and other data on ecosystem dynamics are integrated to provide a better

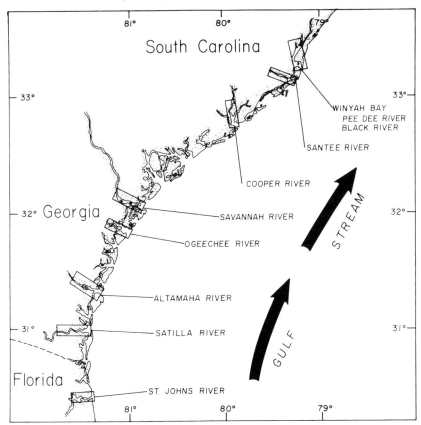

Figure 6-1. Georgia Embayment.

135

understanding of heavy metal transfer across the various biologic and nonbiologic interfaces of this ecosystem.

Cadmium and Mercury Transfer to the Coastal
Marine Ecosystem

Coastal marine ecosystems receive heavy metals from the continents via river
discharge and the atmosphere, and depending on the characteristics of the specific area, either of these two transport pathways may dominate. In the open
ocean the inputs of specific river systems become unimportant. The Georgia
Embayment is characterized by a wide estuarine zone containing extensive salt
marshes which act as a filter for materials passing through them. This is particularly true of suspended sediment which is efficiently trapped in the marshes
(Windom *et al.,* 1971; Meade, 1969). In the case of heavy metals, the estuarine
zone may serve as a temporary or permanent reservoir thus regulating their rate
of input and influencing their biological availability and subsequent transfer. As
in all coastal marine ecosystems, atmospheric inputs to the Georgia Embayment
clearly occur. Because of the greater extent of the air/sea interface; however, it
is more difficult to assess rates of input and influences than is the case with
rivers.

Transfer Through the Estuarine Zone Subsequent to
River Input

The results of studies of cadmium and mercury in the Georgia Embayment serve
to indicate the influence of the estuarine zone on the transfer of these metals.
The annual rate of input of cadmium and mercury by the nine major river systems emptying into the Georgia Embayment (Figure 6-1) was determined by
Windom (1975), based on analyses of river water samples collected bimonthly
and integrating the metal concentrations with flow rates. Once these metals are
delivered to the estuarine zone they follow various pathways within the environment entering major biologic and nonbiologic reservoirs (Figures 6-2 and 6-3).
Knowing sedimentation rates in the salt marshes and metal concentrations
(Table 6-1), Windom (1975) determined the rate of loss of the metals to the
sediment. With information on the production rates of major biological components of the estuarine zone and their metal concentrations, it is possible to determine the importance of the biota in transferring metals through this interface.
For example, the annual production rate of *Spartina alterniflora,* the major pri-
mary producer in the estuarine zone, is approximately 700 g/m^2 dry weight.
With the average concentration of cadmium and mercury observed in *Spartina*
(Table 6-1), this indicates that 3% of the total cadmium and 17% of the total

CADMIUM

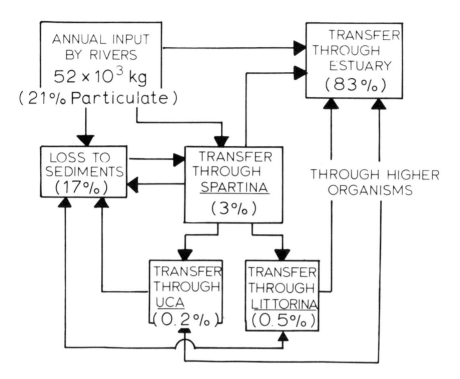

Figure 6-2. Potential transfer pathways of cadmium through the estuarine zone. Estimated percentages of the total input which passes through biological compartments are given in parentheses. An idea of the precision of these values can be obtained from the data in Table 6-1 on which these estimates are based.

mercury annually delivered by rivers is transferred through marsh grass (Windom, 1975). It is clear that a portion of the metal transferred through this compartment may reenter the sediment to be recycled. A similar approach can be taken with the major primary consumers (*Littorina* and *Uca*) since data also exist on their annual production (Odum and Smalley, 1959; Teal, 1962).

Although the importance of different compartments in the movement of cadmium and mercury is evident from the above, the complexities of the various transfer pathways make it difficult to determine their relative importance. The information that we can obtain, however, allows some general observations. For example, for both cadmium and mercury the loss to sediments is roughly equivalent to the total amount of these metals annually delivered by rivers in

MERCURY

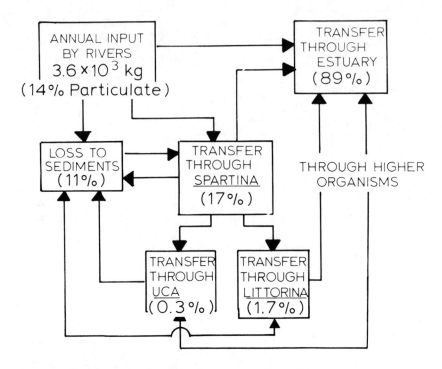

Figure 6-3. Potential transfer pathways of mercury through the estuarine zone. Estimated percentages of the total input which passes through biological compartments are given in parentheses.

Table 6-1
Average Concentration of Cadmium and Mercury in Components of the Estuarine Zone
(On a Dry Weight Basis)

Component	Cd *(ppm)*	Hg *(ppm)*	Reference
Sediments	1.6 ± 1.0	0.07 ± 0.05	Windom (1975)
Spartina alterniflora	0.5 ± 0.3	0.20 ± 0.10	Windom (1975)
Uca sp.	0.2 ± 0.1[a]	0.3 ± 0.1	Windom *et al.* (1976)
Littorina sp.	0.8 ± 0.2[a]	2.6 ± 0.5	Windom *et al.* (1976)

[a]Results of this study based on the analysis of ten samples of *Uca* sp. and six samples of *Littorina* sp.

particulate form (Figures 6-2 and 6-3). This suggests that the soluble fraction is ultimately transferred through the estuarine zone either in solution or in biological compartments. In the case of mercury, cycles within the estuary involving incorporation into various compartments (e.g., *Spartina*) influence its transfer through the region into the coastal marine ecosystem. The importance of similar cycles on cadmium transfer must be considerably less. The transformation of mercury to organic forms such as methylmercury may also determine its transfer through the system. Rahn (1973) has shown that the primary producer, *Spartina alterniflora,* can accumulate both inorganic and methylmercury. Both forms can then be transferred through the food chain to higher organisms. Inorganic mercury may also be released directly to the water from *Spartina* at an annual rate of 35 $\mu g/m^2$. Gardner *et al.* (1975) indicate the greater affinity of methylmercury for animal tissues, especially muscle. Thus, as mercury is transferred up the food chain the efficiency of retention of the methyl form increases, as does the tendency for this form of mercury to remain in the biota rather than to be lost to sediments. If methylmercury is formed in sediments, its mobilization out of this compartment must be rapid since concentrations here are very low (Andren and Harriss, 1973; Windom *et al.,* 1976). The only compartments of the estuarine zone in which methylated forms of mercury are observed are animals.

Transfer Through the Atmosphere

Although no direct determinations of atmospheric input to the Georgia Embayment have been made for cadmium and mercury, an evaluation of the relative importance of this transport mechanism can be obtained indirectly. The Gulf Stream defines the eastern boundary of the Georgia Embayment thus creating a discrete system (Figure 6-1). Gulf Stream eddies and intrusions, and fresh water discharge by the major rivers result in an estimated turnover time for the water in this system of a few months (Blanton, 1971; Webster, 1961).

The Georgia Embayment is approximately 30,000 km^2 in area and has an average depth of about 30 m. The resulting volume of water in the area is approximately 9×10^{11} m^3. Using this volume, existing data on the average concentration of mercury and cadmium in continental shelf waters (Windom and Smith, 1972; Windom *et al.,* 1975, Table 6-2) and their input rates (Windom, 1975), it is possible to estimate a mean residence time for these metals using the following equation:

$$T = \frac{M}{(dM/dt)} \tag{6.1}$$

where T is the residence time of the metal, M is the total amount contained in

Georgia Embayment water and (dM/dt) is the annual input through the estuarine zone (Figures 6-2 and 6-3). The basis for estimates of the mean residence time, assuming steady state conditions of cadmium and mercury is given in Table 6-2;

Although providing only a very crude estimate of residence times, the values obtained should be similar to the turnover time due to the advection of water, if it is assumed that this is the major mechanism for transporting material out of the system and that the only input is through the estuarine zone. The estimated values are clearly longer than expected. Since the uncertainties of estimates given in Table 6-2 are not great, another source of input is necessary to explain this difference. If additional input is due to atmospheric transport or advection from the open ocean, it is clear that these modes of transfer are much more important for mercury than cadmium. This observation is supported by studies of Windom *et al.* (1975) who suggested that observed variations in mercury concentrations in surface waters of the Georgia Embayment can best be explained by atmospheric input. The input of mercury through the atmosphere to this coastal ecosystem may be as much as ten times that supplied by rivers.

Transfer of Cadmium and Mercury Within the Coastal Marine Ecosystem

Once cadmium and mercury have been passed through coastal marshes to the open waters of the Georgia Embayment, transfer from the water to the pelagic food chain becomes important. Some of the metal transferred through the estuarine zone may already be bound in biota (e.g., in rafted detritus, migrating organisms). The greatest portion, however, is delivered to the system in solution and the water thus represents by far the most important reservoir of these metals. The transfer of mercury and cadmium from water into and through the food chain can be schematically diagrammed (Figure 6-4). This presents a simplified model which assumes that uptake from the water column takes place only at the primary producer level. The initial transfer, K_1, is the amount of metal accumulated from the water per unit weight of primary producer which, in this

Table 6-2
Mean Residence Time of Cadmium and Mercury in the Georgia Embayment

	Cd	Hg
Average content of Georgia Embayment waters (ng/1)	100	60
Total content in Georgia Embayment (10^4 kg)	9.0	5.4
Net Input through Estuarine Zone (10^4 kg/yr)	4.3	0.3
Mean Resident Time (yrs)	2	18

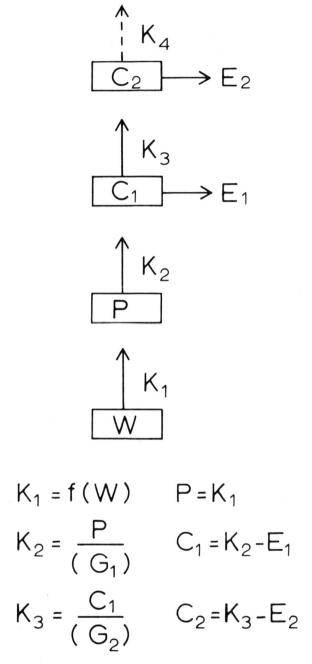

Figure 6-4. Simplified food chain metal transfer model.

case, is the phytoplankton community. As indicated, this amount is a function of the concentration, W, of the metal in the water. Subsequently, metal is transferred to higher trophic levels through food. It is clear, however, that higher trophic levels may also accumulate metals directly from the water though in this model transfer between trophic levels is assumed to be due to feeding only. The transfer to higher trophic levels, K_{1+n}, is a function of the metal concentration of the prey, C_n, and the growth efficiency of the predator, G_n. A portion of the metal taken up by the predator is transferred to the next trophic level while some is eliminated by excretion, molting, etc.

Clearly, the model (Figure 6-4) is oversimplified since it ignores, among other things, direct uptake of metals by organisms other than the primary producers and assumes a simple predator-prey relationship. It can be used, however, to develop a better understanding of food chain transfer if information on feeding habits, metal levels and ecosystem dynamics is also available. The following sections consider the application of the model using such information to simple pelagic food chains occurring in the Georgia Embayment.

Direct Uptake by Organisms

Direct uptake from the water by phytoplankton represents the major transfer pathway of metals into the biota. This is due largely to the fact that these organisms constitute a large portion of the biomass present in pelagic systems. Subsequent transfer through the food chain depends greatly on this uptake and the plants' response to variations in the metal concentration in the water. Preliminary laboratory studies of cadmium uptake by *Skeletonema costatum* in continuous culture (Figure 6-5) show an exponential increase in the cadmium level in the phytoplankton with increasing levels in the medium. Results of batch culture uptake studies for mercury for other species (Sick and Windom, 1975) show similar results (Figure 6-5). The greater biological reactivity of mercury relative to cadmium is clearly evident by comparing metal increases in phytoplankton for given incremental increases in solution. These results are for phytoplankton equilibrated with cadmium and mercury in their growth medium. The time required to reach equilibrium is one to two hours for both cadmium and mercury (Figure 6-6) suggesting that similar uptake mechanisms, possibly adsorption, operate for both metals.

The next most quantitatively important trophic level in coastal and other marine ecosystems consists of the primary consumers represented in the Georgia Embayment by the copepod *Pseudodiaptomus coronatus*. This species is an important food for a number of secondary consumer fin fish (Stickney *et al.*, 1975). Potentially, *P. coronatus* and other marine animals can accumulate metals directly from the water column. Experiments to assess the accumulation directly from the water, however, are difficult to conduct. For such

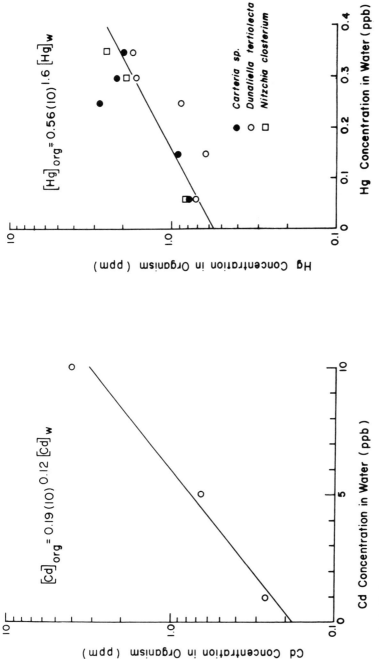

Figure 6-5. Cadmium uptake by *Skeletonema costatum*. Mercury uptake by phytoplankton species. Equations are for the regression lines shown. Metals were added as soluble chlorides.

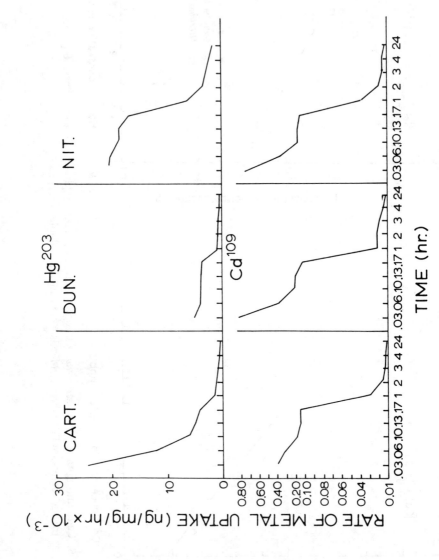

Figure 6-6. Rate of uptake of Cd and Hg by phytoplankton species.

experiments a period of one to two weeks is required during which the animals must be fed. Since their food is phytoplankton and since phytoplankton added to their culture media may accumulate metals very rapidly, it is difficult to determine if the observed metal uptake is direct or is transferred through food. With this difficulty in mind the results of some preliminary studies (Figure 6-7) indicate the *Pseudodiaptomus coronatus* accumulates cadmium at elevated levels in the same manner as *Skeletonema costatum*. During these experiments algae were regularly provided for food and were generally present at concentrations ranging from 1000 to 3000 cells/ml. Since the relationship between cadmium levels in *P. coronatus* and in water is similar to that observed for *S. costatum*, the results may still, however, reflect uptake from the food. Concentrations of cadmium in zooplankton collected from the Georgia Embayment are also similar to those in phytoplankton from the same area (Windom, 1972).

Studies by Stickney *et al.* (1976) on the direct uptake of cadmium by *Micropogon undulatus* (Atlantic croaker), a common fin fish in the Georgia Embayment, indicate little response to increased levels in the water (Table 6-3). The fish were raised for fourteen weeks in three levels of cadmium and were then dissected so that various tissues could be analyzed for the metal. During this experiment the animals were fed a commercial fish food. The results show that there is no difference between the control and experimental fish which were exposed to cadmium. Other workers (Jernelov, 1970; Olson and Fromm, 1973) have shown that mercury can be taken up directly by some fin fish. This is particularly true for the methylated form which appears to be taken up through the gills. This again attests to the observation that mercury is more biologically reactive than cadmium.

Food Chain Transfer

The feeding relationships of a number of important organisms in the Georgia Embayment coastal ecosystem have been determined (Stickney *et al.*, 1974, 1975) and can be schematically diagramed (Figure 6-8) following the simplified model given in Figure 6-4. These relationships were based on analyses of stomach contents of a minimum of one hundred animals of each of the species shown with the exception of *Neomysis americana*, the opossum shrimp. The results of Heinle and Flemer (1975), however, indicate that *N. americana* feeds upon a copepod very similar to *P. coronatus*. All other species listed in trophic levels C_2 and C_3 are fin fish and the indicated prey of these organisms (Figure 6-8) represents 80% or more of the content of stomachs analyzed (Stickney *et al.*, 1974, 1975).

Using the feeding relationships (Figure 6-8) in the simplified model given in Figure 6-4, it is possible to gain a better understanding of food chain transfer of cadmium and mercury in the Georgia Embayment if metal levels in each

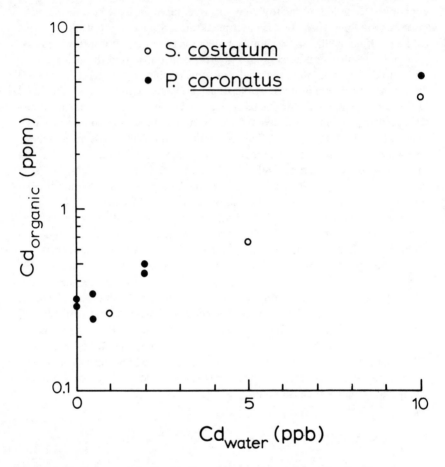

Figure 6-7. Cadmium uptake by *P. coronatus* compared to that by *S. costatum* (metal added as soluble chloride).

Table 6-3
Direct Uptake of Cd by Micropogon undulatus (Atlantic Croaker)

	Cadmium Concentration in Water		
Tissue	*Control*	*0.06 µg/1*	*0.6 µg/1*
Liver	0.10 ± 0.05	0.09 ± 0.04	0.06 ± 0.02
Muscle	0.05 ± 0.04	0.03 ± 0.01	0.02 ± 0.01
Stomach	0.19 ± 0.15	0.23 ± 0.27	0.14 ± 0.16
Gills	0.18 ± 0.15	0.16 ± 0.08	0.27 ± 0.10

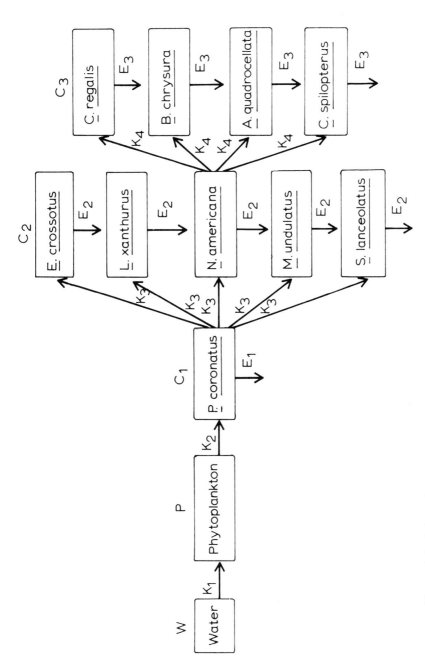

Figure 6–8. Pelagic food chains in the Georgia Embayment.

compartment are known. The following discussion takes this approach employing information on observed metal concentrations and assumed growth efficiencies of the specific members of the food chain.

The direct uptake of metals by phytoplankton from the water has already been discussed above. The concentrations of cadmium and mercury in natural populations collected in the field (Windom, 1972) agree with extrapolated concentrations (Figure 6-5) for the ambient metal levels (Figure 6-9). If it is assumed that the concentration of the metals in *P. coronatus* is due to uptake from its food only, the maximum concentration of the metal it can attain is that of the phytoplankton on which it feeds divided by its growth efficiency. This value is taken to be the transfer coefficient K_2 for the metal exchange between phytoplankton and *P. coronatus* (Figure 6-9). The growth efficiency used (25%) is an average value for juvenile copepods found by Harris and Paffenhofer (1976) for similar species.

The actual concentrations of cadmium and mercury in *P. coronatus* are considerably less than that potentially possible considering the amount of metal available to it in its food. The difference, or the amount of metal that must be eliminated, is taken as E_1. This loss may be due to excretion and/or molting. Fecal pellets and exoskeletal materials of zooplankton are commonly found to contain high levels of metals and could clearly account for the necessary loss (Benayon *et al.*, 1974).

Using a similar approach, the transfer of cadmium and mercury to successive trophic levels (Figure 6-8) can be made. For each step a growth efficiency of 25% is again used as this value appears to be an approximate average for a variety of fin fish (Muller, 1969; Edwards *et al.*, 1969). At each trophic level more cadmium is available in the food than is required to explain its concentration in the consumer (Figures 6-10 and 6-11). Also the excess of cadmium appears to increase the higher the trophic level as does the amount of mercury in the food of secondary consumers. At the tertiary consumer level, however, there is insufficient mercury in the food to explain the levels observed. This suggests that direct uptake from the water may occur or that the growth efficiencies used are too low. Regardless of the uncertainties involved, a clear pattern emerges indicating increased retention efficiencies of mercury at higher trophic levels. This greater efficiency may result from the transformation of inorganic to methylmercury as suggested by Gardner *et al.* (1975). Table 6-4 gives the percent of the total mercury in the methyl form in organisms from trophic levels C_2 and C_3. Generally, tertiary consumers show a greater percentage of methylmercury than do the secondary consumers indicating the likelihood that the higher an organism is in the food chain the greater is the proportion of methylmercury in its food. More methyl than inorganic mercury is retained by the fish as shown schematically in Figure 6-12. This is true regardless of the mode of uptake.

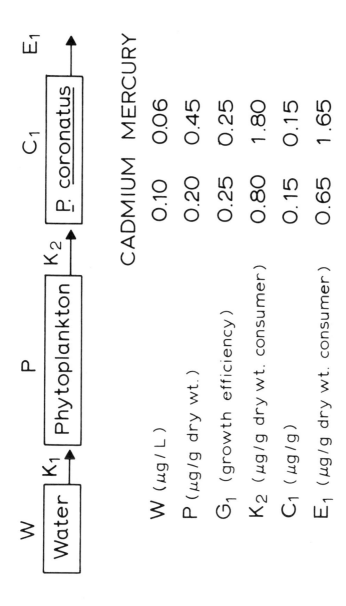

Figure 6-9. Metal transfer to primary consumers. Values for W taken from Windom and Smith (1972) and Windom et al. (1975). Values for P and C_1 taken from Windom (1972) and Windom et al. (1972) and based on samples taken from the field.

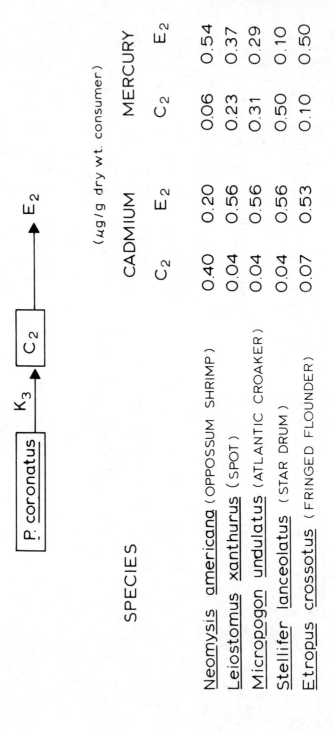

$$P.\ coronatus \xrightarrow{K_3} \boxed{C_2} \longrightarrow E_2$$

(μg/g dry wt. consumer)

SPECIES	CADMIUM		MERCURY	
	C_2	E_2	C_2	E_2
Neomysis americana (OPPOSSUM SHRIMP)	0.40	0.20	0.06	0.54
Leiostomus xanthurus (SPOT)	0.04	0.56	0.23	0.37
Micropogon undulatus (ATLANTIC CROAKER)	0.04	0.56	0.31	0.29
Stellifer lanceolatus (STAR DRUM)	0.04	0.56	0.50	0.10
Etropus crossotus (FRINGED FLOUNDER)	0.07	0.53	0.10	0.50

Figure 6-10. Metal transfer to secondary consumers. Values for C_2 taken from Windom et al. (1973) and Stickney et al. (1976).

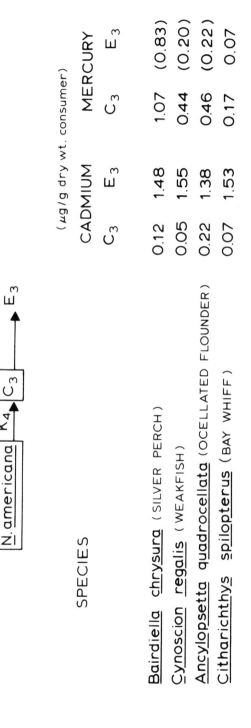

		CADMIUM		MERCURY	
		C_3	E_3	C_3	E_3
SPECIES	(μg/g dry wt. consumer)				
Bairdiella chrysura (SILVER PERCH)		0.12	1.48	1.07	(0.83)
Cynoscion regalis (WEAKFISH)		0.05	1.55	0.44	(0.20)
Ancylopsetta quadrocellata (OCELLATED FLOUNDER)		0.22	1.38	0.46	(0.22)
Citharichthys spilopterus (BAY WHIFF)		0.07	1.53	0.17	0.07

Figure 6-11. Metal transfer to tertiary consumers.

Table 6-4
Percent of Total Mercury in the Methylated Form in Fin Fish from the Georgia Embayment

Trophic Level	Species	Percent Methylated[a]
C_2	*Etropus crossotus* (fringed flounder)	10
	Leiostomus xanthurus (spot)	80
	Micropogon undulatus (Atlantic croaker)	66
	Stellifer lanceolatus (star drum)	54
C_3	*Cynoscion regalis* (weakfish)	100
	Baindiella chrysura (silver perch)	85
	Ancylopsetta quadrocellata (ocellated flounder)	89
	Citharichthes spilopterus (bay whiff)	94

[a]Average of two to six analyses. Methylated mercury was extracted from the tissues by the procedure of Uthe *et al.* (1972) and analyzed by gas chromatography with electron capture detection.

Conclusions

The above discussion is based on a simple model of metal transfer in a coastal marine ecosystem. It ignores direct uptake from the water by organisms other than phytoplankton and uses growth efficiencies obtained by others. Using the approach outlined, however, the possibility of biomagnification of Cd and Hg in food chains can be evaluated. This evaluation is made on the basis of metal transfer rates rather than measured concentrations in the organisms and compares the efficiency of metal transfer (C/K) at each trophic level (Figure 6-13). Results show that the transfer efficiency for cadmium decreases at higher trophic levels while the opposite is true for mercury. It appears then that cadmium is less efficiently retained by higher organisms than mercury.

To predict the impact of increased metal levels in an ecosystem similar to that studied, it is necessary to determine if the trends observed at ambient levels persist at higher levels of metals. If they do, the total impact on the system will also require information on effects at various trophic levels. For example, it is clear that increased metal in the water will eventually decrease primary production. This in turn will affect production at other levels. Also the effect and rate of uptake from the water at the primary producer level will depend on other characteristics of the water such as concentration and composition of dissolved organic matter and levels of nutrients present. Increased levels of metals in the food may have detrimental effects on consumers.

These complexities make it difficult to obtain a complete understanding

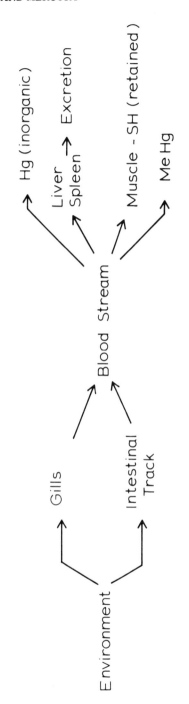

Figure 6-12. Apparent mercury pathways in fin fish.

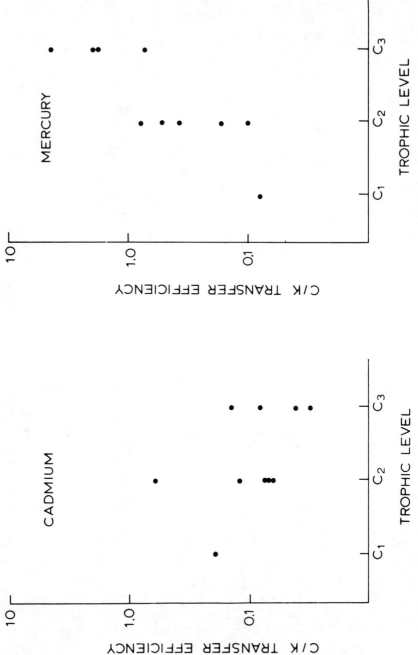

Figure 6-13. Cadmium and mercury transfer efficiencies in Georgia Embayment food chains.

of potential impacts of increased metal levels in the system. By using simplified models, such as that discussed above, however, certain aspects of metal transfer and effects on a system can be better understood. This understanding will form a basis for the construction of more realistic models.

Acknowledgments

This research was supported by a grant from the International Decade of Ocean Exploration Office of the National Science Foundation (IDO76-04243).

References

Andren, A. W. and R. C. Harriss (1973). Methylmercury in estuarine sediments. *Nature, 245,* 256-257.

Benayon, G., S. W. Fowler and B. Oregioni (1974). Flux of cadmium through euphasiids. *Mar. Biol. 27,* 205-212.

Blanton, J. O. (1971). Exchange of Gulf Stream water with North Carolina shelf water in Onslow Bay during stratified conditions. *Deep-Sea Res., 18,* 167-178.

Edwards, R. R. C., D. M. Finlayson and J. H. Steele (1969). The ecology of O-group plaice and dabs in Loch Ewe, II. Experimental studies of metabolism. *J. Exp. Mar. Biol. Ecol. 3,* 1-17.

Gardner, W. S., H. L. Windom, J. A. Stephens, F. E. Taylor and R. R. Stickney, (1975). Concentrations and forms of mercury in fish and other organisms: implications to mercury cycling. In *Mineral Cycling in Southeastern Ecosystems* (F. G. Howell, J. G. Gentry and M. H. Smith, eds.), ERDA Symposium Series (CONF-740513), 268-278.

Harris, R. P. and G.-A. Paffenhöfer, (1976) The effects of food concentration on cumulative ingestion and growth efficiency of two small marine planktonic copepods. *J. Mar. Biol. Assoc. U. K.,* in press.

Heinle, D. R. and D. A. Flemer (1975). Carbon requirements of a population of estuarine copepod *Eurytemora affinis. Mar. Biol., 31,* 235-247.

Jernelöv, A. (1970). Release of methylmercury from sediments with layers containing mercury at different depths. *Limnol. Oceanogr., 15,* 958-960.

Meade, R. H. (1969). Landward transport of bottom sediments in estuaries of the Atlantic Coastal Plain. *J. Sed. Pet., 39,* 222-234.

Müller A. (1969) Körpergewicht and gewichtszunahme junger Platt-fische in Nord-und Ostee. *Ber. dt. Wiss Kommn Meeresforsch, 20,* 198-207.

Odum, E. P. and A. E. Smalley (1959). Comparison of population energy flow of a herbivorous and deposit-feeding invertebrate in a salt marsh ecosystem. *Proceeding Nat. Acad. Sci., 45,* 617-622.

Olson, K. R. and P. O. Fromm (1973). Mercury uptake and ion distribution in gills of rainbow trout (*Salmo gairdneri*): Tissue scans with an electron microprobe. *J. Fish. Res. Bd. Can., 30,* 1575-1578.

Rahn, W. R. (1973). *The role of Spartina alterniflora in the transfer of mercury in a salt marsh environment.* Georgia Institute of Technology, MS Dissertation, 61 pp.

Sick, L. V. and H. L. Windom (1975). Effects of environmental levels of mercury and cadmium on rates of metal uptake and growth physiology of selected genera of marine phytoplankton. In *Mineral Cycling in Southeastern Ecosystems* (F. G. Howell, J. B. Gentry and M. H. Smith, eds), ERDA Symposium Series (CONF-740513), 279-308.

Stickney, R. R., G. L. Taylor and R. W. Heard (1974). Food habits of Georgia estuarine fishes. I. Four species of flounder (Pleuroneoti formes: Bothidae). *Fish. Bull. Natn. Fish. Serv., 72,* 515-525.

Stickney, R. R., H. L. Windom, D. B. White and F. E. Taylor (1975). Interrelationships between heavy metal concentration and the food habits of selected Georgia estuarine fishes. In *Mineral Cycling in Southeastern Ecosystems.* (F. G. Howell, J. B. Gentry and M. H. Smith, eds.), ERDA Symposium Series (CONF-740513), 256-167.

Stickney, R. R., H. L. Windom, S. Knowles and R. Smith (1976). Tissue response in croaker, *Micropogon undulatus,* exposed to low levels of cadmium. Submitted to *J. Fish. Res. Bd. Can.*

Teal, J. M. (1962). Energy flow in the salt marsh ecosystem of Georgia. *Ecology, 53,* 614-624.

Uthe, J. F., J. Solomon and B. Grift (1972). Rapid semimicro method for the determination of methylmercury in fish tissue. *J. Assoc. Offic. Anal. Chem., 55,* 583-589.

Webster, F. (1961). A description of Gulf Stream meanders off Onslow Bay. *Deep-Sea Res., 8,* 130-143.

Windom, H. L. (1972). Arsenic, cadmium, copper, lead, mercury and zinc in marine biota — North Atlantic Ocean. In *Baseline Studies of Pollutants in the Marine Environment,* IDOE-NSF, Brookhaven National Lab., 121-148.

Windom, H. L., F. E. Taylor and R. R. Stickney (1972). Mercury in North Atlantic plankton. *Jour. Cons. Int. Explor. Mer., 35,* 18-21.

Windom, H. L. (1975). Heavy metal fluxes through salt marsh estuaries. In *Estuarine Research* (L. E. Cronin, ed.), Vol. 1, 137-152, Academic Press, New York, San Francisco, London.

Windom, H. L., W. Gardner, J. Stephens and F. Taylor (1976). The role of methylmercury production in the transfer of mercury in a salt marsh ecosystem. *Estuarine Coastal Mar. Res.* (in press).

Windom, H. L., W. J. Neal and K. C. Beck (1971). Mineralogy of sediments in three Georgia estuaries. *J. Sed. Pet., 41,* 497-504.

Windom, H. L. and R. G. Smith (1972). Distribution of cadmium, cobalt, nickel and zinc in southeastern United States Continental Shelf waters. *Deep-Sea Res., 19,* 727-730.

Windom, H. L., R. R. Stickney, R. G. Smith, D. B. White and F. E. Taylor, (1973). Arsenic, cadmium, copper, mercury and zinc in some species of North Atlantic finfish. *J. Fish. Bd. Can., 30,* 275-279.

Windom, H. L., F. E. Taylor and E. M. Waiters (1975). Possible influence of atmospheric transport on the total mercury content of southeastern Atlantic Continental Shelf surface waters. *Deep-Sea Res., 22,* 629-633.

7

Cadmium Transport in the California Current

J. H. Martin, K. W. Bruland and W. W. Broenkow

Introduction

Each year large quantities of potentially toxic heavy metals are introduced into the world's oceans from both natural and anthropogenic sources. The major portions of these metals are thought to be removed to the sediments soon after entering the marine environment by natural sedimentation processes (e.g., Chow et al., 1973; Bruland et al., 1974). Nevertheless, in some cases a significant soluble fraction may become available for vertical and horizontal transport. Much of this soluble material ultimately becomes associated with planktonic organisms where it can affect their physiology and/or be transported across mixing barriers via active or passive vertical transport.

Our research is primarily aimed at establishing the importance of various transport routes after an element enters the marine environment, with special emphasis on its removal from the surface waters via the sinking of planktonic remains. We are presently concentrating on cadmium, a metal well known for its toxicity and one that appears to be more apt to stay in solution than others (Galloway, 1972; Bruland et al., 1974). Our study area is located off Southern and Baja California. This region was chosen because (1) the California Current is typical of Eastern Boundary current systems (Wooster and Reid, 1963); (2) it is a typical upwelling area with associated high rates of primary productivity (Ryther, 1969); (3) Southern California is a well-studied area where anthropogenic input rates are high (SCCWRP 1973, 1974, 1975); and (4) elevated Cd concentrations have been found in plankton off Baja California (Martin and Broenkow, 1975).

We hope to estimate vertical and horizontal cadmium transport rates based upon our measurements of Cd and other selected elements in the water column and plankton using representative velocity fields. A simplified model (Figure 7-1) illustrates transport pathways of interest. Box I represents the mixed layer and, depending on distance from the coast, will be the California Current or the waters lying inshore or offshore of it. Box II represents the waters immediately underlying the mixed layer beneath the pycnocline. Along the shoreline this box contains the northward flowing, subsurface California Countercurrent. Box III represents the water column lying beneath these layers. The six transport vectors shown in Figure 7-1 represent (1) the southerly flowing California Current; (2) the northerly flowing California Countercurrent; (3) metals introduced via atmospheric fallout on the air/sea interface; (4) metals introduced via

159

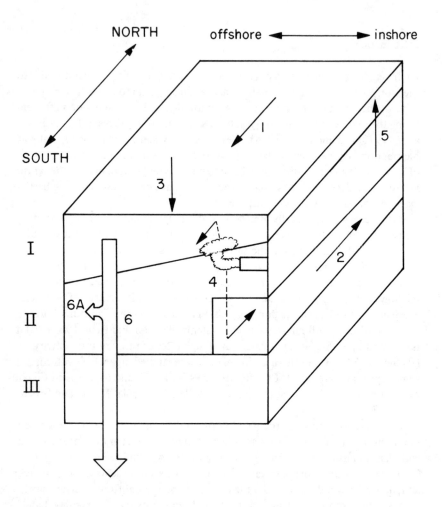

Figure 7-1. Block model of cadmium transport in the California Current; see
text for explanation.

submerged outfalls and storm channels; (5) metals transported into the surface layer by upwelling; (6) metals removed from the water column by sinking particulates; and (6A) metals removed from surface water and released at mid-depths.

To understand pollutant fluxes in the total system depicted in Figure 7-1 and to discern the amounts transported to the open Pacific Ocean via currents, it is necessary to measure the elemental concentrations and to estimate representative volume or mass transports that apply in the various compartments. The purpose of this paper is to describe our progress toward this goal. In the sections that follow we will describe methods we have developed and the general results we have obtained using them. We will also discuss preliminary findings pertaining to the six transport vectors shown in Figure 7-1.

Methods

Water and plankton samples were collected on a cruise of the *R/V Cayuse* to Southern and Baja California in February and March of 1975 (Figure 7-2). In addition to the specialized water and plankton collections for trace element analysis, routine hydrographic data were also collected: temperature and salinity were measured with a CTD on stations 41 to 70 and by hydrocast on later stations. Nutrient samples were collected at stations 71 to 95, filtered, and stored frozen until analysis. Phosphate and nitrate analyses were performed on a Technicon Autoanalyzer using the methods of Murphy and Riley (1962) and Wood *et al.* (1967) with minor modifications described in the Technicon manual. Samples for trace element analyses were collected as follows.

Plankton

Because of paint, rust, and other particles falling off all research vessels, plankton samples are especially easy to contaminate. We have developed the following methods that, we believe, insure the collection of clean samples. The basic sampling equipment consisted of a 3/4-m diameter brass net ring which was completely enclosed in polyethylene tubing; a 3/4-m diameter, 64-μm aperture phytoplankton net constructed of canvas and nylon with no metal grommets; a cod end which was a 600 ml polyethylene beaker secured to the end of the net with nylon line; and a nylon bridle and tow line. All of this equipment was stored in a polyethylene bag to prevent contamination between collections. Sample processing equipment included two polyethylene buckets with tight fitting covers; a polypropylene buchner funnel; a piece of 64-μm aperture netting; a plastic spoon; Assembly Wipes; and acid-cleaned polypropylene scintillation vials.

With this equipment, the following procedures were used: a rubber raft

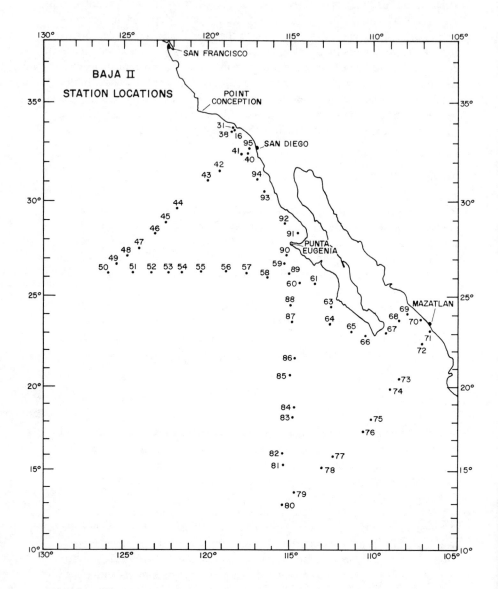

Figure 7-2. Station locations and cruise track, *R/V Cayuse,* 17 February to
30 March 1975.

was rowed with unpainted wooden oars several hundred meters upwind from the research vessel. The plankton net was removed from its plastic bag and fished up and down through the mixed layer for approximately one hour. When the tow was completed, the cod end was placed in one of the buckets which had been half-filled with clean seawater. Aboard ship the cod end was removed from the net; its contents were poured through the piece of 64-μm netting which was supported by the buchner funnel; the end of the net was washed in the remaining bucket also half-filled with clean seawater; and the contents were concentrated on the netting. After all the water had passed through the netting, the plankton was formed into a pellet with a plastic spoon. The netting and pellet were placed on top of a wad of Assembly Wipes, and excess seawater was removed by pressing the pellet with the plastic spoon; the processed pellet was transferred to an acid-cleaned scintillation vial which was placed in a plastic bag. The sample was frozen until laboratory processing. Polyethylene gloves were worn while performing the operations described above.

Water Collection

Contamination problems associated with the collection of seawater samples for trace element analyses are severe (see Patterson and Settle, 1976). In an attempt to overcome these problems, we have developed noncontaminating sampling methods.

Seawater samples were collected from the raft at the air/sea interface and approximately 0.3 meters below the interface, during the plankton sampling. A polyethylene screen (30 cm $\times$ 62 cm, 1.0 mm aperture), supported on a polypropylene frame, was passed through the interface (see Garrett, 1965); and the entrapped water was collected in a 1-liter, acid-cleaned polyethylene bottle. This procedure was repeated until the bottle was full. The bottle was then capped, placed in two polyethylene bags, and returned to the research vessel. The collecting screen was stored between stations in a dilute HCl bath constructed of welded polypropylene.

Subsurface water samples were collected by holding a 4-liter, acid-cleaned polyethylene bottle at arm's length beneath the sea surface. Polyethylene gloves were worn during this procedure, and the bottle was submerged through the interface while still enclosed in a polyethylene bag. The bottle was uncapped, filled, and recapped while held well beneath the surface. The collected sample was returned in a polyethylene bag to the research vessel.

The remainder of the water samples was collected using two tube pumps connected to 100 meters of 1.9 cm ID $\times$ 3.2 cm OD tygon tubing. Pump tubing was 1.3 cm ID $\times$ 1.9 cm OD tygon, joined to the larger sampling tubing by means of a PVC joint. The large tubing was weighted near the inlet with a completely sealed, welded-polypropylene cube filled with lead shot. After passing the pumps, the water flowed directly into a corner of the ship's laboratory enclosed in polyethylene sheeting. The sample bottles were flushed three times and filled with the effluent from the pump system. Polyethylene gloves were worn during these operations. The water flow rate was approximately eight liters per minute, and the tube flushing time was four minutes. Additional 250-ml samples were collected along with the larger volume samples. The smaller samples were acidified with 2 ml of 6 N redistilled HCl and stored in two polyethylene bags for later analysis using organic extraction techniques.

A beam transmissometer/temperature/depth profiler was attached to the tygon tube two meters above the inlet which allowed water samples to be taken relative to turbidity extrema, the mixed layer, and the thermocline. Half of the water was pumped through a fluorometer to measure *in vivo* chlorophyll. At each station, the water sampling tube and transmissometer were lowered slowly through the water column, and the location of the thermocline and turbidity features were recorded on an XYY' recorder. The tube was then brought back up to depths of interest; and, after a six-minute wait, the sample was collected. The pump system was kept clean by pumping 20% HCl through the tubing for approximately 15 minutes. The acid was left in the system between stations.

Laboratory Methods

Plankton

The frozen plankton samples were weighed in the closed scintillation vials to obtain approximate wet weights. Holes were then punched through the vial cap with a clean syringe needle, and the vial and its frozen contents were lyophilized. In a clean, filtered-air room, the dried samples were weighed, removed from the scintillation vial, and placed in a nylon sieve (366-μm aperture). The pellets were broken up, and the pieces were rubbed against the netting using the bulb end of an all polyethylene dropping pipet, resulting in complete sample homogenization. A teflon-coated magnetic stirring bar, covered with a small plastic bag, was passed over a thin layer of the homogenized plankton powder several times. This procedure removed shiny black magnetic particles that were frequently found in association with plankton collected off Baja. (We are convinced that these particles were in the water column at the time of sampling and were not added by contamination during any of our procedures.) The powdered plankton was then ready for digestion.

Digestion procedures were the same as those described in Martin and Knauer (1973) with the following exceptions. All procedures were performed in an isolated, filtered-air room, and all-quartz redistilled nitric acid and water were used in the digestion procedure. (The stills are the same as those used by NBS and are described in Kuehner et al. (1972).) Commercial reagent grade hydrogen peroxide was used, as we have been unable to locate a source of the ultrapure reagent or to produce our own. After the plankton were dissolved and diluted, the solution and silica residue were transferred to an acid-cleaned polypropylene centrifuge tube with a polypropylene cap. The silica was spun down, and the soluble fraction was decanted back into the original digestion beaker. The silica pellet was then resuspended, washed, and centrifuged twice. The washings were added to the first soluble fraction, and this solution was evaporated down until the desired dilution was obtained. The solution was weighed and transferred to an acid-cleaned scintillation vial in which the sample was kept until analyses were completed.

The silica pellet was transferred to a polyethylene beaker, dried, and weighed. The silica was dissolved with HF; and, after evaporating it to dryness to drive off SiF_4, the residue was reweighed and then taken up in 5 ml of 5% HNO_3. This solution was weighed and placed in a scintillation vial where it was stored until analyses were completed.

Almost all of the analyses were done by atomic absorption. Standard flame techniques were used for Na, K, Mg, Ca, Sr, Ba, Al, Fe, Mn, Cu, and Zn. A graphite furnace was used for Ag, Ni, and Pb analyses. Cadmium was analyzed by flame and then diluted approximately 100:1 to confirm the results with flameless atomization. Matrix problems can be severe when using graphite furnaces; and, for this reason, standard additions were used when warranted (e.g., Pb). This was accomplished by loading an aliquot of standard into the graphite tube after loading the sample, then atomizing as usual. Standard additions were not required for Cd because of the large dilutions prior to analysis. Matrix effects while analyzing the samples for Ag and Ni appeared to be negligible. A deuterium background corrector was used during all graphite furnace work and while analyzing Cd and Zn by flame. Silica was estimated gravimetrically. A 0.05 ml aliquot of the digested plankton was diluted to 50 ml and analyzed for phosphorous colorimetrically (Wood *et al.*, 1967). With the exception of Si, the accuracy of these techniques has been checked by analyzing NBS orchard leaves and bovine liver; good agreement was obtained. With the use of subboiling, quartz distilled water and acid in the filtered-air room, we were unable to detect Ag, Cd, Ni, or Pb in our blanks when analyzing with the graphite furnace. No detectable reagent blanks were found for the other elements analyzed by flame.

Seawater

Cadmium, along with other trace metals, was preconcentrated from seawater by two methods: chelex-100 ion exchange (Riley and Taylor, 1968), and a chelation-organic extraction technique. The chelex-100 ion exchange resin was purchased in the sodium form (Biorad, 100-200 mesh). The chelex resin was initially stored with 50% HCl for approximately one month. This was followed by daily shaking, decanting, and adding fresh 2 N HNO_3 to a large batch of resin for a period of one week. The resin was then recharged with ammonia, and 7.5 ml of the resin was loaded into acid-cleaned Biorad polypropylene disposable columns. The excess ammonia was rinsed from the columns with 30 ml of distilled, deionized water to an initial pH ≤9. The columns were then stoppered with 1 cm of distilled, deionized water over the resin bed and stored individually wrapped in plastic bags.

Chelex extractions were performed at sea immediately after sample collection. Sample water was peristaltically pumped from 4-liter polyethylene sample bottles via teflon and tygon tubing through the prepared chelex columns. The tubing passed through a polyethylene stopper which was inserted tightly into the top of the columns allowing the multichannel peristaltic pump to maintain a flow rate of 5 ml/min. The effluent passed through each column was measured volumetrically.

Special precautions to avoid contamination included: (1) all personnel wore nontalced polyethylene gloves while handling sample containers and other pieces of the apparatus; (2) the 4-liter bottles were cleaned in a hot 6 N HCl bath for one week prior to their initial use; (3) the bottles were cleaned between stations with HCl rinses; (4) dilute HCl was pumped through the tubing with empty columns in place between each sample; (5) sample bottles were rinsed at least twice with seawater from the selected depth prior to filling; (6) approximately 30 ml of sample seawater was pumped through the tubing and the pH checked prior to placing the prepared column in line; (7) the ion exchange system, which was mounted in a polypropylene rack, was enclosed in a corner of the ship's laboratory by polyethylene sheeting.

After these precautions, approximately 3.5 liters of sample were pumped through each column. The columns were then removed, capped at both ends, placed in polyethylene bags, and frozen until return to the laboratory. The columns were never allowed to go dry during preparation, processing, or storage. Samples were not filtered since we did not have time for the development of a clean filtering technique prior to our first cruise.

After returning to the laboratory, the columns were thawed and then washed with 60 ml of distilled, deionized water. Trace elements were eluted off the columns with 30 ml of 2 N HNO_3 into acid-cleaned polyethylene bottles. Blanks and efficiency have only been checked in detail for Cd thus far. Cadmium retention and recovery was 100% when 7 ml of resin in the ammonium form was used. The total Cd blank caused by our apparatus and reagents was consistent and corresponded to 0.8 ng Cd/liter. Preliminary studies on Pb, Zn, and Cu also indicate 100% retention and recovery under conditions identical to those described.

The amounts of Cd in the eluate were determined by flameless atomic absorption. Severe matrix problems, due to the large amounts of sea salts retained on the columns, markedly suppressed Cd absorbance. To overcome this problem, we measured the amounts of major ions in sample eluates. Large amounts of Ca (2800 μg/g) and Mg (1700 μg/g) were found, in conjunction with smaller amounts of Na (150 μg/g), K (40 μg/g), and Sr (10 μg/g). Since amounts of these elements were constant from column to column, a matching matrix of precleaned alkali and alkaline earth salts was used in preparation of standards. The sensitivity for the standards prepared in this manner was identical to that obtained from standards made up in stripped seawater and processed like the samples. Cadmium analyses were performed using a Varian AA-6 (with a Model 63 carbon rod analyzer and BC-6 background corrector) and confirmed using a Perkin Elmer 305B with a Model 2100 graphite furnace and deuterium background corrector.

Several samples were also checked by an organic extraction technique. This involved chelation with ammonium-l-pyrrolidine dithiocarbamate (APDC) and diethylammonium diethyldithiocarbamate (DDDC), a double extraction into chloroform, and back extraction into nitric acid. The acidified sample was transferred to a 250-ml teflon separatory funnel and buffered to pH 4 with ammonium acetate. One ml of 1% APDC, 1% DDDC and 10 ml of chloroform were added, and the solution was shaken vigorously for one minute. After five minutes for complete phase separation, the chloroform was dispensed into a 125-ml teflon separatory funnel containing 2 ml of 6 N HNO_3. An additional 10 ml of chloroform was added to the original seawater sample to complete the extraction of the chelated species and effectively rinse the solution and funnel. This chloroform layer was then combined with the first fraction and shaken for at least 15 minutes. The phases were allowed to separate for five minutes, after which the chloroform was discarded and the 2 ml acid phase, containing the redissolved cadmium and other trace metals, was run into a polyethylene vial and subsequently analyzed by flameless atomic absorption spectrophotometry. Detailed Cd $^{115\,m}$ radio tracer studies gave a total extraction efficiency of 98.9 ±0.7 (1 SD; n = 10) percent.

The comparison of the two methods resulted in a regression coefficient of 1.00 ±0.08. However, the blank of the organic extraction technique was 3 ng/ liter, a factor of four higher than the blank associated with the chelex ion exchange method. Thus, for low levels of cadmium, the chelex method was preferred; and the results presented are from the ion exchange technique.

Results

Seawater Analyses

Results of the seawater cadmium analyses together with ancillary temperature, salinity, and (for stations 72-95, Figure 7-2) phosphate and nitrate data are presented in Table 7-1. The data can be broken down into two general groups:

Table 7-1

Seawater Cadmium Concentrations and Ancillary Water Column Data, R/V Cayuse, 17 February to 30 March 1975

(Station locations are shown in Figure 7-2.)

Station	Depth (m)	Temperature (°C)	Salinity (ppt)	Cd (ng/l)	Transmission/m (%)
16-9	20	11.84		62	70
-8	25	11.55		101	60
-7	26	10.72		112	39
-6	32	10.45		124	48
31-5	5	12.1		46	57
-4	13.5	11.7		52	94
-3	16	10.6		147	33
-2	24.5	10.5		74	96
-1	38.5	10.1		116	54
38-10	interface			156	
-8	12	12.8		16	75
-7	28	12.6		16	89
-6	45	11.7		39	99
-5	60	11.1		51	100
-4	90	10.2		73	100
-2	0.3 pump	13.0		14	
-1	0.3 raft	13.0		21	
40-8	0.3 raft	14.1	33.51	18	
-4	9	13.9	33.62	17	
-3	13	13.7	33.60	15	
-2	20	13.2	33.53	18	
-1	62	10.9	33.63	52	
44-10	interface			60	
-9	0.3 raft	14.5	33.48	12	
-4	30	13.8	33.47	8.6	
-3	60	13.5	33.45	8.2	
-2	80	12.7	33.45	12	
-1	95	12.4	33.46	21	
52-7	20	16.9	33.88	6.0	
-6	40	16.8	33.87	5.3	
-5	60	16.7	33.87	5.1	

Station	Depth (m)	Temperature (°C)	Salinity (ppt)	Cd (ng/l)	PO_4 (µmole/l)	NO_3 (µmole/l)
72-2	0.3 raft			7.2		
-3	20			6.3	.55	.02
73-1	interface			26		
-2	0.3 raft	22.4	34.81	6.3		
-3	22	22.4	34.81	6.6	.47	.02
-8	47	20.0	34.60	21	.98	5.7
-9	80	15.5	34.53	67	2.13	21.4

Station	Depth (m)	Temperature (°C)	Salinity (ppt)	Cd (ng/l)	PO_4 (µmole/l)	NO_3 (µmole/l)
74-2	0.3 raft	22.6	34.83	6.9		
-3	20	22.4	34.83	7.2		
-8	44	21.2	34.73	8.9	.63	1.3
-9	70	18.2	34.62	31	1.35	11.6
75-1	interface			21		
-2	0.3 raft	24.2	34.54	6.7		
-3	25	24.2	34.54	5.0	.31	.00
-8	76	21.5	34.57	19	.87	6.9
-10	43	24.0	34.54	8.8		
76-1	interface			31		
-2	0.3 raft	24.1	34.12	12		
-4	30	23.8	34.09	3.9	.23	.04
-8	65	18.4	34.39	16	.62	3.0
-9	91	14.9	34.57	54	1.98	21.3
77-10	interface			25		
-1b	interface			25		
-2	0.3 raft	23.9	33.91	4.9		
-4	26	23.9	33.91	4.5	.31	.42
-8	44	22.8	34.23	7.0	.49	3.0
-9	67	19.0	34.40	10.4	1.13	9.5
78-1	interface			29		
-2	0.3 raft		34.04	8.2		
-4	36		34.39	5.9	.35	1.7
-9	90		34.68	66	1.93	17.6
79-1	interface			16		
80-1	interface			18		
-2	0.3 raft	26.4	33.56	4.8		
-3	28	26.1	33.56	4.8	.21	.24
-4	60	26.1	33.56	4.5	.23	.28
81-1	interface			25		
-2	0.3 raft	25.7	33.76	5.3		
-3	20	25.7	33.77	4.2	.29	.18
-8a	25	25.5	33.90	5.4	.39	.91
-8b	31	24.9	34.20	4.8	.32	.31
-9	84	18.7	34.11	63	2.22	25.6
-12	45	22.5	34.15	6.2	.34	.31
82-2	0.3 raft	25.1	33.73	4.3		
-3	28	24.9	33.73	4.0	.38	.25
84-1	interface	23.3		15		
-8	60	19.7	34.23	12	.92	6.0
-9	85	15.8	34.04	31	1.32	12.0

Table 7-1 — continued

Station	Depth (m)	Temperature (°C)	Salinity (ppt)	Cd (ng/l)	PO_4 (µmole/l)	NO_3 (µmole/l)
89-1	interface			33		
-2	0.3 raft	16.0	33.88	9.1		
-3	30	15.9	33.85	4.6	.27	.01
-8	68	15.2	33.75	5.1	.30	.15
90-1	interface			29		
-2	0.3 raft	15.1	33.71	6.9		
-3a	27	14.7	33.71	7.2	.28	.15
-3b	27	14.7	33.71	6.7		
-8	40	14.4	33.70	9.5	.32	.48
-9	60	14.0	33.67	13	.61	4.5
91-1	interface			43		
-2	0.3 raft	13.3	33.72	12		
-8	30	13.2	33.72	11	.31	.65
-9	70	11.0	33.75	24	1.04	8.4
-12	40	12.6	33.67	11	.72	3.8
92-3a	30	14.7	33.60	9.2		
-3b	30	14.7	33.60	7.6	.30	.22
93-1	interface			29		
-2	0.3 raft	13.6	33.61	8.5		
94-1	interface			38		
95-1	interface			35		
-2	0.3 raft	12.9	33.58	27		
-8a	40	10.3	33.82	45	.74	5.9
-8b	40	10.3	33.82	44		
-9a	80	9.7	33.98	72	1.17	12.8
-9b	80	9.7	33.98	68		

open ocean stations away from coastal influence and anthropogenic input sources and coastal stations near input sources (e.g., stations 16, 31, 38, 40, and 95).

Distributions at offshore stations were similar and, thus, only the data for one typical station (Station 73) are shown (Figure 7-3). At this station the 30-m deep mixed layer was underlain by a strong thermocline. Phosphate and nitrate levels were typically low in the upper layer and increased markedly through the thermocline. Cadmium concentrations were high at the air/sea interface (26 ng/liter); low in the mixed layer (6.3, 6.6 ng/liter); and, then, as was the case with nitrate and phosphate, Cd levels increased in the pycnocline to a maximum of 67 ng/liter at 80 meters. At this and all other open ocean stations, a remarkable relationship was found between cadmium, phosphate, and nitrate. A scatter diagram of Cd vs. PO_4 (Figure 7-4) shows a linear relationship

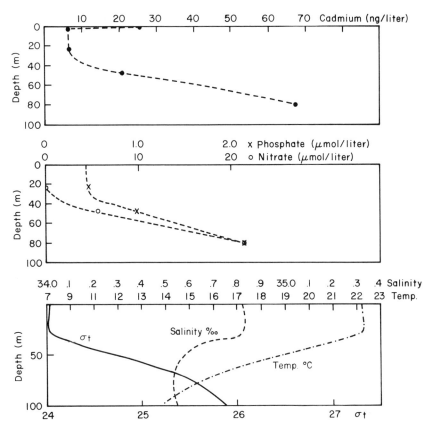

Figure 7-3. Vertical distribution of properties at station 73, 20° 29.0' N, 108° 27.2' W, 17 March 1975.

for which the correlation coefficient is 0.957, indicating a highly significant correlation ($P < 0.01$). Similar results are obtained between Cd and NO_3 (Figure 7-5), and complete regression statistics are given in Table 7-2.

Samples were also collected at five nearshore stations located in the Southern California Bight. Higher Cd concentrations were consistently observed; i.e., the average Cd value for the mixed layer at the open ocean stations was 5.6 ±1.4 (1 SD) ng Cd/liter, while off San Diego at stations 40 and 95 mixed layer Cd values were 17 ($n = 4$) and 27 ($n = 1$) ng/liter; 15 km west of Los Angeles, the surface Cd concentration was 16 ng/liter, and, in the immediate vicinity of the White's Point Sewer Outfall, surface values were 62 (station 16, $n = 1$) and 49 (station 31, $n = 2$) ng Cd/liter.

We also have some evidence that Cd levels are elevated at subsurface depths near shore. At station 95, PO_4 and NO_3 concentrations were 0.74 and 5.9

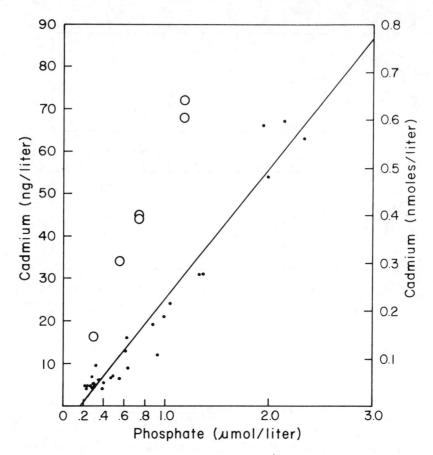

Figure 7-4. Cadmium-phosphate relation in seawater, data in Table 7-1. Line
shows linear least-square regression relations given in Table 7-2.
The open circles represent stations in the Southern California Bight
and were not included in the linear regression calculations.

μmole/liter at 40 m and 1.17 and 12.8 μmole/liter at 80 m. From the regression
equations obtained from the offshore data (Table 7-2) these nutrient levels indicate that the Cd concentrations should be 17–19 ng/liter at 40 m, and 30–36
ng/liter at 80 m. However, in duplicate samples from these depths, we found 44
and 45 ng Cd/liter at 40 meters and 68 and 72 ng/liter at 80 m. An anomalously
high Cd value of 38 ng/liter was also observed at station 94-1 compared to 11–16
ng Cd/liter calculated from the PO_4 and NO_3 values. Thus, these nearshore
values were approximately twice those predicted by the observed Cd-nutrient
relationship (see Figures 7-4 and 7-5).

Elevated Cd concentrations were observed consistently in the water collected

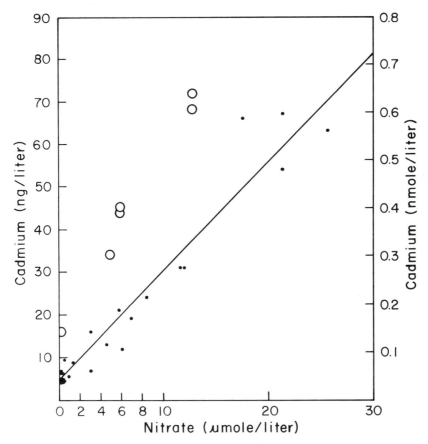

Figure 7-5. Cadmium-nitrate relation in seawater, data in Table 7-1. Line shows linear least-square regression relations given in Table 7-2. The open circles represent stations in the Southern California Bight and were not included in the linear regression calculations.

at the air/sea interface. For example, an average of 24 ±6 (1 SD; $n = 11$) ng/liter was obtained for the samples collected west and south of Baja California, in comparison with an average of 6.9 ±2.1 (1 SD; $n = 12$) ng/liter found just beneath the surface and 5.6 ±1.4 (1 SD, $n = 19$) ng/liter observed in the mixed layer. As expected, higher concentrations were observed in the vicinity of the Southern California Bight; e.g., values for the interface, subsurface, and mixed layer samples collect at station 44 were 60, 12, and 8.4 ng Cd/liter; within the Bight, comparable depths had 156, 21, and 16 ng Cd/liter (see Station 38, Table 7-1).

We also obtained Cd data from the water column in the immediate vicinity

Table 7-2
Linear Least-Square Cadmium/Nutrient Statistics: $Y = a + bX$

	n	a	b	r	$S_{Y/X}$	S_a	S_b
Plankton							
Y: Cd μg/g X: P mg/g	22	–1.8	1.8	0.709	±4.3	±3.1	±0.4
Water							
Y: Cd ng/l X: PO$_4$ μmole/l	35	–5.5	30	0.957	±5.4	±1.5	±1.6
Y: Cd ng/l X: NO$_3$ μmole/l	35	4.1	2.5	0.957	±5.4	±1.1	±0.1

NOTE: r = coefficient of correlation, $S_{Y/X}$ = standard error of estimate, S_b = standard error of regression coefficient, S_a = standard error of intercept.

of the Los Angeles County outfall at White's Point (371 mgd flow) at stations 16 and 31. The beam transmissometer-temperature record at station 31 (Figure 7-6) shows clear evidence of three turbid layers: the near-surface layer to about 3 m showed high *in vivo* chlorophyll which indicates that the particles were predominantly phytoplankton; a well-defined, 8-m-thick lens centered at 15 m which was most likely sewage effluent; and the near-bottom effluent below 26 m. The mid-depth lens was trapped below a strong thermocline centered at 12 m, and the near-bottom plume lay below a weaker thermocline at 26 m.

Water samples were collected by tube pump in each of these water layers (Figures 7-6 and 7-7). Light transmission was converted to attenuation to facilitate comparison between the trace element and turbidity data. Cadmium values were relatively low (46 and 52 ng/liter) in the near-surface layers. A marked increase was observed in the mid-depth plume (150 ng/liter) followed by a decrease (72 ng/liter) in the clear water and another increase (120 ng/liter) in the near-bottom plume. Copper, lead, zinc, manganese, and nickel analyses were also done for these samples (Figure 7-7). With the exception of Mn, the distributions of these elements were very similar to that observed for Cd, with minima occurring in the clear and phytoplankton-rich waters and maxima occurring in the turbid plume waters. Manganese also followed this pattern except that high values also occurred in the near-surface waters. Elevated Cd levels were also observed at station 16. Three samples collected in a sewer plume contained 101, 112, and 124 ng Cd/liter while the water above the plume contained 62 ng Cd/liter (see Table 7-1).

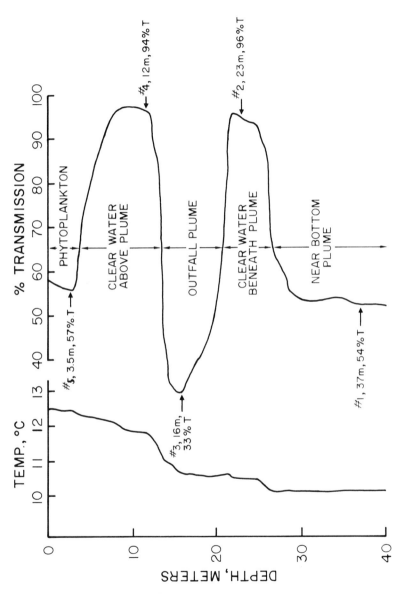

Figure 7-6. Vertical temperature and percent light transmission per meter at station 31 near Los Angeles County outfall at White's Point, 25 February 1975. Sample numbers for data in Table 7-1 are indicated.

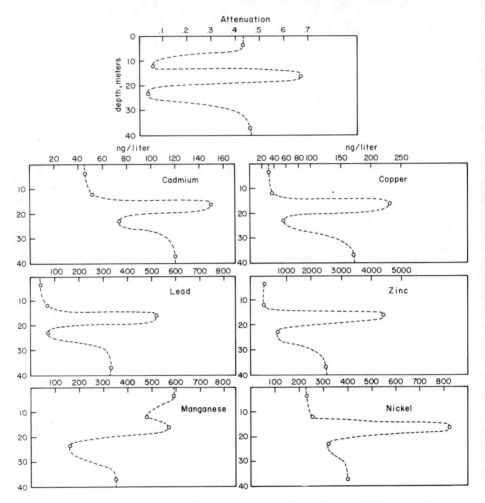

Figure 7-7. Vertical profiles of light attenuation ($A = -\log \% T/100$), and selected trace elements at station 31 (see Figure 7-6).

Plankton

Thus far, 22 of the 44 plankton samples collected during the last cruise (Figure 7-2) have been analyzed; the results are presented in Table 7-3. As was the case on two previous cruises (Martin and Broenkow, 1975), elevated Cd concentrations were found again near Baja California (Figure 7-8). On this cruise, the highest levels (up to 24.7 μg/g dry weight) were just off Mazatlan and in the waters to the south of Baja California. All of these samples had at least 14 μg

Table 7-3
Concentrations of Trace and Major Elements in Plankton Collected at Stations Shown in Figure 7-2

Station	Cd	Zn	Ag	Pb	Cu	Ni	Mn	Fe	Al	Ba	Si	P	Sr	Ca	Mg	K	Na
					μg/g dry weight										mg/g dry weight		
38	8.2	65.3	0.20	4.4	5.0	3.4	17.9	460	375	58	24	7.7	4.4	14.6	7.4	11.7	56.8
42	11.0	56.8	0.17	13.5	17.7	5.0	18.3	392	290	27	17	6.7	2.1	26.7	10.8	10.4	84.3
44	14.1	68.2	0.10	6.2	6.0	6.2	9.1	240	156	122	23	9.9	16.8	15.0	7.8	10.7	63.0
47	10.2	67.6	0.13	12.5	14.8	18.5	7.5	259	167	174	17	9.7	7.4	18.4	10.4	10.9	90.2
51	8.4	70.4	0.06	8.3	7.8	16.5	11.5	208	NA	90	24	9.0	10.5	21.5	11.8	10.1	94.0
54	4.4	11.4	0.08	4.3	6.0	4.7	4.7	187	93	103	14	7.2	7.4	7.8	10.5	3.6	89.7
57	14.1	45.2	0.05	13.6	5.2	11.3	6.8	407	NA	24	10	9.1	6.3	31.7	11.1	11.3	95.0
63	8.7	NA	0.03	6.8	NA	2.1	NA	NA	NA	676	46	4.7	NA	NA	NA	NA	NA
66	8.2	47.8	0.06	5.8	8.6	9.5	9.2	104	181	64	82	6.1	4.7	15.4	11.3	13.5	93.2
69	19.4	39.7	0.18	2.4	6.4	3.7	8.5	79	31	22	9	8.3	5.8	28.1	9.5	12.1	81.2
71	24.7	40.6	0.51	2.5	8.7	4.0	9.3	158	126	3	6	10.4	4.0	16.1	6.1	9.8	50.9
73	16.7	28.4	0.06	1.4	7.0	3.8	5.6	59	39	7	7	10.1	3.0	33.7	11.4	6.5	73.2
75	15.2	40.4	0.07	0.8	6.8	6.3	4.5	75	51	36	10	9.0	10.9	17.8	8.8	8.7	70.5
77	18.0	16.5	0.08	1.0	7.6	6.3	5.9	37	33	80	13	10.8	7.4	13.9	10.0	12.8	82.3
78	17.9	15.1	0.05	1.7	6.7	9.8	5.8	36	72	79	45	8.6	4.1	13.7	12.6	9.4	105
81	2.8	3.2	0.05	1.5	3.0	2.8	3.5	16	13	<3	3	1.5	1.6	16.1	26.4	8.0	250
83	14.3	17.1	0.06	1.7	5.6	4.8	7.6	187	118	281	25	7.9	6.7	16.4	12.0	11.2	97.1
85	18.5	72.3	0.10	2.4	8.9	5.3	4.8	65	41	46	11	8.8	7.3	22.1	9.6	13.8	74.7
88	4.4	24.5	0.04	2.0	2.9	2.8	4.3	57	80	737	242	2.7	3.7	17.9	17.8	8.3	155
90	8.7	30.3	0.05	2.6	4.4	3.3	8.3	175	134	99	47	5.7	6.6	19.4	14.4	9.9	126
93	7.8	86.9	0.13	11.6	4.4	4.3	12.8	362	279	261	69	6.4	1.5	15.1	11.6	11.1	95.8
95	2.2	85.2	0.24	11.4	7.3	7.2	35.9	3870	3710	93	47	7.5	1.0	17.1	12.5	13.5	92.5

NOTE: NA = Not analyzed because of insufficient sample.

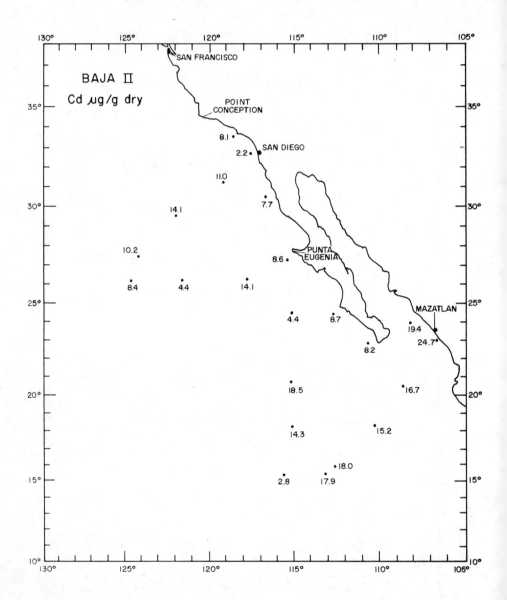

Figure 7-8. Distribution of cadmium (μg/g dry weight) in plankton collected off Baja California. Data in Table 7-3.

Cd/g except that collected at station 81. This sample was practically pure sea salt (see Na and Mg data) which is the reason for the low Cd value (2.8 μg/g). Cadmium concentrations in the samples collected west of Baja California were low or intermediate depending on station location. Once again, relatively low Cd levels were observed in the vicinity of the Southern California Bight (2.2-8.2 μg/g; stations 38, 93, 95).

Because of the Cd-P relationship we observed in seawater, the plankton samples were also analyzed for phosphorus. Almost all of the samples had at least 4.7 mg P/g dry weight. The only exceptions were at stations 81 and 88 where 1.5 and 2.7 mg P/g were found. At both stations 81 and 88, samples contained large amounts of sea salt, and that collected at station 88 also contained 242 mg Si/g. Thus the small P concentrations are due to the limited amounts of organic matter in these samples. The highest P concentrations were found in the samples collected immediately southwest of Mazatlan (up to 10.4 mg P at stations 71, 73, and 77).

The cadmium-phosphorus relation in plankton shows a significant correlation ($r = 0.709$), but the regression coefficient for the plankton, $b = 1.77 \pm 0.39$ (1 SD) μg Cd/mg P (i.e., 54.8 $\pm$12 ng Cd/μmole P), is larger than that observed in seawater (Table 7-2). Thus, relative to seawater, plankton appear to concentrate Cd more than P.

Low concentrations were observed south of Baja California for Pb (0.8-2.5 μg/g, Zn (3.2 - 40.6 μg/g), Cu (3.0 - 8.7 μg/g), and Ni (2.8 - 9.8 μg/g). Low to intermediate values for these elements were observed at the remaining stations. Silver values were highest (0.13 - 0.51 μg/g) at the nearshore stations (38, 42, 47, 69, 71, 93, and 95), while almost all of the offshore samples contained less than 0.10 ppm.

The low amounts of Al, Fe, and Mn suggest that only minimal quantities of terrigenous materials were present in the samples; the only exception was that collected at station 95 which contained 3710 μg/g Al, 3870 μg/g Fe, and 35.9 μg/g Mn. Major element concentrations were typical of those observed in plankton samples; the elevated Sr concentrations were also similar to those found in previous offshore microplankton collections (Martin and Knauer, 1973).

Discussion

Surface depletion and deep water enrichment has been recognized previously for a variety of elements including P, N, C, Si, Ca, Sr, Ba, and Ra[226] caused by incorporation into the biogeochemical cycle (RIME Report, 1971). However, it was not until recently that any trace-transition or heavy metals have been shown to exhibit such distributions. Boyle and Edmond (1975) found copper varying by greater than a factor of three, from 0.98 nmol/kg (56 ng/kg) to 3.25 nmol/kg (185 ng/kg), in the surface waters across the Antarctic Circumpolar current south of New Zealand. This was in an area of marked horizontal gradients in chemical properties, and it was found that the copper appeared to be correlated with nitrate with a correlation coefficient of 0.88 and a molar ratio Cu:N of 1 : 9200.

Boyle and Edmond postulated that if the correlation with nitrate is valid, then the levels in the nutrient poor surface waters of low and mid latitudes should be less than 0.1 nmole/kg (6 ng/kg) and the maximum values in the deep Pacific would be close to 4 nmole/kg (230 ng/kg). However, because of contamination problems associated with obtaining clean deep-water samples, the investigators have yet to verify this extrapolation.

Until this time, only the nutrients P (as PO_4^{3-}), N (as NO_3^-), and Si (as $Si(OH)_4^{3-}$) have been shown to have deep-water to surface-water enrichments of greater than a factor of 10. Radium-226 and barium exhibit roughly a fourfold enrichment, whereas calcium and strontium exhibit variations of only a few percent or less. The cadmium data from the northeast Pacific off Baja California vary from 4 ng/liter to roughly 70 ng/liter and, by extrapolation, would approach maximum values close to 100 ng/liter in the deep Pacific. This yields deep Pacific to surface water enrichment of roughly 25-fold.

Cadmium, like the nutrients, appears to be taken up by phytoplankton in the surface waters of the oceans and transported to, and released at, depths where solution or oxidation of the detritus takes place. The net result is a depletion of these elements in the surface waters and an enrichment in deeper waters. Thus, cadmium appears to behave similarly to nutrients, being an extremely reactive element and having a short residence time in the surface waters.

The Cd : P correlations in both seawater and plankton are of interest especially with regard to transport processes across the sea/biosphere interface and to the detection of anthropogenic input. At this point we can only speculate, but our present findings suggest that either (1) P and Cd are taken up together – that is, when phosphate is actively concentrated by the phytoplankton, Cd is also taken up, in approximately a ratio of 1 mg P for each 1 μg of Cd; and/or (2) PO_4 is taken up during bloom conditions, and after growth ceases, Cd is taken up passively by adsorption; and eventually an equilibrium is reached. We believe that the second process or some approximation to this process is more likely. In Monterey Bay, when large phytoplankton blooms occur, Cd levels in the plankton are always low, <2.0 μg/g Cd (Martin and Knauer, 1973; Knauer and Martin, 1973). Although P was not measured in these samples, large concentrations would be expected since actively growing phytoplankton usually contain larger amounts of this element than senescent cells (Parsons and Takahashi, 1973). If we assume that these samples contained about 10 mg of P compared to 2 μg of Cd, a weight ratio of 5000 P : 1 Cd would be obtained. This is five times higher than the ratio that we have observed for waters at the top of the pycnocline which would eventually reach the surface during upwelling.

If Cd is taken up passively after active P uptake ceases, measurements of Cd and P in seawater during a plankton bloom would result in an excess of Cd in the water column in relation to the P that has been taken up by the phytoplankton. This finding thus should not be misinterpreted as evidence of anthropogenic input. On the other hand, if an excess of Cd in relation to phosphate and nitrate is found below the thermocline where plankton have not depleted these nutrients, there is good reason to suspect anthropogenic input. Thus, the Cd-nutrient relationship may prove to be a useful tool in differentiating natural versus anthropogenic Cd in mid-depth waters.

Additional speculation is, of course, unwarranted. We must now determine if the Cd : P relationship exists ocean wide at both nearshore and open ocean locations. We must also determine if the Cd remains correlated with the P throughout the open ocean water column or if it is correlated just in the zone beneath the mixed layer where rates of biological oxidation are greatest. We must also determine if Cd in phytoplankton is correlated with organic carbon which would be a better indication of standing crop than phosphorus.

The information obtained thus far allows us to make initial rate estimates for some of the transport mechanisms shown in Figure 7-1. For example, the California Current (arrow 1) has a total transport rate of about 12×10^6 m^3/sec (Wooster and Reid, 1963). We estimate that average Cd concentration for the mixed layer plus the upper portion of the pycnocline is about 10 $\mu g/m^3$, thus the total transport past a fixed point would be 10,000 kg Cd/day. A portion of the California Current breaks off and flows northward through the Southern California Bight. The volume transport for this small, shallow countercurrent is about 1/100 of the main current, providing a surface current transport of about 100 kg of Cd per day into the Bight. This is the same order of magnitude as that introduced via outfalls (arrow 4). Based on estimates for 1974 (55.4 metric tons/year) (SCCWRP, 1975), the daily sewage discharge is approximately 150 kg of Cd.

Because of the observed increase of Cd with depth, upwelling (arrow 5) is an important transport mechanism. During upwelling periods, subsurface waters will move upward at a rate of 0.1 to 1.0 meters/day. We estimate that the average Cd concentration in these waters will be about 40 ng/liter based on a phosphate content of 1.5 μmoles/liter. Total transport will depend, of course, on both the intensity of upwelling and the size of the area considered; for example, in an area the size of the Southern California Bight (500 km $\times$ 10 km), approximately 20 to 200 kg Cd/day would be transported to the surface.

Although the northerly flowing subsurface undercurrent (arrow 2) carries only about one fourth of the water transported by the surface California Current, it is located at a depth of 200–400 m (Wickham, 1975) where Cd levels should be high (about 75 ng/liter, based on our present understanding). Thus, total Cd transport to the north is approximately 10,000 to 20,000 kg/day, an amount similar to that flowing south.

We are presently unable to make estimates for the remaining fluxes (arrows

3 and 6). We know that Cd concentrations are elevated at the air/sea interface by a factor of 5 to 10 depending on the area sampled. However, we cannot yet estimate the fallout rate, nor have we estimated the amounts of Cd leaving the surface waters by the sinking of particulates.

Nevertheless, it now appears that we have a reasonable hypothesis for the high Cd levels observed in the plankton off Baja California. Elevated quantities of Cd are brought to the surface by the pronounced upwelling in this area, and large amounts of plankton bloom in these nutrient-rich waters. A significant portion of the plankton remains in the mixed layer for sufficient time to permit maximum Cd adsorption to take place. As a result, the plankton end up with high Cd concentrations, and the surface waters are depleted of this element. Ultimately, these Cd-rich particles sink through the pycnocline, and Cd is released as the organic matter is oxidized. This is, of course, an entirely natural process and remarkably similar to that predicted by Schutz and Turekian (1965). With the knowledge we have gained of the natural biogeochemical cycling of Cd off Baja California, we feel that we are now in a better position to understand and discern anthropogenic impingement on this cycle in other parts of the oceans.

Acknowledgments

The authors gratefully acknowledge Oregon State University for the use of the R/V Cayuse. We are also indebted to Pat Elliott, Lisa Nelson, David Seielstad, Mark Stephenson, Keith Skaug, and Mike Gordon for their help with the collection and analyses of these samples. This research was supported by the National Science Foundation, International Decade of Ocean Exploration Pollutant Transport Program, Grant No. IDO 75-01303.

References

Boyle, E. and J. M. Edmond (1975). Copper in surface waters south of New Zealand. *Nature, 253,* 107–109.

Bruland, K. W., K. Bertine, M. Koide and E. D. Goldberg (1974). History of metal pollution in Southern California coastal zone. *Environ. Sci. Technol., 8,* 425–432.

Chow, T. J., K. W. Bruland, K. K. Bertine, A. Soutar, M. Koide and E. D. Goldberg (1973). Lead pollution: Records in Southern California coastal sediments. *Science, 181,* 551–552.

Galloway, J. N. (1972). *Man's alteration of the natural geochemical cycle of selected trace metals.* Ph.D. Thesis, University of California, San Diego, 143 pp.

Garrett, W. D. (1965). Collection of slick-forming materials from the sea surface. *Limnol. Oceanogr., 10,* 602–605.

Knauer, G. A., and J. H. Martin (1973). Seasonal variations of cadmium, copper, manganese, lead, and zinc in water and phytoplankton in Monterey Bay, California. *Limnol. Oceanogr., 18,* 597–604.

Kuehner, E. C., R. Alvarez, P. J. Paulsen and T. J. Murphy (1972). Production and analysis of special high-purity acids purified by sub-boiling distillation. *Analyt. Chem., 44,* 2050–2056.

Martin, J. H. and W. W. Broenkow (1975). Cadmium in plankton: Elevated concentrations off Baja California. *Science, 190,* 884–885.

Martin, J. H., and G. A. Knauer (1973). The elemental composition of plankton. *Geochim. Cosmochim. Acta, 37,* 1639–1653.

Murphy, J. and J. P. Riley (1962). A modified single solution method for the determination of phosphate in natural waters. *Anal. Chim. Acta, 27,* 31–36.

Parsons, T. and M. Takahashi (1973). Biological oceanographic processes. Pergamon Press, New York, 186 pp.

Patterson, C. C. and D. M. Settle (1976). The reduction of orders of magnitude errors in lead analyses of biological materials and natural waters by evaluating and controlling the extent and sources of industrial lead contamination introduced during sample collecting and analysis. To be published in the National Bureau of Standards special publication: *Accuracy in Trace Analysis, Proc. of the Seventh Materials Research Symp.* (P. LeFleur, ed.).

Riley, J. P. and D. Taylor (1968). Chelating resins for the concentration of trace elements from sea water and their analytical use in conjunction with atomic absorption spectrophotometry. *Anal. Chim. Acta, 40,* 479–485.

RIME Report (1971). Radioactivity in the Marine Environment. National Academy of Sciences, Washington, D. C., 272 pp.

Ryther, J. H. (1969). Photosynthesis and fish production in the sea. *Science, 166,* 72–76.

SCCWRP (1973). The Ecology of the Southern California Bight: Implications for water quality management. Southern California Coastal Water Research Project, El Segundo, Ca., 531 pp.

SCCWRP (1974). Coastal water research project: 1974 annual report. Southern California Coastal Water Research Project, El Segundo, Ca., 197 pp.

SCCWRP (1975). Coastal water research project: 1975 annual report. Southern California Coastal Water Research Project, El Segundo, Ca., 211 pp.

Schutz, D. F. and K. K. Turekian (1965). The investigation of the geographical and vertical distribution of several trace elements in sea water using neutron activation analysis. *Geochim. Cosmochim. Acta, 29,* 259–313.

Wickham, J. B. (1975). Observations of the California countercurrent. *J. Mar. Res., 33,* 325–340.

Wood, E. D., F. A. J. Armstrong and F. A. Richards (1967). Determination of nitrate in sea water by cadmium-copper reduction to nitrite. *J. Mar. Biol. Ass. U. K., 47,* 23-31.

Wooster, W. S. and J. L. Reid, Jr. (1963). Eastern boundary currents. In *The Sea,* Vol. 2. (M. N. Hill, ed.). Interscience Publ., New York, pp. 253-280.

8 The Flux of Light Hydrocarbons into the Gulf of Mexico via Runoff

J. M. Brooks

Introduction

The light gaseous hydrocarbons, methane through pentanes, are sensitive indicators of petroleum pollution. They are the most mobile fraction of petroleum. Their high solubility and vapor pressures makes them susceptible both to solution into seawater where they can be transported by mixing processes and to evaporation to the atmosphere. Brooks and Sackett (1973) reported that the important inputs of light hydrocarbons to the Gulf from man-related sources were (1) ports and rivers with their associated shipping and industrial activity, (2) offshore production operations, and (3) open ocean shipping activity. The important natural sources include seepage from oil and gas reservoirs and anaerobic production of methane. High light hydrocarbon concentrations have been found in the water column several miles distant from each of these sources (Brooks and Sackett, 1973; Sackett and Brooks, 1974; and Brooks and Sackett, 1976).

The primary processes for the production of light hydrocarbons in nature are (1) bacterial catalysis in anoxic environments yielding principally methane and (2) cracking, either thermal or catalytic, yielding a large spectrum of saturated and unsaturated products. Since neither of these processes is operative in the aerobic portion of the ocean, the light hydrocarbons present in the Gulf of Mexico must enter chiefly from man-derived sources, river runoff, and/or across the sea-air or sea-sediment interface. This paper will focus on the flux of light hydrocarbons in Gulf of Mexico rivers and estuaries with emphasis on the Mississippi River and Delta region.

Experimental Procedures

Light Hydrocarbons

Light hydrocarbons are determined according to the methods described by Brooks and Sackett (1973). Discrete analysis of hydrocarbon samples is performed by McAuliffe's (1971) method of multiple phase equilibrium. This involves equilibrating a 1 : 1 mixture of seawater and helium in a gas-tight glass syringe. Since more than 95% of the light hydrocarbons partition into the helium phase, a sample of the helium phase is injected into a gas chromatograph with a FID detector. The sensitivity of this method is 5×10^{-9} liters of gas/liter of seawater (nl/L).

185

Continuous determinations of light hydrocarbons in surface waters (3 meters below the sea surface) are performed using a hydrocarbon "sniffer". This technique is described by Brooks and Sackett (1973). Since the concentrations reported by the "sniffer" are relative values, the values are normalized relative to open ocean equilibrium concentrations. Thus, the relative value of one for methane, ethane plus ethene and propane are equal to open ocean surface concentrations of 45, 3 and 1 nannoliters per liter, respectively.

Isotopic Composition

The isotopic composition of the methane dissolved in river or seawater was determined by a method that will be described in detail in a forthcoming paper. The method basically involves vacuum extraction of the total gases from solution, removal of CO_2 by ascarite, combustion of methane to CO_2 in a hot combustion tube filled with cupric oxide and collection of the methane derived CO_2 by bubbling through potassium hydroxide solution. The KOH is sealed in ampoules and brought back to the laboratory. The isotopic composition of the CO_2 derived from the dissolved methane is determined according to methods described by Sackett et al. (1970).

Results and Discussion

Light hydrocarbons found in runoff are important since they may be a major source of contamination in many coastal areas. Brooks and Sackett (1973) reported that higher than normal light hydrocarbon concentrations can be seen over 50 miles south of the Mississippi Delta and typically up to 10 to 20 miles off ports and estuaries such as Galveston, Port Arthur and Coatzacoalcos, Mexico. Since light hydrocarbons are sensitive indicators of higher hydrocarbon pollution, it also becomes important to examine the sources of the light hydrocarbons in riverine systems. This paper contains data on Gulf coast rivers collected over the last five years.

The Gulf of Mexico receives a total annual runoff volume of approximately 1.1×10^{15} liters (Moody, 1967). Three quarters of this comes from runoff from approximately two thirds of the land area of the U.S., and the remaining amount enters via Mexican and Cuban Rivers. This runoff carries in many instances large concentrations of both natural and man-derived hydrocarbons. The nonmethane hydrocarbons, however, in coastal rivers are probably almost exclusively man-derived. Methane in many heavily polluted rivers receiving sewage and industrial wastes is derived from the anaerobic breakdown of organic matter. An example of this is the high concentrations of methane found in the Houston Ship Channel (Figure 8-1 and Table 8-1). Methane and other higher hydrocarbons can also originate from industrial and petroleum-derived pollution.

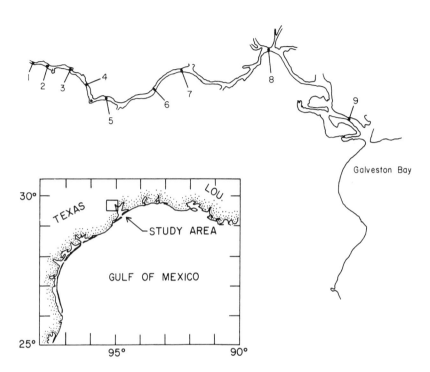

Figure 8-1. Locations of sampling stations in the Houston Ship Channel. (Hydrocarbon concentrations tabulated in Table 8-1.)

Mississippi River

The Mississippi–Atchafalaya River system largely controls the riverine inputs of light hydrocarbons to the Gulf of Mexico since it constitutes 60 to 65% of the total runoff. The Mississippi River passes (South and Southwest Pass to Head of Passes) have been surveyed four times over the last five years using the "sniffer" which measures the light hydrocarbons concentrations at 3 meters depth every few minutes while underway. Figure 8-2 shows a typical survey taken during May 1974 in South and Southwest Passes. Figure 8-3 shows a survey taken during July 1975 from South Pass to New Orleans. In addition to hydrocarbon surveys, discrete samples for laboratory analysis have been obtained during several cruises. Table 8-2 and 8-3 show light hydrocarbons, nutrient and hydrographic data obtained during two cruises in 1975. Although these data represent hundreds of analyses, only crude estimates of hydrocarbon sources and concentrations in the Mississippi Delta are possible.

Table 8-1
Light Hydrocarbons in the Houston Ship Channel (September 11, 1971)

Station	Methane[a]	Ethene[a] + Ethene	Propane[a]	Propene[a]	Iso-Butane[a]	n-Butane[a]	$NO_3 =$ (μg-at/L)	$SO_4 =$ (μg-at/L)	$PO_4 \equiv$ (μg-at/L)	O_2 (ml/L)
1	1200	<4	—[b]	—	—	—	20.2	290	15.9	0.04
2	800	14	—	3	—	110	26.1	170	22.9	—
3	8400	9	—	—	—	—	24.5	230	46.0	—
4	11000	—	5.6	<1	—	30	11.6	130	37.5	—
5	22000	6	7.0	6	—	30	0.75	120	51.5	—
6	21000	17	22	13	90	130	0.65	120	58.0	0.2
7	14000	9	17	11	50	100	1.7	150	57.5	0.3
8	530	12	14	6	40	160	4.2	130	47.5	0.4
9	200	—	2.8	—	10	30	41.3	125	47.5	0.8

[a]Concentrations expressed as nl l[-1].
[b]Dash indicates analysis was not obtained.

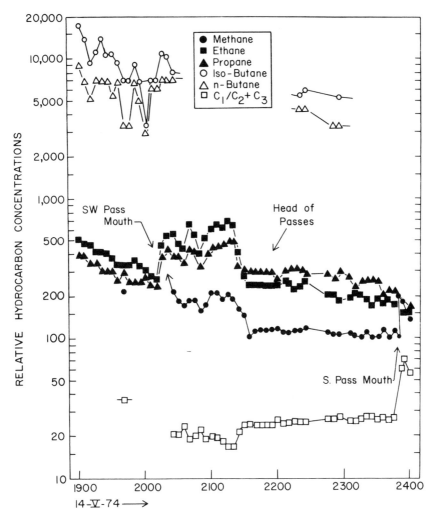

Figure 8-2. Relative hydrocarbon concentrations and $C_1/(C_2+C_3)$ ratios in the Mississippi River Delta (14 May 74).

The four hydrocarbon surveys in the Mississippi Passes showed similar trends, although the concentrations of the light hydrocarbons in the river varied widely. Typically, the highest methane concentrations are found in Southwest Pass where most of the shipping and refinery activity is centered. Brooks and Sackett (1973) earlier attributed this primarily to shipping activity, but it now appears it is chiefly the result of dumping of oil field brines. Sport (1969) reported Shell Oil Company dumps 290,000 bbls/day of water into this pass. These inputs from refineries along Southwest Pass are no doubt largely

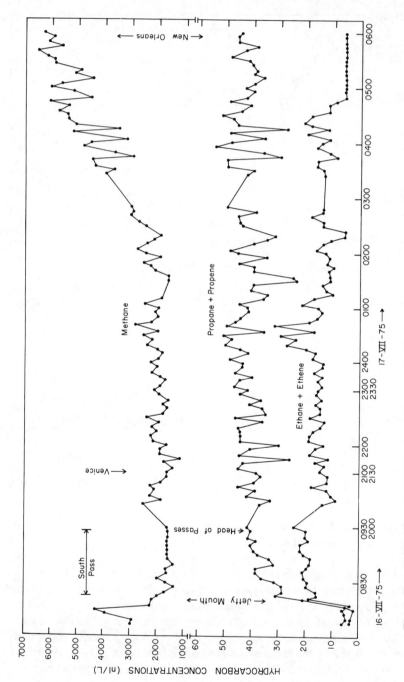

Figure 8-3. Variations in light hydrocarbons in the Mississippi River between south pass and New Orleans.

Table 8-2
Light Hydrocarbons in the Mississippi River and Delta During Cruise 75-G-1 (February 26-March 1, 1975).

Mississippi River at Head of Passes (March 1, 1975)

Time	Concentrations (nl/L)				
	Methane	*Ethene*	*Ethane*	*Propene*	*Propane*
0930	8800	22	63	2200	810
1030	5400	<20	45	1600	670
1130	3900	<20	34	1200	440
1230	4600	<20	48	1200	300
1330	4800	<20	52	1400	440
1430	5600	<20	55	1300	370

Mississippi Delta Samples[a]

Location	Depth	Concentration (nl/L)				
		Methane	*Ethene*	*Ethane*	*Propene*	*Propane*
28° 54.1'N	Sfc	6570	7	49	295	150
89° 29.4'W	20 m	680	1.8	12	nd	7
28° 52.6'N	Sfc	11200	22	163	960	480
89° 28.8'W	18 m	2200	1.5	20	19	15
28° 59.3'N	Sfc	13700	22	100	1330	630
89° 07.1'W	9 m	4530	nd	4.7	5	5.5
29° 03.4'N 89° 12.3'W[b]	Sfc	10300	350	908	2220	960
Miss. River @ Pilot Town	Sfc	12000	20	40	2360	1110

nd: not detected

[a]River flow 750,000 cfs during this period.

[b]South Pass at Oysterville.

responsible for the higher concentrations. In general, methane concentrations decline as one moves up the Mississippi River from its outflows. There is an anomaly in this trend in Southwest Pass that is attributed to brine discharges. This same decline is seen in the other light hydrocarbons with the exception of propane. In several of the surveys, propane increased toward Head of Passes indicating an upstream source of propane. High concentrations of propene and propane were observed during cruise 75-G-1 (Table 8-2). The C_3's (propene and propane) are considerably higher in the river than the C_2's (ethene and ethane) which is opposite of what is found on the Louisiana Shelf.

Table 8-3
Mississippi River and Delta Data During Cruise 75-G-8 (July 10-14, 1975)

Station	Depth	CH_4 (nl/L)	C_2H_6[a] (nl/L)	C_3H_8[a] (nl/L)	Salinity (ppt)	Temperature (°C)	O_2 (ml/L)	DOC (mgC/L)	POC (mgC/L)	Phosphate (µg-at/L)	Silicate (µg-at/L)
New Orleans[b]											
29° 55.3′N & 89° 55.9′W[b]		6200	6	46	0	—	—	3.68	2.95	—	—
29° 51.4′N & 89° 58.5′W[b]		6110	6	42	0	—	—	3.41	2.68	—	—
29° 44.9′N & 90° 00.7′W[b]		4500	14	47	0	—	—	3.37	—	—	—
29° 36.7′N & 89° 53.3′W[b]		3040	14	46	0	—	—	3.40	—	—	—
29° 32.1′N & 89° 45.2′W[b]		2500	12	46	0	—	—	3.40	2.06	—	—
29° 26.8′N & 89° 35.2′W[b]		2040	16	43	0	—	—	3.43	1.69	—	—
		2200	15	43	0	—	—	—	—	—	—
29° 22.1′N 89° 33.1′W	Sfc	2480	29	110	0	—	—	3.55	1.81	—	—
	8 m	2070	18	70	0	—	—	3.41	3.73	—	—
Venice[c]	Sfc	1750	13	43	0	—	—	3.68	1.60	—	—
	15 m	2250	20	110	0	—	—	3.55	2.85	—	—
Pilot Town	Sfc	2540	21	47	0	—	4.03	3.19	1.43	72.7	—
	2 m	2560	18	47	0	—	4.06	3.24	1.84	61.3	—
	4 m	2580	21	44	0	—	4.03	3.15	1.78	81.7	—
	6 m	2630	21	47	0	—	3.98	3.15	1.73	96.2	—
S.W. Pass[d]	Sfc	1630	30	—	0	—	—	3.24	3.95	—	—
	7 m	1750	15	78	0	—	—	—	—	—	—
	14 m	2060	60	93	0	—	—	3.13	1.18	—	—
~1 mi. off South Pass (Station 6)	Sfc	3660	15	30	6	—	5.36	2.73	0.39	9.5	110
	2 m	4050	15	7	32	—	4.07	1.51	0.27	1.5	23
	4 m	4050	15	<10	34	—	3.68	1.01	0.25	0.9	13.2
	6 m	4160	15	<10	36	—	3.44	1.05	0.22	0.7	8.0
~1 mi. off S.W. Pass (Station 5)	Sfc	6860	190	160	4	27.14	4.45	2.93	0.33	11.8	106
	2 m	13700	145	130	32	26.08	3.15	1.39	0.17	1.2	13.8
	4 m	45900	670	740	35	25.40	2.87	1.03	0.17	1.2	13.7
	6 m	8330	—	60	36	25.04	3.16	0.98	0.19	1.2	13.7
	8 m	4450	20	15	36	23.75	3.30	0.97	0.20	1.0	11.2
~2 mi. off S.W. Pass (Station 4)	Sfc	4720	43	35	14	27.54	4.80	2.29	0.37	9.8	76.0
	2 m	3860	36	14	20	27.77	4.41	2.18	0.34	13.8	45.6
	5 m	5160	50	12	28	26.47	2.48	1.85	0.26	0.9	17.3
	10 m	2730	36	14	34	22.10	2.85	1.15	0.15	1.3	19.5
	30 m	730	<15	<10	36	20.36	3.25	0.79	0.07	0.6	7.1

[a] Includes the unsaturated analogue, ethene and ethane and propane and propene are not separated.

[b] Hydrocarbon concentrations obtained by hydrocarbon "sniffing" at three meters depth.

[c] River flow 470,000 cfs during this period.

[d] Station was halfway between the Head of Passes and Southwest Pass outlet.

The light hydrocarbons are quite variable in the river. Table 8-4 shows methane concentrations in the Mississippi River (near Head of Passes) taken sporadically over the last several years. Although one would expect concentration differences seasonally because of differences in temperature, nutrients, river flow and man-derived additions, the concentrations are quite variable even on a hourly basis. Table 8-2 shows that hydrocarbon values taken hourly on March 1, 1975, varied by as much as a factor of two. Figure 8-3 showing hydrocarbon levels in the Mississippi River between South Pass and New Orleans indicates that although the concentrations are quite variable there is a steady increase in methane upstream to New Orleans. Although propane plus propene are high, they do not show a discernible trend in this part of the river. The pattern of light hydrocarbons in the Mississippi River is complex because of the heavy use of the Lower Mississippi for transportation, petrochemical and petroleum activity, and/or sewage outfalls.

Off the Mississippi River outflows in the Gulf of Mexico, there is a large increase in dissolved methane (Figure 8-2 and Tables 8-2 and 8-3). This may be a result of the large amount of suspended material brought into the Delta region. Figure 8-2 shows the $C_1/(C_2+C_3)$ ratios in the Mississippi River and just outside of the passes. The $C_1/(C_2+C_3)$ ratios in the river are generally in the 15 to 30 range which is characteristic of light hydrocarbons derived from oil and gas (Moore and Shrewsbury, 1966). The $C_1/(C_2+C_3)$ ratios off the passes, however, can climb to over 100. The high ratios are indicative of a biogenic source of methane.

Table 8-4
Methane in the Mississippi River (Near Head of Passes)

Date	Methane (nl/L)	Temperature (°C)	O_2 (ml/L)	River Flow (cfs)	Source
1/3/75	3900-8000[a]	10	–	750,000	1
13/7/75	2540-2630[b]	28	4.0	470,000	1
14/5/74	5,600	23[c]	7.4[c]	750,000	1
17/2/73	1,350	7[c]	10.5[c]	840,000	1
9/6/71	2,430	24[d]	–	250,000	2
25/10/71	2,320	22[d]	–	180,000	2

Source: (1) Samples analyzed by McAullife's (1971) method.
              (2) Samples taken by the author and analyzed by Swinnerton et al. (unpublished data).

[a]Represents six samples taken over a six-hour period.

[b]Represents four samples taken every two meters of water depth.

[c]U.S.G.S. data at Venice.

[d]U.S.G.S. data at New Orleans.

Isotopic Analysis

Table 8-5 shows the $\delta^{13}C$ value of the dissolved methane in the Mississippi River and Delta. To the author's knowledge, these are the first isotopic carbon compositions obtained on dissolved methane in river and estuarine systems. The isotope analyses of dissolved methane at Pilottown just above the Head of Passes and outside of the passes confirms the conclusions reached from $C_1/(C_2+C_3)$ ratios about the origins of methane in the delta. The methane in the river had a carbon isotope value of -41±1 $^O/oo$ on both sampling occasions. This value which is indicative of the carbon isotopic composition of petrogenic methane, suggests that the methane at this location near the outflows of the Mississippi River is essentially man-derived. Outside of the Mississippi River outflows the $\delta^{13}C$ of the dissolved methane decreases dramatically to -59±2$^O/oo$. This represents a significant dilution of the petrogenic methane brought into the Delta by river outflow with isotopically lighter methane from the bacterial reduction of organic matter either in the carbon-rich sediments or water column.

An estimate of the natural and man-derived component of the dissolved methane can be made by making several assumptions. It is necessary to estimate the $\delta^{13}C$ content of the methane derived from petrogenic and from biogenic sources. The man-derived methane in the river from petroleum and petrochemical sources should have values in the -35 to -50 $^O/oo$ range. This is characteristic

Table 8-5
$\delta^{13}C$ of DISSOLVED METHANE in the Mississippi River and Delta

Station[a]	Date	$\delta^{13}C$[b] $^O/oo$	Vol. Collected[c] (cc)
Cruise 75-G-11			
Pilottown	8 Sept 1975	–40.4	10.2
Cruise 75-G-8			
Pilottown	13 July 1975	–41.9	15.7
~2 mi. off S.W. Pass (Sta. 4)	11 July 1975	–57.1	3.8
~1 mi. off S.W. Pass (Sta. 5)	12 July 1975	–61.9	3.9
~1 mi. off S. Pass (Sta. 6)	13 July 1975	–57.9	4.5

[a]All samples represent dissolved methane extracted from water taken three meters below the river or sea surface.

[b]$\delta^{13}C = [(R/Rstd)-1] \times 1000$ where R and $Rstd$ represent the ratio of ^{13}C to ^{12}C in samples and standard respectively. The standard is PDB_1 — the Chicago belemnite standard.

[c]Volume of CO_2 collected (methane is combusted to CO_2) for isotope analysis.

of methane derived from thermal-catalytic cracking of organic matter. Since the methane in the river has δ^{13}C values near -41 ^{O}/oo, it is assumed that the man-derived component of the dissolved methane in the river has values predominately in the -35 to -40 ^{O}/oo range. Secondly, biogenic methane is assumed to have a δ^{13}C value in the -65 to -80 ^{O}/oo range. In a core at Pilottown, Bernard and Brooks (1976) found that methane in interstitial water of river sediments had -66.7 and -70.7 ^{O}/oo isotopic values at the 10-20 and 20-30 cm core intervals, respectively. They found δ^{13}C values for dissolved methane in anaerobic sediments of the Delta region as high as -85 ^{O}/oo. Using these assumptions, it can be concluded that the methane in the Mississippi River is almost exclusively petrogenic (80 to 98%) and the methane just off the Passes is mostly biogenic (48 to 83%).

Mississippi Delta

Although there is considerable petroleum production in the Mississippi Delta region, it appears from molecular and isotopic compositions that the methane within a few miles of the Mississippi River outflows is predominantly biogenic in origin. This means that the majority of the methane off the passes does not originate directly from the Mississippi River outflows, but rather from "in situ" production either in the water column and/or in the sediments. Since the sediments in much of the Mississippi Delta region are anaerobic, there has to be some biogenic methane seepage out of the sediments. However, if this was the major source of the biogenic methane found in the water column, one would expect to observe a methane gradient increasing towards the sediments in vertical profiles. Although in many stations (Tables 8-2, 8-3 and 8-6) there is a slight increase in dissolved methane off the bottom, in no station is there a significant gradient above the sediments. The high methane maximum is always found at some mid-depth suggesting that the majority of the methan originates from "in situ" production in the water column.

Vertical profiles at stations 4 and 5 of cruise 75-G-8 (Table 8-3) show methane maxima in the upper 10 meters directly off the Southwest Pass outflow. Since the maxima exist at mid-depths and are biogenic in origin according to δ^{13}C values (Table 8-5), they have to originate "in situ" in the water column or possibly from mixing of biogenic methane from seepage out of the sedimentary column in some other source area. No profiles taken in this region support a biogenic source for methane from seepage out of sediments that could account for these water column maxima. The methane maxima in stations 5 and 6 are accompanied by oxygen minima. Although the data suggest an "in situ" production of methane in the water column, theoretically bacteria do not produce methane in the presence of oxygen. One explanation for this occurrence is the possible production of methane in small reducing micro-environments in the particulate material.

Table 8-6
Light Hydrocarbons and Hydographic Data in the Mississippi Delta Region (Cruise 75-G-8)

Depth (meters)	Methane (nl/L)	Ethane[a] (nl/L)	Temperature (°C)	Salinity (ppt)	Oxygen (ml/L)
Station 26	July 9, 1975		28° 47.0'N & 89° 31.5'W		
Sfc	4,030	–	23.53	21.511	4.951
10	6,700	–	27.86	34.361	4.499
20	5,600	–	26.65	35.823	4.799
30	1,100	–	23.08	35.973	4.451
40	790	–	22.36	36.166	4.850
50	500	–	21.05	36.177	3.910
60	270	–	20.28	36.324	3.389
70	250	–	19.73	36.342	3.127
80	290	–	19.65	36.351	3.092
Station 25	July 8, 1975		28° 30.0'N & 89° 53.0'W		
Sfc	970	90	29.50	26.978	4.825
10	1,100	100	29.06	34.405	4.727
20	380	50	29.69	35.680	4.592
30	1,300	60	28.27	35.995	4.746
40	9,660	110	24.28	35.833	4.942
50	2,400	50	23.20	36.112	4.940
60	3,100	50	21.52	36.079	4.487
70	850	50	21.16	36.098	4.288
80	1,300	50	20.46	36.198	3.734
90	330	50	18.89	36.340	3.091
100	380	50	18.01	36.326	3.029
Station 24	July 8, 1975		28° 11.5'N & 90° 28.1'W		
Sfc	2,500	190	29.18	29.226	4.907
10	4,530	290	29.36	33.532	4.734
20	7,660	550	29.50	34.420	4.794
30	3,530	80	25.72	35.442	–
40	1,940	<70	22.33	35.857	4.665
50	2,110	<70	21.94	36.095	4.662
60	1,350	<70	21.23	36.104	4.053
70	710	<70	20.71	36.162	3.519
80	330	<70	20.32	36.259	3.015

[a]Represents ethene plus ethane.

Table 8-6 shows three profiles in the delta region taken 7, 35 and 70 miles southwest of Southwest Pass. It is thought that the methane maxima in these profiles also originates from similar in situ production in the water column as the water moves off the shelf and sinks. The patterns are so complex in the delta region, however, that one can not correlate satisfactorily the maxima found in Table 8-6 with sinking of water from stations 5 and 6 along thermosteric anomalies. It appears probable, however, that "in situ" production of methane continues as the water moves away from the delta region, since some of the methane maxima are higher in stations away from the delta than just off the Mississippi River passes.

Other Rivers

Sampling for methane has also been done in other rivers. Table 8-7 shows methane, river flow, nutrients and hydrographic data for stations taken in the Brazos River, Texas. This river was sampled at one station periodically over a three-month period during the spring of 1975. Methane values were more stable in this river than the Mississippi, although there were some fluctuations that could be correlated roughly with river flow. There appears to be an inverse relationship between methane content and river flow in this river. Unlike the Mississippi River, methane is the only light hydrocarbon detectable (limit of detection 5nl/L) in the Brazos River. It appears from the high $C_1/(C_2+C_3)$ ratios in the river that the methane originates from the breakdown of natural and domestic organic matter. The Coatzacoalcos River, Mexico has also been sampled in this study. It has methane concentrations around 4000 nl/L and C_2-C_4 hydrocarbon levels comparable to the Mississippi River (Brooks, 1975).

Inputs to the Gulf of Mexico

Assuming a runoff of 8.4×10^{14} l/yr from the Mississippi–Atchafalaya River system and other U.S. Gulf coast rivers (Moody, 1967), and assuming that measured values for the Mississippi River are typical for the total runoff, the following inputs into the Gulf of Mexico can be estimated:

Input of methane:	2.5×10^9 l/yr (assume 3000 nl/l)
Input of C_2-C_4:	1.3×10^8 l/yr (assume 150 nl/l)

Assuming the input of the Coatzacoalcos River is typical of the 2.2×10^{14} l/yr runoff from Mexican rivers into the Gulf, the following inputs can be estimated:

Input of methane:	0.9×10^9 l/yr (assume 4000 nl/l)
Input of C_2-C_4:	4.4×10^7 l/yr (assume 200 nl/l)

Thus the input of methane from rivers is estimated at 3.4×10^9 l/yr. This compares to a coastal Gulf of Mexico methane reservoir of 13×10^9 liters that has a residence time of approximately 10 days on the continental shelf (Brooks, 1975).

Table 8-7
Methane, Nutrient and Hydrographic Data in the Brazos River, Texas

Station	Date	CH$_4$ (nl/L)	River Flow (cfs)	Temp. (°C)	O$_2$ (ml/L)	DOC (mgC/L)	POC (mgC/L)	DOC/POC	Nitrate (μg-at/L)	Phosphate (μg-at/L)	Silicate (μg-at/L)
Bryan U.S.G.S. Guage	3/11/75	5650	7440	14.3	—	3.82	1.02	3.8	207	200	1428
Bryan U.S.G.S. Guage	3/12/75	3760	7050	15.8	—	3.43	1.20	2.9	290	192	1170
Bryan U.S.G.S. Guage	3/13/75	5030	5620	12.9	—	3.25	1.19	2.7	274	192	936
Bryan U.S.G.S. Guage	3/14/75	6140	5460	12.6	—	3.34	1.31	2.6	390	394	996
Bryan U.S.G.S. Guage	3/15/75	4870	6400	12.3	—	3.16	1.27	2.5	371	264	876
Bryan U.S.G.S. Guage	3/16/75	5030	6130	13.7	—	3.46	1.66	2.1	423	366	750
Bryan U.S.G.S. Guage	3/17/75	5590	5680	14.0	—	3.16	1.19	2.7	441	264	990
Bryan U.S.G.S. Guage	3/18/75	4113	6100	14.9	—	3.82	1.75	2.2	549	228	1122
Bryan U.S.G.S. Guage	3/19/75	3800	6340	15.5	—	3.40	1.14	3.0	402	229	1014
Bryan U.S.G.S. Guage	3/20/75	5500	4670	17.3	—	3.27	1.13	2.9	438	194	780
Bryan U.S.G.S. Guage	3/21/75	6090	3700	17.8	—	3.55	1.15	3.1	468	195	1200
Bryan U.S.G.S. Guage	3/22/75	6450	3290	20.3	—	3.19	1.15	2.8	522	174	1152
Bryan U.S.G.S. Guage	3/23/75	5160	3050	20.7	—	3.28	1.14	2.9	537	246	924
Bryan U.S.G.S. Guage	3/24/75	4890	2850	19.9	—	3.10	1.36	2.3	420	216	1242
Bryan U.S.G.S. Guage	3/27/75	2860	4430	20.2	—	3.46	1.77	2.0	291	194	750
Bryan U.S.G.S. Guage	3/28/75	2360	4610	19.8	6.19	—	—	—	672	194	936
Bryan U.S.G.S. Guage	3/29/75	2790	3460	14.8	6.86	3.46	0.90	3.8	396	186	1196
Bryan U.S.G.S. Guage	3/30/75	3570	3310	14.0	6.98	3.40	1.06	3.2	462	206	1011
Bryan U.S.G.S. Guage	3/31/75	4180	3270	13.7	7.34	3.33	0.80	4.2	444	224	780
Bryan U.S.G.S. Guage	4/4/75	3420	3100	16.5	—	3.22	0.84	3.8	405	188	1062
Bryan U.S.G.S. Guage	4/29/75	2470	8120	22.5	—	4.91	4.68	1.1	570	316	1124
Bryan U.S.G.S. Guage	5/7/75	2640	13300	23.0	—	4.97	4.97	0.7	663	332	1668
Bryan U.S.G.S. Guage	5/11/75	—	9300	23.5	—	—	—	—	368	430	1266
115 mi.[a]	4/5/75	7150		15.9	6.70	3.34	0.91	3.7	500	234	1158
Navosota R.	4/5/75	21500		15.5	6.28	6.55	1.80	3.6	126	806	2600
110 mi. to outflow[a]	4/5/75	6150		15.9	6.48	2.79	0.81	3.4	540	254	1422
105 mi. to outflow[a]	4/5/75	23700		16.6	6.87	3.94	1.45	2.7	350	270	1188
85 mi. to outflow[a]	4/5/75	12200		16.8	7.48	3.63	1.48	2.5	275	222	1422
75 mi. to outflow[a]	4/5/75	11300		16.9	7.29	3.46	1.33	2.6	250	197	1302
50 mi. to outflow[a]	4/5/75	8500		16.8	7.08	3.82	1.31	2.9	231	258	1446
35 mi. to outflow[a]	4/5/75	4590		17.0	7.36	3.70	1.59	2.3	254	196	1698
25 mi. to outflow[a]	4/5/75	8800		17.3	7.27	3.94	1.46	2.7	395	302	1410
1.2 mi. to outflow[a]	4/5/75	9000		18.4	6.23	3.88	1.00	3.9	164	236	756
0.8 mi. to outflow[a]	4/5/75	12000		18.4	5.92	3.88	1.15	3.4	230	416	1050

[a]Miles from the Brazos River outflow represent straight line distances, not river miles.

Conclusions

Light hydrocarbons in rivers originate from both natural and man-derived sources. It appears from molecular and isotopic data that the light gaseous hydrocarbons (C_1 to C_4) in the Mississippi River originate predominately from man-related sources. This is the result of the extensive use of the lower Mississippi for petroleum and petrochemical activities. In more agrarian rivers, such as the Brazos River, Texas, the methane originates principally from the bacterial breakdown of natural and man-derived organic matter.

The light hydrocarbons discharged into the Gulf of Mexico by rivers have significant impact on the coastal waters. Hydrocarbon anomalies are seen typically from 10 to 20 miles off port and estuaries and as much as 50 miles off the Mississippi River. The light hydrocarbons introduced into the surface layer of the ocean are rapidly lost to the atmosphere. The residence time of methane and other gaseous hydrocarbons in the mixed layer of the ocean is on the order of days (Brooks, 1975).

Rivers also have an influence on the light hydrocarbon concentrations in coastal waters because of the suspended material they carry. Some of the organic matter in the suspended material appears to be reduced slowly to methane possibly in small micro-reducing environments. This methane formation seems to occur *in situ* in the water column forming a maximum at some depth in the upper hundred meters in the Mississippi Delta region. As the water in the delta region spreads across the shelf, the methane maximum in the water column may increase as the water moves away from the delta. The extent of methane formation in the methane maximum and the movement and fate of the maximum in the Gulf of Mexico are poorly understood.

Acknowledgments

The help of W.M. Sackett and B.B. Bernard in collection and analyses of some samples is gratefully acknowledged. This work was supported by NSF Grant Nos. GX-37344 and GX-42576.

References

Bernard, B.B. and J.M. Brooks (1976). Methane in anaerobic marine sediments (unpublished data).

Brooks, J.M. (1975). Sources, sinks, concentrations and sublethal effects of light aliphatic and aromatic hydrocarbons in the Gulf of Mexico. Technical Report 75-3-T, Dept. of Oceanography, Texas A&M University, 342 pp.

Brooks, J.M. and W.M. Sackett (1976). Significance of low-molecular-weight hydrocarbons in coastal waters. *Advances in Organic Geochemistry, 1975* (in press).

Brooks, J.M. and W.M. Sackett (1973). Sources, sinks and concentrations of light hydrocarbons in the Gulf of Mexico. *J. Geophys. Res., 78,* 5248-58.

McAullife, C. (1971). Gas chromatographic determination of solutes by multiple phase equilibrium. *Chemical Technology, 1,* 46.

Moody, G.L. (1967). Gulf of Mexico distributive province. *Am. Assoc. Petroleum Geol. Bull., 51,* 179-99.

Moore, B.J., R.D. Miller and R.D. Shrewsbury (1966). Analysis of natural gases of the United States, 1964. U.S. Bur. Mines, Information Circular 8302.

Sackett, W.M. and J.M. Brooks (1974). Use of low-molecular-weight hydrocarbon concentrations as indicators of marine pollution. NBS Spec. Publ. 409, *Marine Pollution Monitoring (Petroleum),* Proceedings of a Symposium and Workshop held at NBS, Gaithersburg, Maryland, May 13-17.

Sackett, W.M., S. Nakaparksin and D. Dalrymple (1970). Carbon isotope effects in methane production by thermal cracking. *Advances in Organic Geochemistry,* (G.D. Hobson, ed.), New York: Pergamon Press, pp. 37-53.

Sport, M.C. (1969). Design and operation of gas flotation equipment for the treatment of oil field produced brines. Paper #OTC 1015, presented at the First Offshore Technology Conference, Houston, Texas.

**Transfer of Petroleum Residues
from Sea to Air: Evaporative
Weathering**

J. N. Butler

Introduction

The age and weathering processes of pelagic tar (petroleum residues found on
the open ocean) have been at issue since the first observations were published
(Horn *et al.*, 1970). Combining estimates of the standing stock of tar with estimates of the amount of oil spilled annually and estimates of the fraction of this
spilled oil which remains as pelagic tar, a residence time of the order of a year
was obtained (Morris, 1971; Butler *et al.*, 1973). Is there any more direct way
of predicting the rates of hydrocarbon pollutant transfer to the atmosphere and
water column? This paper develops a simple semiquantitative model of evaporative weathering which predicts the transport of petroleum components to the
atmosphere from the sea surface. An attempt to determine the age of individual
tar lumps is described elsewhere (Butler, 1975; Ehrhardt and Derenbach, 1976).

The various modes of dispersion and decomposition of oil spilled at sea have
been discussed elsewhere (Blumer *et al.*, 1973; National Academy of Sciences,
1975) and may be briefly summarized as evaporation, dissolution, emulsification, photochemical oxidation, and microbial degradation. Evaporation and dissolution are closely related, since for nonpolar compounds, at least, the equilibrium solubility is so low that most of the material partitions into the atmosphere. Emulsification, should it occur, would retain the petroleum in the water
column but would not lead to the formation of pelagic tar. Photochemical oxidation and microbial degradation are complex and will not be considered here
except in passing.

Derivation of the Model

Let x_0 be the amount of a particular component of a crude oil present at the
beginning of weathering ($t = 0$), and let x be the amount of that same compound at time t. Assume that the weathering rate is proportional to the equilibrium vapor pressure P of the compound and to the fraction remaining:

$$dx/dt = -kP\,(x/x_0) \tag{9.1}$$

Since petroleum is a complicated mixture of compounds, one would not expect
P to be precisely equal to the vapor pressure of the pure compound, but neither

would one expect large variations in the activity coefficient as the weathering process occurs. For this reason, the activity coefficients can be subsumed in the empirical rate coefficient k, which depends primarily on evaporation rates, concentration gradients, diffusion coefficients, and sample geometry.

Furthermore, if P and k are independent of the amount x remaining over a fairly wide range of x, the equation can be directly integrated to give the fraction of the original compound remaining after weathering:

$$x/x_0 = \exp(-ktP/x_0) \tag{9.2}$$

On continued evaporation of volatile material from the surface of a tar lump, the surface layers would become depleted in that material, and eventually the rate of disappearance of that component could become limited by diffusion from the center portion. If x were taken to be the average amount in the entire lump (instead of the amount at the surface) then diffusion limitation would result in a decrease in k, and hence a slower decay with time for long times and for larger values of x_0. For well-defined geometries, the differential equations of this more complex model are straightforward but will not be developed in detail here.

Although in principle P could be obtained for each compound of interest resolved by the chromatogram, it is useful to obtain a simple correlation of P with carbon number so as to facilitate interpretation of chromatograms which may be poorly resolved.

For example, Blumer et al. (1973) employed a number of parameters to describe the gas chromatograms of crude oil in various stages of weathering. In particular, the "equivalent n-paraffin carbon numbers where environmental losses reach 50% (C_n^{50}) and 90% (C_n^{90})", express the qualitative observation that the initial slope of either the unresolved background or the envelope of normal paraffin peaks rises quite steeply, and that the position of this steep rise would be expected to retreat to higher carbon numbers as weathering proceeds.

Some vapor pressure data (Weast, 1971) for hydrocarbons containing 6 to 18 carbon atoms (N) are displayed in Figure 9-1. A regression line was fitted to the data for normal paraffins (P is in torr):

$$\log_{10}P = 4.75 - 0.46\,N \tag{9.3}$$

or

$$P = \exp(10.94 - 1.06\,N) \tag{9.4}$$

For these coefficients, the 95% confidence limits on $\log_{10}P$ are ±0.08 logarithmic units or about ±20% on P. Other hydrocarbons, including branched alkanes, cyclo-alkanes, alkyl benzenes, alkyl naphthalenes, and other polycyclic

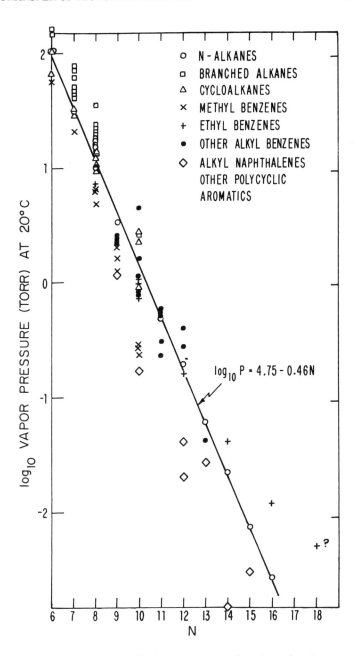

Figure 9-1. Vapor pressures of hydrocarbons as a function of carbon number N. The regression line predicts $\log_{10}P$ for normal alkanes within 95% confidence limits of ±0.08

compounds in this molecular weight range cluster about the normal alkane line with confidence limits of the order of ±0.4 logarithmic units or a factor of 2.5. This implies that the effective vapor pressure of the unresolved compounds of the usual gas chromatogram will be similar to the normal paraffin of the same carbon number. In fact, it is likely that the correlation is even more precise, since a nonpolar silicone or Apiezon column will tend to show retention times which correlate more precisely with vapor pressure at the column temperature than with carbon number *per se*.

Thus, if the normal alkane peaks of the chromatogram are used as calibration points and the retention times of other components referenced to that scale, one may use this model to predict the shape of the chromatogram obtained after varying times of weathering. Combining Eqs. 9.2 and 9.4 yields

$$x/x_0 = \exp[-(kt/x_0)\exp(10.94 - 1.06\,N)] \tag{9.5}$$

This predicts that the fraction weathered at a given time decreases more than exponentially with increasing carbon number. If the initial distribution of compounds is essentially uniform (x_0 independent of N), then Eq. 9.5 predicts that the carbon number where a constant fraction (e.g. half) the initial amount has been lost ($x = 0.5\,x_0$) is a logarithmic function of the time of weathering:

$$N_{1/2} = 10.66 + 2.17\log_{10}(kt/x_0) \tag{9.6}$$

If the amount of material initially present x_0 and the coefficient of the evaporation rate k (actually, their ratio k/x_0) can be approximately evaluated, this equation provides a one-parameter semiquantitative estimate of the age of a weathered petroleum residue from its chromatogram. Even if neither the parameters nor their ratio can be evaluated absolutely, Eq. 9.6 predicts the slope $dN_{1/2}/d(\log t) = 2.17$ for constant k/x_0.

Test of the Model

The experiments of Kreider (1971) on simultaed weathering, carried out by "pouring a pint of crude oil onto the surface of a barrel of seawater exposed to the weather", provide some data to test Eqs. 9.5 or 9.6 at times from 1 to 21 days.

His observations on the weathering of a 0.5-mm film over a period of 21 days can be used to test the time-dependence as well as the dependence on N predicted by Eq. 9.5. This comparison is displayed in Figure 9-2. The points are experimental peak heights (x_0 was taken to be the peak height in the initial sample) and the curves are calculated using $k/x_0 = 35$ torr^{-1} day^{-1}; a value arbitrarily chosen to make a central point (on the four-day curve, $x/x_0 = 0.38$

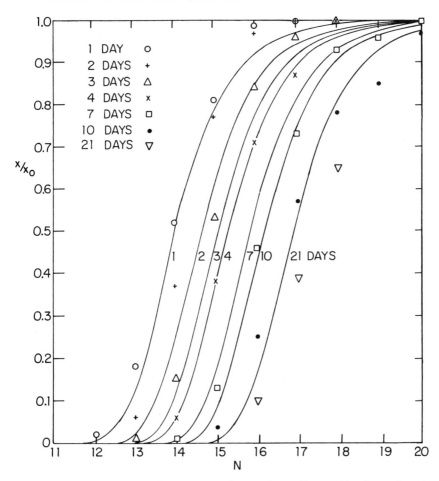

Figure 9-2. Comparison of experimental normal paraffin profiles (points) with curves calculated from Eq. 9.5 using $k/x_0 = 35$.

at $N = 15$) coincide with the family of calculated curves. No other adjustments were made.

A test of the model covering a longer time span can be made using the data of Blumer *et al.* (1973) on the weathering of crude oil stranded on rocky shores of Martha's Vineyard and Bermuda. Here, unfortunately, the age of the sample is relatively uncertain at short times, since the exact date of the original spill was unknown. However, by analogy with the weathering data of Kreider (1971), the samples were probably not more than two weeks old at the time they were first observed, and thus the relative uncertainty in their age is small at times of the order of many months.

Since well-resolved chromatograms were available, Eq. 9.5 was used directly

with individual components to obtain k/x_0. Since a chromatogram of the unweathered crude oil was not available, x/x_0 was taken to be the ratio of a di-
minished normal alkane peak ($N = 17$ to 20) to an undiminished peak ($N > 21$).
Figure 9-3 shows values of k/x_0 obtained from the heights of normal alkane
peaks on the same chromatograms (of Bermuda crude oil residue) used by
Blumer et al. (1973). (These chromatograms were published by Butler et al.,
(1973).) In these calculations, the zero of the time scale was taken to be at the
time the oil was first stranded; this might lead to an underestimation of the age
by 4 to 14%. Other errors may result if the initial peak heights were not independent of N. The values of k/x_0 fall within a factor of two on either side of
10, with little dependence on N. Note that at the longest time (388 days) there
is a relatively sudden qualitative change in the chromatograms which results in
much larger values of k/x_0 and a strong dependence on N. This effect is probably due to the onset of microbial attack, since it coincides with a rapid decrease
in the ratio of normal alkane peaks to the unresolved envelope and to the more
resistant isoprenoids (Blumer et al., 1973).

It should be noted here that if diffusion were an important factor limiting
the rate of disappearance of the C_{17} to C_{20} paraffins, one would expect that the
value of k/x_0 would be effectively smaller for the more volatile components at
the longest times. This would show up on Figure 9-3 as a systematic deviation
at longer times. Within experimental error, there is no such deviation observed.

Although the geometry of the evaporating oil film in Kreider's experiment
is simple, the corresponding geometries of the crude oil stranded on rocky shores
in Bermuda and Martha's Vineyard are less well defined. The Bermuda sample
was "25 cm dia and 2.5 cm thick"; the Martha's Vineyard sample was "less than
5 cm dia." In addition, different portions of the sample appear to weather at
different rates. This was pointed out by Blumer et al. (1973) in their designa-
tions of analyses as being from the "main," "crust," or "wax" of the Bermuda
sample.

Thus the little data available on weathering of crude oil in the marine environment where chronological age can be assigned to samples analyzed by gas
chromatography tends to confirm the simple evaporative weathering model
proposed. Additional experiments conducted under more precisely defined conditions should be conducted, of course, particularly at long times of weathering.

Comparison with Dissolution

A recent study of experimental ($1 m^3$) oil spills spiked with cumene (Harrison
et al., 1975) illuminated quantitatively the distinction between evaporation and
dissolution as modes of weathering. Even though cumene has a solubility of the
order of 50 mg/l as compared with 0.2 mg/l for n-nonane (which has a similar
vapor pressure) cumene is lost principally by evaporation. The model developed

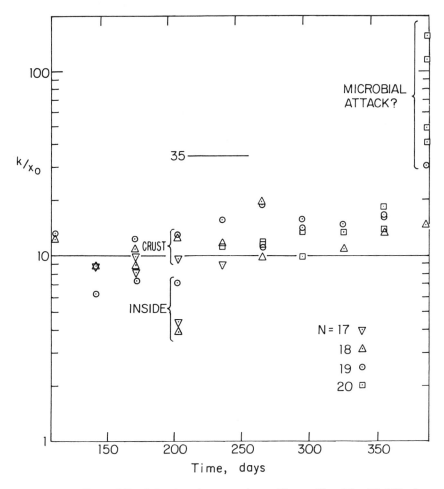

Figure 9-3. Test of Eq. 9.5 using the normal paraffin profiles ($N = 17$–20) of the crude oil weathering on a rocky shore in Bermuda. Perfect agreement would result in a value of k/x_0 independent of both N and t. The sudden breakdown of this consistency at 400 days may indicate the onset of rapid microbial attack on the normal paraffins of the sample.

by Harrison et al. is slightly more elaborate than the one presented above, but is based on the same principles. Especially for the n-paraffins and related compounds, which have even lower solubilities in seawater: 0.1 µg/l for C_{26} in seawater as compared to 1.7 µg/l in distilled water (Sutton and Calder, 1974), their new results confirm that *evaporation will be the principal mode of weathering compared to dissolution.* Only highly polar materials (e.g., those containing nitrogen, sulfur, and oxygen) would tend to be preferentially dissolved.

Discussion

Although the evaporative weathering model does not take account of all the complexities of the weathering process, it still can estimate the rate of transfer of petroleum residues from the ocean surface to the atmosphere. Critical factors in this estimate are the initial volatility distribution of the petroleum input and factors which influence mass transport such as the geometry of the residue (slick or globule? how large?) and the state of the sea. In particular, if the input is a refined fuel product, one would expect a much smaller fraction of it to remain as residue than if the input were a crude oil sludge.

If the parameters derived from the experiments discussed above are used with volumes typical of pelagic tar lumps, the normal paraffin profile predicted by the evaporative weathering model shows much more degradation than is actually observed (Butler, 1975; Ehrhardt and Derenbach, 1976). One possibility is that the diffusion coefficients in the almost perfectly isothermal pelagic tar lumps are so much lower than in either the laboratory or the rocky-shore situation (where the oil is heated by the sun part of the time) that k/x_0 for lumps about 1 cm^3 volume may be of the order of 1 torr^{-1} day^{-1} cm^3 (Ehrhardt and Derenbach, 1976). An alternative explanation (Butler, 1975) is that the pelagic tar lumps collected are but fragments of the original masses which began to weather. Undoubtedly, both these explanations are qualitatively correct, and we will present data in another paper in this volume regarding the particulate hydrocarbons in the water column which are probably the product of disintegrating pelagic tar lumps.

Steady State Model

The evaporation rate for petroleum from a region of the ocean surface results from a distribution of partly weathered residues, which may be approximated by a steady state model. Equations 9.1 and 9.4 give the rate of evaporation of any normal paraffin of chain length N (or compound of similar volatility) and if a given area of the ocean is considered a continuously stirred reactor, then the rate of change with time for each component is just equal to the evaporation rate for each component plus a source term R_n:

$$dx_n/dt = -A(x_n/x_0)\exp(10.94 - 1.06n) + R_n = 0 \tag{9.7}$$

However, petroleum is a complex mixture, and for the sake of simplicity, assume that the initial mixture consists of equal amounts x_0 of normal paraffins from one to 40 carbon atoms. Then the total input rate R (sum of the R_n) depends on the profile (values of the various x_n) at steady state:

$$R = A \sum_{n=1}^{40} (x_n/x_0)\exp(10.94 - 1.06\,n) \tag{9.8}$$

If $x_n = x_0$ for all n, R must be extremely large; if $x_n = 0$ for $n<12$, R is a factor of 10^5 smaller. Suppose the profile at steady state is the median for the pelagic tar lumps collected at Station S near Bermuda (Butler *et al.*, 1973; Butler, 1975) with $N_{1/2} = 16.7$. We can model this by setting $x_n = 0$ for $n<16$, $x_{16} = 0.1\,x_0$, $x_{17} = 0.5\,x_0$, $x_{18} = 0.9\,x_0$ and $x_n = 1$ for $n>18$ (cf. Figure 9-2.). This leads to the input rate required to maintain this profile: $R = 10^{-4}\,(A/x_0)$. The coefficient A in this expression is not the same as k in Eq. 9.1, but applies to a *distribution* of pelagic tar lumps of varying sizes disposed on the ocean surface, and thus A/x_0 would not be easily predictable either from Kreider's experiments (1971) or the rocky shore field studies of Blumer, Ehrhardt and Jones (1973) which were discussed earlier.

However, the amount of petroleum spilled into the currents which feed the Sargasso Sea has been estimated (NAS, 1975); and a rough value for the steady input rate R is 70 mg/m^2 yr. From the profile steady state model discussed above ($N_{1/2} = 17$) approximately 40% of the steady input must evaporate, and the rest forms pelagic tar. (Actually, since it appears as if much of the pelagic tar is the residue of tanker sludge, which contains relatively more of the high-boiling components, one might expect that an even smaller fraction of the input evaporates.)

To the extent that A/x_0 is constant, this steady state model can be used to estimate the fluxes of petroleum residues in parts of the world oceans where the inputs are not well known. This requires analysis of a representative set of pelagic tar samples for their normal paraffin profiles. For example, if the median of $N_{1/2}$ is 15 instead of 17, the flux of crude oil and the flux of evaporation from the sea surface would be approximately 10 times as great (this follows from Eq. 9.8). Of course, a different source of crude oil, a different type of sea conditions, a different temperature, all will affect the coefficient A/x_0; but perhaps the differences are not too large. As more analyses of pelagic tar from the world oceans become available, it will be possible to verify this idea quantitatively.

Residence Time of Pelagic Tar

Even though the amount of pelagic tar in the Sargasso Sea may fluctuate from place to place and from time to time, the average amount at Station S near Bermuda (Morris, 1971; Butler *et al.,* 1973; NAS, 1975) is approximately 10 mg/m^2 of ocean surface, and values from other parts of the Sargasso Sea fall within the range of values obtained near Bermuda. If the rate of production of residue is 40 mg/m^2 yr (see above), the residence time of pelagic tar at steady state is only one fourth of a year. This is lower than our previous estimates but not so low as the one to four months estimated by McAuliffe (private communication) and of course depends critically on the estimate of the input rate.

Since tar lumps have been collected with organisms such as goose barnacles (*Lepas pectinata*) growing on them which appear from their size to be up to four months old (Horn *et al.,* 1970; Ehrhardt and Derenbach, 1976; Morris, unpublished) these observations too are not inconsistent with the above estimates.

The global fluxes are of course much more uncertain, but if the total petroleum input to the marine environment is estimated (NAS, 1975) to be 5 × 10^{12} g/yr, then evaporation from the sea would be expected to be roughly half that amount. Naturally, the distribution of this input is not uniform, and the marine environment is not the only source of atmospheric hydrocarbons. Far more significant than the global flux would be the local input and removal rates in sensitive ecological areas, but modelling these is a real challenge. The Sargasso Sea may be large, but it is conceptually simple compared to a shallow estuary.

In summary, evaporation of petroleum residues appears to be an important mechanism for transfer; certainly in the initial phases of weathering it is much more rapid than any other process. The little data that are available for longer times indicate that evaporation is probably also the most important weathering process for at least several months of the life of a petroleum residue on the surface of the open ocean.

Acknowledgments

This work was supported by the National Science Foundation, International Decade of Ocean Exploration Grant No. IDO 72-06411 and a grant from the Zemurray Foundation. Part of the material in this paper was previously published in *Marine Chemistry,* Volume 3, pp. 9-21, © 1975 by the Elsevier Scientific Publishing Co., Amsterdam, and is reprinted by permission.

References

Blumer, M., M. Ehrhardt and J. H. Jones (1973). The environmental fate of stranded crude oil. *Deep-Sea Res., 20,* 239-259.

Butler, J. N., (1975). Evaporative weathering of petroleum residues: the age of pelagic tar. *Marine Chemistry, 3,* 9-21.

Butler, J. N., B. F. Morris and J. Sass (1973). Pelagic tar from Bermuda and the Sargasso Sea. Bermuda Biological Station for Research, St. George's West, Spec. Publ., *10*, 346 pp.

Ehrhardt, M. and J. Derenbach (1976). Composition and weight per area of pelagic tar collected between Portugal and south of the Canary Islands. *Meteor. Forschungsber., Reihe A.* (in press).

Harrison, W., M. A. Winnick, P. T. Y. Kwong and D. McKay (1975). Crude oil spills: disappearance of aromatic and aliphatic components from small sea-surface slicks. *Environmental Science and Technology, 9,* 231-234.

Horn, M. H., J. M. Teal and R. H. Backus (1970). Petroleum lumps on the surface of the sea. *Science, 168,* 245-246.

Kreider, R. E. (1971). Identification of oil leaks and spills. *Proc. Joint Conf. Prevention and Control of Oil Spills,* American Petroleum Institute, pp. 119-124.

Morris, B. F. (1971). Petroleum: Tar quantities floating in the northwestern Atlantic taken with a new quantitative neuston net. *Science, 173,* 430-432.

National Academy of Sciences–National Research Council, (1975). Petroleum in the Marine Environment. Washington, D. C.

Sutton, C. and J. A. Calder (1974). Solubility of higher molecular weight *n*-paraffins in distilled water and seawater. *Environmental Science and Technology, 8,* 654-657.

Weast, R. C., Ed. (1971). *CRC Handbook of Chemistry and Physics.* Chemical Rubber Co., Cleveland, 52nd ed., pp. D-151 – D-170.

10 Transfer of Particulate Hydrocarbon Material from the Ocean Surface to the Water Column

B. F. Morris, J. N. Butler,
T. D. Sleeter and J. Cadwallader

Introduction

Long-lived petroleum residues ("pelagic tar") released on the surface of the sea by crude oil tankers in the process of tank cleaning and deballasting now commonly occur in many oceanic regions (Butler *et al.,* 1973; NAS, 1975). Highest concentrations occur in the Sargasso and Mediterranean Seas. In the Sargasso Sea during 1970–72, the concentration of tar lumps (several mm diameter or larger) collected with neuston nets averaged 9.4 mg/m^2 of ocean surface (Morris and Butler, 1973); in the Mediterranean Sea during December 1974–January 1975, the average amount of these larger tar lumps was 9.7 mg/m^2 (Morris *et al.,* 1975).

What Is The Fate of Pelagic Tar?

Thus far, the fate of pelagic tar lumps such as those collected by neuston nets at the ocean surface has been largely unknown. In the open ocean, especially in central oceanic gyres such as the Sargasso Sea, stranding on beaches cannot be a major removal mechanism. Evaporation is limited to the more volatile fractions (below C_{17}).

What happens beyond the initial evaporation and the possible microbial action (which has, incidentally, never been observed under open-ocean field conditions) depends to a large extent on how long pelagic tar lumps remain at the ocean surface. They may eventually sink because of the accumulation of calcareous organisms such as barnacles (Horn *et al.,* 1970). Photo-oxidation and microbial degradation appear *not* to be rapid on the open ocean (NAS, 1975). The one field study (Blumer *et al.,* 1973) of degradation under conditions (rocky shores, Bermuda) which might approximate the open ocean, has clearly indicated that evaporative weathering (Butler, 1975) is the important process for over a year, before microbial action appears to become important.

Age of Pelagic Tar

An evaporative weathering model (Butler, 1975) has been used to relate the age of a petroleum residue to the major features of its gas chromatogram. However,

213

the theory requires values for the evaporation rate coefficient k and the initial amount of each component x_0, neither of which is known accurately for pelagic tar lumps.

Gas chromatograms as well as sample volumes of 40 tar lumps collected at Station S near Bermuda from May to October 1972 (Butler, *et al.*, 1973) were measured (Butler, 1975, Table II). The volumes varied over a factor of 100 (from 0.06 cm^3 to 8 cm^3) and the degree of weathering as measured by $N_{1/2}$ (carbon number of normal paraffin where weathering has reduced the peak to half the height of a relatively unweathered peak) varied from 14 to 23, with median 16.7. However, relation of this parameter to chronological age is not a trivial problem.

Organisms provide a lower limit to the age of some samples. Horn *et al.*, (1970) collected tar lumps with barnacles (*Lepas pectinata*) up to four months old growing on them, so the lumps must have been at least that old, if not older.

Ehrhardt and Derenbach (1976) collected tar lumps with a volume less than 1 cm^3 which had barnacles attached to them and therefore must have been at least several weeks old, but also had $N_{1/2}$ as low as 15. Although such an observation does not invalidate evaporative weathering as a model, it does imply that the rate of hydrocarbon loss for these pelagic samples is slower than would be predicted using the constants derived from larger crude oil samples in the laboratory or on rocky shores (Butler, 1975). Ehrhardt and Derenbach proposed, for example, that because of the high viscosity of petroleum, the diffusion coefficient of n-tetradecane e.g. would be of the order of 10^{-8} cm^2/sec, which in turn implies that the age of tar lumps about 1 cm^3 with $N_{1/2} = 14$ could be as long as 250 days. Those with higher $N_{1/2}$ could be even older.

On the other hand, the few large tar lumps which have been sectioned and analyzed by M. Blumer and J. Sass (data given in Butler *et al.*, 1973) have shown relatively high values of $N_{1/2}$ and little variation in $N_{1/2}$ between the surface and the interior of the lump. This implies that evaporation, not diffusion within the lump, is rate-controlling. For example, a 50-g fragment of a 155-g lump (Series 125) had $N_{1/2} = 21.0$ to 21.5 on the outer crust and 20.3 to 21.4 on the fracture face. Applying Eq. 8 (of Butler 1975) with $N_{1/2} = 21$ one obtains an apparent age of 450 days. This large tar lump was apparently still nearly the same size as when it was formed, or it would be many years old at the time it was collected. Of course, by this model, with the same parameters, a sample with $N_{1/2} = 14 - 15$ would have to be formed in a very short time from a very much larger sample, and this is not consistent with Ehrhardt and Derenbach's (1976) observations.

Physical Disintegration

Until recently, it has been tacitly assumed that the size of pelagic tar lumps collected at sea was a substantial fraction (10–30%) of the original globule of crude oil or lump of crude oil sludge from which the tar lumps were formed by pri-marily evaporative weathering. From the previously published estimates (Mor-ris, 1971; Butler *et al.,* 1973; McAuliffe, 1976) of the mean residence time (standing stock/input rate) of pelagic tar—a few months to a year—it seemed reasonable to assume also that the median age of the samples collected is also in the same range. Since the median of $t \cdot f$ (age times fraction of the original) is only 0.02 days for the Station S samples (Butler, 1975), the median of f could be as small as $5 \cdot 10^{-5}$. Even allowing for a smaller diffusion coefficient and shorter residence time, this argument thus implies that most of the pelagic tar lumps of the size range represented by the samples we measured are *extremely small fragments* of the original globule or lump from which they were formed by weathering.

Thus the unsuccessful attempt to obtain the age of pelagic tar lumps direct-ly from their gas chromatographic profiles has suggested that most pelagic tar lumps collected in neuston tows near Bermuda are only a small fraction of their original size, and that physical disintegration is an important process in removing tar lumps from the ocean surface.

Since material which flakes off a tar lump is not completely decomposed, it remains for some time in the form of particles which may become suspended deeper in the water column, and should be detectable there. Some preliminary sam-ples of seawater (Morris, 1974) showed small black tar-like particles when filtered. Gas chromatographic analyses of similar particles (Quinn, private communication, 1974; Wade and Quinn, 1975) suggested that their composition was roughly the same as that of the larger pelagic tar lumps

Hydrocarbons in the Water Column

Measurements of the hydrocarbon content of seawater often concern only the "dissolved" fraction, i.e., compounds that pass through a fine porosity filter with the seawater sample and are later partitioned into an organic solvent. Workers using gravimetric methods of hydrocarbon quantification (e.g., Bar-bier *et al.,* 1973; Illife and Calder, 1974) have filtered the seawater before ex-traction, because particles might introduce gravimetric errors. Others, using gas chromatography or UV spectrophotometry or fluorescence (Keizer and Gor-don, 1973; Gordon and Keizer, 1974; Brown *et al.,* 1974; Zsolnay, 1972) have extracted unfiltered seawater with solvent, since they felt that particles did not interfere with their methods.

Of course, pelagic tar is not the only source of hydrocarbon-bearing parti-culate material in the water column. Natural or petroleum -derived hydrocar-bons contained in organisms, or detritus, hydrocarbons adsorbed from the water on to particulate material, or microscopic emulsified liquid oil droplets, all can be suspended in the water column.

The concentration of hydrocarbons bound in a particulate form is not measured in filtered seawater, but emulsified liquid oil may pass through some filters and be adsorbed by others. The efficiency of solvent extraction for removing hydrocarbons from the particulate fraction in unfiltered seawater has not been assessed quantitatively, although some laboratory experiments have been conducted (Gordon *et al.,* 1973).

Our project of sampling specifically for particulate hydrocarbon material in subsurface water was initiated in November 1972 to quantify the amounts of such material, to characterize its composition, and to determine its contribution to the solvent-extractable fraction of unfiltered water samples. Further details of this work are presented elsewhere (Morris *et al.,* 1976) and only the highlights are given here.

Sample Collection

Seawater samples from various depths (typically 5 to 100m) were collected between 28 Nov. 1972 and 7 June 1975 aboard the *R/V Panulirus II* at Station S (32°10'N 64°30'W), located in 3000 m of water about 25 km SE of Bermuda. (Schroeder and Stommel, 1969; Morris and Schroeder, 1973) Station numbers, dates and depths sampled are given by Morris *et al.* (1976).

Niskin bottles (5 or 8 liter volume) were used to collect water samples. To minimize shipboard contamination, the samplers were kept closed at all times except after mounting on the hydrowire, and were opened just prior to lowering. The bottles were rinsed only by the seawater flowing through them while being lowered to depth.

The full samples (5 to 8 l) of seawater were gravity filtered directly from the Niskin bottle. Several types of fine porosity filters were tested, including glass fibre (Whatman GFC #3), polyethylene filter discs (SGA Scientific #F2332), and Millipore (0.45μ) filters. Because of their smooth flat surface, the Millipore filters were superior for particle counts. After filtration, the filters were stored in Millipore Petri-slides, and dried under cover for 24 to 48 hours.

Filter blanks, when subjected to extraction and gas chromatrography of the extracted materials, showed characteristic patterns of peaks but these were easily distinguishable from the pattern characteristic of the particles (see below). All filters were preextracted (before filtration of the seawater samples) with pentane; the polyethylene and glass fibre filters were also preextracted with benzene. Attempts to preextract the Millipore filters with benzene damaged the filters, and so these were extracted only with pentane.

One sediment sample (top 10 cm of core) was collected by Professor R. Berner on 6 August 1975 at 2800 m depth on the slope of the Bermuda platform. It was extracted and analyzed by the same methods as were used for the biological samples from the Sargassum Community (see Morris, Cadwallader, Geiselman and Butler, paper in this volume).

Physical Characteristics of Particles

The possibility that fragments of pelagic tar lumps were suspended in the water column was first investigated by counting all the "tar-like" particles visible in microscopic examination of the filters, both before and after solvent extractions. Gas chromatographic analysis of the extracts is discussed later.

Visual Particle Counts

Particles visible on the filters under a dissecting microscope at 50X were counted while still within the petri-slide holder. Characteristics used to categorize the particles were color, texture, and shape. Such particles varied in appearance from glossy to dull, in color from brown to grey to black, and in texture from solid to porous. Shapes varied from thin flakes to spheroid granules. Figure 10-1 shows scanning electron micrographs of three of these particles. No bacterial colonies, diatoms, forams, or fragments of organisms are recognizable.

All tar-like particles (black, brown or grey fragments, nonmetallic in appearance) were tallied according to size (mean diameter of the particle as viewed if its cross-section were not round) determined by an ocular micrometer. Metallic or other inorganic-looking particles were not counted.

This visual distinction was not always reliable. For example, light-colored particles that may have been paraffinic wax were difficult to distinguish from other organic particles of similar color, and were not counted. Similarly, not all particles counted proved to be soluble in either pentane or benzene. Many that had every superficial appearance of being a petroleum speck were not removed by solvent. Others were only partially soluble, suggesting that perhaps the hydrocarbon material was bound to some less soluble matrix. After extraction, some of these complex particles left a brownish halo on the filter. No attempt was made to determine the chemical composition of the brownish halo; it may have been inorganic (e.g., ferric oxide, which has been found in larger pelagic tar lumps) or a relatively polar organic residue.

The count data for each sampling depth are summarized in Figure 10-2. Detailed data are given by Morris *et al.* (1976). The total for all depths of solvent-soluble particles averaged from 1.4 to 6.3 particles per liter. Highest average counts were at the 50 m depth and lowest just beneath the surface (1 m).

The decrease in number of particles on extraction was statistically significant for most particle sizes. Using a paired observation t-test on the counts before and after extraction we obtained $p < 0.05$ to 0.001 that extraction did not result in solution of particles. (The largest size class did not show a significant decrease because of the infrequency of such particles: generally less than one particle per hundred liters.) The largest particles were the most soluble; indeed, a strong correlation was observed between size class and the fraction which dissolved (Morris *et al.,* 1976).

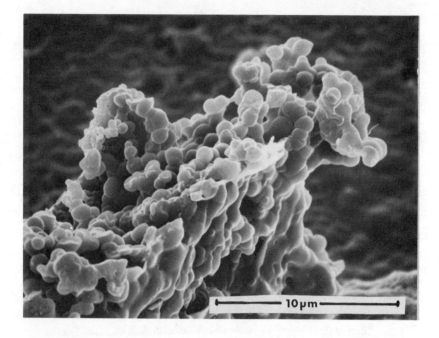

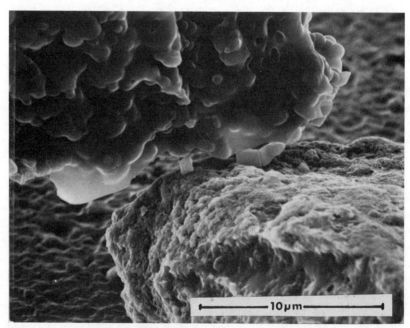

Figure 10-1. Scanning electron micrographs of tar-like particles retained on
Millipore filters. Sampling depth 1 meter, 20 August 1975.

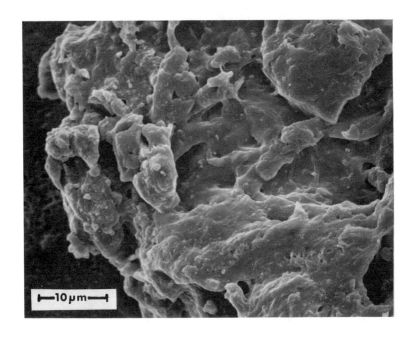

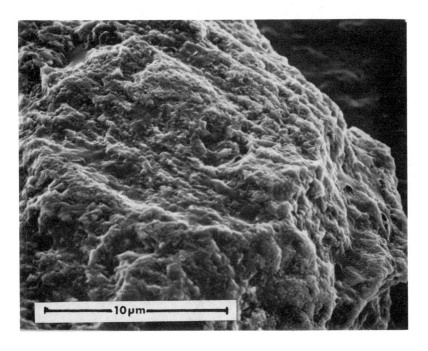

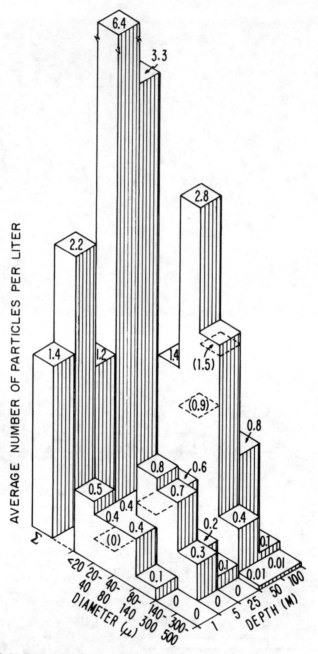

Figure 10-2. Isometric three-variable graph of number of particles vs. depth and
size class shows the total of all size classes.

Conversion of the Counts to Weight

To compare with the amount of hydrocarbon found in the gas chromatographic analyses, the weight of particles was required. Simple weighing of the filters was too imprecise to give usable data. Because of the varying shapes and porosity of the particles we have obtained only approximate values from geometric shape. Most particles are oblong, rectangular or round flakes or lumps whose product of length and width can be approximated by D^2. The thickness of the particles was less variable than diameter; the larger ones tended to be flakes rather than lumps; we assumed it to be approximately $D^{1/2}$. Thus the approximate weight Q (in μg) of tar particles on the filter is estimated by the expression:

$$Q = \sum_i 10^{-6} N_i D_i^{2.5} \tag{10.1}$$

where N_i is the number of particles in each size class dissolved from filters by pentane-benzene, and D_i is the average diameter in μm of each size class i. The specific gravity of the particles was assumed to be 1.0, and so the factor 10^{-6} has the units μg μm$^{-2.5}$.

Estimated weights of soluble particles in the water column derived from this equation are given in Figure 10-3. The average at 1 m may be too low because only five samples were available.

Comparing the probable number of tar particles per liter of water at each depth (Figure 10-2) with the weight of the same particles (Figure 10-3) shows that although the most common size is in the 40-80 μm size class (0.4 to 3.3 particles per liter), by weight the particles >80 μm were most important and the presence of only one particle >300 μm could contribute as much as half the total weight. However, both the greatest number of particles (6.4 per liter) and the greatest weight (1.0 μg/l) occurred at the 50 m depth. This is near the bottom of the seasonal thermocline and tar particles (as well as other detritus) could accumulate there during part of the year. Just below the surface (where one might expect a large number of particles from disintegration of surface tar lumps) only small particles (<140 μm) were found, and the least concentration by weight (0.03 μg/l) also occurred.

Chemical Characteristics of Particles

Changes in Particle Counts After Extraction

After the initial count, the filter was placed in a flat glass funnel and extracted (at 22-25°C) with 25 ml of pentane. This procedure was about 90% efficient, since a second portion of pentane gave only about 10% additional *n*-alkanes.

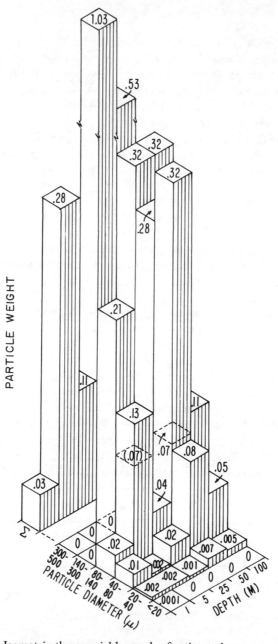

Figure 10-3. Isometric three-variable graph of estimated average weight of particles per liter vs. depth and size class. Note reversed order of size
classes compared to Figure 10-2.

An extraction with benzene was also made to dissolve any aromatic compounds. For the 1972–73 samples, only initial counts and final counts following both the pentane and benzene extractions were made (Stations 372–386). Later (Stations 397–402), counts were also made after the pentane extraction. The number of soluble particles was determined in all samples as the difference in counts between the initial filter and the benzene-extracted filter.

Table 10–1 presents two examples of data on the loss in particles of each size class between the initial count and after extraction. Normally, the number of particles decreased on extraction, but occasionally (e.g. 5 and 25 m samples of S-382) the number increased; and this resulted from fragmentation of the original particles under the action of the solvent. In many cases, particles of a larger size class did not dissolve completely, and thereby increased the count in a smaller size class.

Gas Chromatography

The extracts were evaporated to near-dryness and kept frozen until analyzed by gas-liquid chromatography. Pentane extracts were analyzed for nonpolar hydrocarbon compounds by redissolving the residue in 100 μl carbon disulfide and injecting 1 to 10 μl of this solution into the gas chromatograph.

The chromatograph used was a Varian Model 1440 with a 10 ft. $\times$ 1/16 inch stainless steel column of 3% SE-30 on Varaport 30 100/120 mesh, or a 6 ft $\times$ 1/16 inch column of 3% SP 2100 on Supelcoport 100/120 mesh. Operating conditions were: N_2 carrier gas at 30 ml/min; temperature programming from 75 to 300°C at 6°/min with a final temperature of 300° maintained until all peaks were eluted.

Quantitative standards of n-alkanes (C_{14} to C_{36}) were regularly run and amounts of each resolved component determined by comparison of peak areas. Unresolved peaks below the principal alkanes formed an "envelope" which was measured by planimetry and quantified by assuming the calibration factor was the same as for the alkane of comparable retention time.

A typical chromatogram of a pentane extract is shown in Figure 10–4. In 43% of the chromatograms, an unresolved envelope was detected; in the remainder, only alkane peaks appeared, with an occasional additional peak probably corresponding to a branched-chain or olefinic hydrocarbon eluting between normal C_{25} and C_{26}.

The hydrocarbons retained on the filter and extractable by pentane contained normal alkanes primarily in the range from C_{25} to C_{36} with trace amounts down to C_{14} and up to C_{40}. This pattern was essentially independent of the sampling depth (Figure 10–5).

(Chromatograms of the benzene extracts were complex, containing not only hydrocarbons but fatty acids, and their analysis is too incomplete to be reported at this time.)

Table 10-1
Examples of Counts of Tar-Like Particles on Filters and the Effect of Solvent Extraction

Depth (m)	Size Class (μm)					
	<20	20-40	40-80	80-140	140-300	All
Station 382	**6 June 1973** — *pentane extraction not counted*					
	I B	*I B*	*I B*	*I B*	*I B*	*I B*
5	14 - 52	8 - 15	6 - 3	4 - 0	1 - 0	33 - 70
25	24 - 34	14 - 14	8 - 7	3 - 0	1 - 0	50 - 55
50	29 - 34	30 - 16	33 - 7	16 - 2	2 - 0	110 - 59
100	38 - 70	39 - 49	58 - 19	32 - 3	6 - 0	172 - 141
Station 402	**7 June 1975** — *pentane extraction counted*					
	I P B	*I P B*	*I P B*	*I P B*	*I P B*	*I P B*
1	19-17-16	11-10- 4	0 - 1-1	3 - 2-1	0-0-0	33-30-22
5	27-22-25	13-18-10	5 - 6-4	1 - 1-0	0-0-0	46-47-39
25	17-11-19	18-13- 6	7 - 3-2	3 - 2-1	1 - 1-0	46-30-28
50	24-15-20	17-14- 8	8 - 3-1	3 - 3-1	0-0-0	52-35-30
100	6- 5- 6	5- 7- 4	2 - 2-3	2 - 2-1	1-0-0	16-16-14

NOTE: The counts give the total number of particles at each stage—initial (I), after pentane extraction (P), and after benzene extraction (B). The difference between counts represents dissolution of particles.

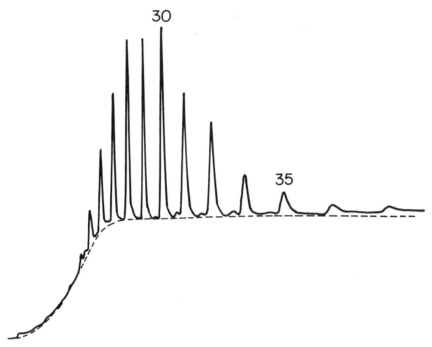

30

35

Figure 10-4. Chromatogram of sample with medium *n*-alkane concentration:
Sample 388 – 50m, lacking unresolved envelope. The broken line
represents the baseline obtained in a blank run due to column
bleed.

The total concentrations of pentane-extractable hydrocarbons removed
from the filters are given in Table 10-2. Highest concentrations occurred at the
1 m sampling depth, with slightly lower concentrations occurring at greater
depths.

At Station S-385a, we sampled the water column to a depth of 800 m. Despite the considerable variability in the upper layer, a pronounced decrease in
concentration occurred beneath the 100 m depth, corresponding to the bottom
of the seasonal pycnocline. (See Morris *et al.,* 1976, for more details.)

The pentane extract of the surface sediment sample from 2800 m depth on
the slope of the Bermuda platform gave a gas chromatogram with more light
hydrocarbons than the filter extracts: normal alkanes from C_{17} to C_{33}, with C_{21}
and C_{24} being the highest peaks, and phytane distinctly resolved from C_{18}. The
presence of phytane, the monotonic change of peak height with carbon number
below C_{24}, and the presence of an unresolved envelope implies a petroleum
source, but the alternating heights of the peaks beyond C_{24} implies some biogenic contribution as well. The hydrocarbon contribution from the *n*-alkane

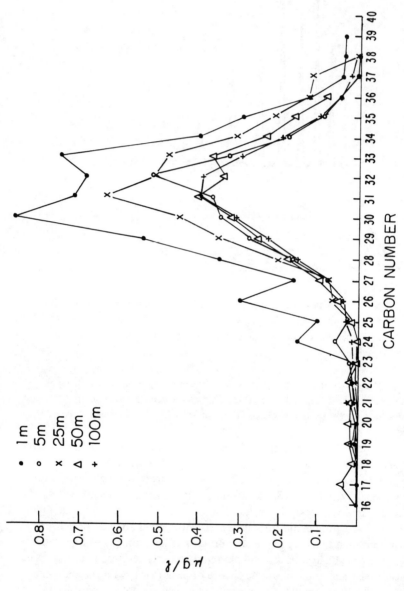

Figure 10-5. Average distribution (all stations) of *n*-alkane concentrations from pentane extracts of filterable material from five depths in the Sargasso Sea off Berumda.

Table 10-2

Estimates by Gas Chromatography of Nonpolar Hydrocarbon Concentrations ($\mu g/1$) at Different Sampling Depths Retained on Filters and Recovered by Pentane Extraction

Depth (m)	Number of samples	n-Alkanes median (range)	Unresolved Envelope median (range)	Total median (range)
1	5	4.38 (0.67-12.7)	0.39 (0.0-4.8)	6.15 (0.67-13.1)
5	21	1.13 (0.0-18.8)	0.13 (0.0-7.6)	1.23 (0.0-23.0)
25	21	0.65 (0.0-18.2)	0.0 (0.0-3.0)	0.98 (0.0-18.6)
50	21	1.15 (0.0-14.1)	0.0 (0.0-5.3)	1.69 (0.0-25.9)
100	22	0.89 (0.0-9.9)	0.0 (0.0-91.3)	1.78 (0.0-96.2)

peaks was 0.18 $\mu g/g$ dry weight and with the unresolved envelope, a total of 0.60 $\mu g/g$ dry weight was obtained. This is comparable to the few other known values (1 to 4 $\mu g/g$) reported for open ocean sediments (NAS, 1975; Farrington and Tripp, 1975) and very much lower than typical coastal sediments (100 to 1000 $\mu g/g$; NAS, 1975).

Discussion

Similarity to Pelagic Tar

A distinct similarity exists between our profiles of *n*-alkanes from the filter extracts, and the profiles obtained from wax inclusions in pelagic tar lumps (Blumer, Ehrhardt and Jones, 1973; Butler, Morris, and Sass, 1973). These may be a source of some of the C_{25} to C_{37} alkanes. Because of their light color, they would have been overlooked in the counts of black or brown "tar" particles

Wade and Quinn (1975) have also reported finding black particles in surface and subsurface samples from the Sargasso Sea. By analogy with a laboratory experiment in which large pelagic tar lumps were shaken with filtered sea water, they concur that these particles were formed by the disintegration of pelagic tar.

Peculiarly, our normal alkane distributions (e.g., Figure 10-5) are also similar to those reported by Barbier *et al.* (1973) from the *dissolved* fraction. They used Millipore 0.45 μm filters to remove particulates before extraction, 100-liter samples of seawater, and found in the extract a range of n-alkanes from C_{14} to C_{37} with maxima between C_{27} and C_{30}. No odd-carbon predominance occurred and the unresolved envelope was small. Our samples showed a somewhat higher n-alkane maximum (between n-C_{30} and n-C_{33}) but were similar in the frequent lack of an unresolved envelope, and an odd/even carbon-preference index between 0.9 and 1.1.

Biogenic vs. Petroleum Sources

Identifying the source of the hydrocarbons in the water column is well known to be a difficult problem. Three criteria (NAS, 1975) for distinguishing petroleum derived from biogenic hydrocarbons are (1) lack of odd-even carbon preference in the paraffins, (2) the presence of the C_{18} to C_{20} isoprenoids, with pristane/phytane ratios of the order of 1/2, and (3) the presence of a complex mixture of naphthenic aromatic and heterocyclic compounds in an unresolved envelope of the chromatogram.

Our chromatograms (Figure 10-4) lacked significant amounts of pristane, phytane, or normal paraffins below about C_{25}. The unresolved envelope was undetectable in about half the samples. In a few samples there were small peaks other than normal paraffins between C_{18} and C_{19} (but not phytane), between C_{19} and C_{20}, and a large peak between C_{25} and C_{26}.

Biogenic hydrocarbons extracted from unialgal phytoplankton cultures generally show only a few specific alkanes (Blumer, Guillard and Chase, 1971) and those are often branched or cyclic. By comparison, the extracts of mixed plankton (Clark and Blumer, 1967; Blumer, Guillard and Chase, 1971) usually show compounds most strongly in the C_{15} to C_{22} region. Although some mixtures of plankton show equal amounts of odd and even normal paraffins, odd-carbon chain preference is more typical. In Figure 10-5, only the 1 m average shows any indication of an alternation in amounts of successive n-paraffins.

The absence in our samples of these lower alkanes, or the isoprenoids, the carbon predominance index near 1.0, the presence of a few dominant nonalkanes, and the partial presence of an unresolved envelope, suggest both a petroleum and a biogenic origin for the material extracted from the filters. Quantification of the relative amounts is not possible without a more detailed analysis of individual particles.

Filter Extracts vs. Particle Counts

Measurement by gas chromatography of the amount of hydrocarbon material obtained from pentane-extracts of filters gives much higher concentrations than estimates from particle counts. The values differ by an order of magnitude for most depths (Figure 10-6). The greatest discrepancy is at the one-meter depth where the highest gas chromatography (6.2 $\mu g/l$) and lowest count (0.03 $\mu g/l$) estimates occurred. Although the variation of either estimate is large, the differences are statistically significant at the 99% confidence level. Hydrocarbons in addition to those contained in the tar particles counted visually must be present on the filter, and it is possible that some of the hydrocarbons on the filters came from the adsorption of dissolved or colloidally dispersed compounds.

Brown et al. (1974) and Gordon and Keizer (1974) reported concentrations similar to ours by extracting *unfiltered* sea water. Higher values because of contribution from the dissolved fraction would be expected, unless our filters contained also most of the dissolved fraction. Similarly, Wade and Quinn (1975) found that filtering seawater resulted in retention of most of the hydrocarbons on the filter. On the other hand, one must reconcile the high values for total hydrocarbons and the similar gas chromatographic profiles of the extracts from *filtered* seawater (Barbier et al., 1973; Illife and Calder, 1974) with our data and with the studies just mentioned. This is not obvious. One possibility is that Barbier et al., for example, because they filtered such large (100-liter) samples of water, fully saturated whatever capacity their Millipore filters had for adsorbing dissolved hydrocarbons, whereas our smaller (5- to 8-liter) samples did not.

Another puzzling question is why so few compounds of lower molecular weight were found in the filter extracts. Simple evaporative weathering (Butler, 1975), even for long periods of time, would not be expected to deplete the hydrocarbon profile above about C_{20}. Microbial attack normally degrades paraffins at approximately the same rate over a wide range of chain lengths, but branched or cyclic compounds are attacked much more slowly (Liu, 1973; Kator 1973; National Academy of Sciences, 1975 and references cited therein). A microbially degraded sample of petroleum-derived hydrocarbons might be expected to show an unresolved envelope with a few distinctive isoprenoid or other peaks, but not a regular succession of higher paraffins only, although new evidence exists (M. Ehrhardt, private communication) that normal paraffins above C_{25} are more resistant to some microbial cultures.

Since longer chain n-alkanes are very insoluble (e.g. 0.1 $\mu g/l$ for C_{26}) in seawater (Sutton and Calder, 1974) and must be dispersed as micelles if at all, these longer molecules may be preferentially adsorbed on to the filter or suspended particulate material while the lower molecular weight hydrocarbons pass through in the dissolved or colloidal state. Analysis of the filtered water could test this hypothesis. However, the data of Barbier et al. (1973) from *filtered* water do not show any low molecular weight hydrocarbons, either.

The large excess of extractable hydrocarbons of uncertain origin on the filter unfortunately masked the chromatographic analysis of the composition of the smaller fraction of "tar" particles. Since their small size precluded the analysis of individual particles, their true composition is still unresolved. At present,

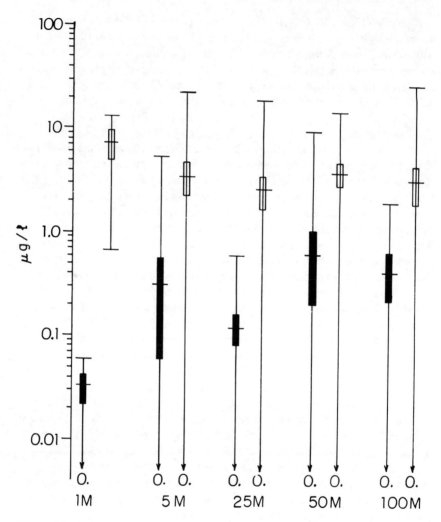

Figure 10-6. Comparison of estimated concentrations (μg/l) of particulate material on filters based on particle counts (dark bars) and on gas
chromatographic analysis (white bars) at five sampling depths.
Range of values indicated by end lines, mean by central horizontal line, and one standard error either side of the mean by the wide
vertical bar.

the basic evidence we have for assuming the dark particles on the filter are "tar"
is (1) the knowledge that tar lumps do disintegrate into small particles, (2) the
presence in the ocean of particles of similar appearance that are soluble in pentane and benzene, and (3) their physical appearance under the dissecting and

scanning electron microscopes. Other more sensitive and selective methods of analysis, such as gas chromatography—mass spectrometry (Giger and Blumer, 1974) or high performance liquid chromatography (Zsolnay, 1973), should provide further information.

Steady State Fluxes for the Sargasso Sea

Integrating our estimated concentrations of tar particles (from visual counts) within the upper 100 m gives an estimate of 40 mg of tar particles beneath each square meter of ocean surface (the filter extract values give 300 mg). The surface concentration of tar lumps in the Sargasso Sea is approximately 10 mg/m^2, with a residence time of several months to a year (see discussion earlier in this paper). Tar particles in the water column, although derived from the surface tar lumps, may exceed the quantity of lumps floating on the surface at any given time simply by being more resistant to degradation, and thus having a longer residence time in the water column than the residence time of the parent lumps on the surface. A schematic diagram (based on data from NAS 1975 and the paper by Butler in this volume) of the fluxes of petroleum hydrocarbons near the surface of the Sargasso Sea is shown in Figure 10-7.

Considering the inaccuracies (up to a factor of ten — see Figure 10-6) inherent in the tar particle estimates and the variability encountered during surface tar sampling (a factor of two for the geometric mean of 16 samples) a steady-state model implies a residence time for the suspended particles of the order of years. Ultimately, of course, they are oxidized, probably by microorganisms but also by larger organisms, to carbon dioxide or else are accumulated in the sediments.

Extension of this model requires quantitative data on the amount of petroleum and biogenic hydrocarbons in the particulate, dissolved, and emulsified states in the water column, quantitative data on the rate of oxidation under actual field conditions, and the profile of hydrocarbons in the first 10 cm of sediments, including sufficiently detailed chemical analysis to distinguish between petroleum, biogenic, and other sources (such as atmospheric fallout of combustion products) of hydrocarbons in the sediments.

Acknowledgments

This work was supported by the National Science Foundation, International Decade of Ocean Exploration, Grant No. IDO-72-06411, and a grant from the Zemurray Foundation. The assistance of the Captain and crew of the R/V Panulirus II is gratefully acknowledged. The authors thank Professor R. Berner of Yale University for supplying the sediment sample for analysis; and

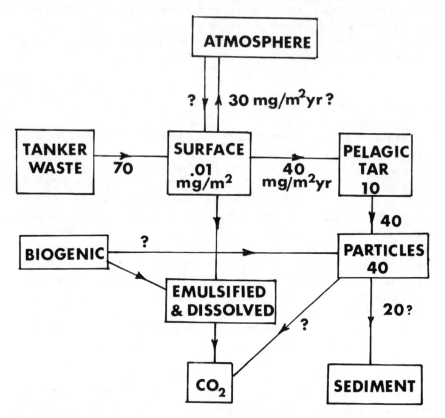

Figure 10-7. Schematic diagram of petroleum hydrocarbon fluxes and contents of the Sargasso Sea environment.

C. Bradford Calloway and Ed Seling of the Harvard Museum of Comparative Zoology for the scanning electron microscopy. Byron F. Morris presented much of this information at the Workshop on Petroleum Hydrocarbons in the Marine Environment, sponsored by the International Council for the Exploration of the Sea, at Aberdeen, Scotland, U.K., September 1975.

References

Barbier, M., D. Joly, A. Saliot and D. Tourres (1973). Hydrocarbons from sea-water. *Deep-Sea Res., 20,* 305-314.

Blumer, M., M. Ehrhardt and J. H. Jones (1973). The environmental fate of stranded crude oil. *Deep-Sea Res., 20,* 239–259.

Blumer, M., R. R. L. Guillard and T. Chase (1971). Hydrocarbons of marine phytoplankton. *Mar. Biol. 8,* 183-189.

Brown, R. A., J. J. Elliott and T. D. Searl (1974). Measurement and characterization of nonvolatile hydrocarbons in ocean water. In *Marine Pollution Monitoring (Petroleum).* Symposium and Workshop, National Bureau of Standards. NBS Spec. Publ. *409,* 131-133.

Butler, J. N., B. F. Morris and J. Sass (1973). Pelagic tar from Bermuda and the Sargasso Sea. Spec. Publ. No. 10, Bermuda Biological Station for Research, St. George's West.

Butler, J. N. (1975). Evaporative weathering of petroleum residues: the age of pelagic tar. *Marine Chemistry, 3,* 1-7. See also chapter in this volume.

Clark, R. C., Jr., and M. Blumer (1967). Distribution of paraffins in marine organisms and sediment. *Limnol. Oceanogr. 12,* 79-87.

Ehrhardt, M., and J. Derenbach (1976). Composition and weight per area of pelagic tar collected between Portugal and south of the Canary Islands. *Meteor. Forschungsber., Reihe A.* (in press).

Farrington, J. W., and B. W. Tripp (1975). A comparison of analysis methods for hydrocarbons in surface sediments. In *Marine Chemistry in the Coastal Environment* (T.M. Church, ed.), Amer. Chem. Soc. Symp. Series No. 18.

Giger, W., and M. Blumer (1974). Polycyclic aromatic hydrocarbons in the environment: Isolation and characterization by chromatography, visible, ultraviolet, and mass spectrometry. *Anal. Chem., 46,* 1663-1671.

Gordon, D. C., Jr., P. D. Keizer and N. J. Prouse (1973). Laboratory studies of the accomodation of some crude and residual fuel oils in seawater. *J. Fish Res. Bd., Can., 30,* 1611-1618.

Gordon, D. C., Jr., and P. D. Keizer (1974). Hydrocarbon concentrations in seawater along the Halifax-Bermuda section: Lessons learned regarding sampling and some results. In *Marine Pollution Monitoring (Petroleum).* National Bureau of Standards, Spec. Publ. *409,* 113-115.

Iliffe, T. M., and J. A. Calder (1974). Dissolved hydrocarbons in the eastern Gulf of Mexico Loop Current and the Caribbean Sea. *Deep-Sea Res., 21,* 481-488.

Kator, H. (1973). Utilization of crude oil hydrocarbons by mixed cultures of marine bacteria. In *Microbial Degradation of Oil Pollutants* (D. G. Ahearn and S. P. Meyers, eds.), Louisiana State University Center for Wetland Resources, pp. 47-66.

Keizer, P. D., and D. C. Gordon, Jr. (1973). Detection of trace amounts of oil in seawater by fluorescence spectroscopy. *J. Fish. Res. Bd., Can., 30,* 1039-1046.

Liu, D. L. S. (1973). Microbial degradation of crude oil and the various hydrocarbon derivatives. In *Microbial Degradation of Oil Pollutants,*

(D. G. Ahearn and S. P. Meyers, eds.), Louisiana State University Center for Wetland Resources, pp. 95-104.

McAuliffe, C. (1976). *Surveillance of the Marine Environment for Hydrocarbon.* Unpublished manuscript, 30 pp.

Morris, B. F. (1974). Unpublished data cited in *Pollutant Transfer to the Marine Environment,* National Science Foundation, p. 34.

Morris, B. F., and J. N. Butler (1973). Petroleum residues in the Sargasso Sea and on Bermuda beaches. In *Proceedings Joint Conference on Prevention and Control of Oil Spills,* American Petroleum Institute, Washington, D.C., pp. 521-530.

Morris, B. F., J. N. Butler and A. Zsolnay (1975). Pelagic tar in the Mediterranean Sea. *Environmental Conservation, 2,* 275-281.

Morris, B. F., J. N. Butler, T. D. Sleeter and J. Cadwallader (1976). Particulate hydrocarbon material in oceanic waters. *J. du Consiel* (in press).

Morris, B., and E. Schroeder (1973). Hydrographic observations in the Sargasso Sea off Bermuda; May, 1967 - February, 1973. Bermuda Biological Station Spec. Publ. *12.*

NAS (National Academy of Sciences) (1975). *Petroleum in the Marine Environment. Workshop on Inputs, Fates, and Effects of Petroleum in the Marine Environment,* May 21-25, 1973. Nat. Acad. Sci., Wash. D. C., 107 pp.

Schroeder, E., and H. Stommel (1969). How representative is the series of Panulirus stations of mean monthly conditions off Bermuda? *Prog. Oceanogr., 5,* 31-40.

Sutton, C., and J. A. Calder (1974). Solubility of higher molecular weight *n*-paraffin in distilled water and seawater. *Environ. Sci. Tech., 8,* 654-657.

Wade, T. L., and J. G. Quinn (1975). Hydrocarbons in the Sargasso Sea surface microlayer. *Mar. Pollution Bull., 6,* 54-57.

Zsolnay, A. (1972). Preliminary study of the dissolved hydrocarbons and hydrocarbons on particulate material in the Götland Deep of the Baltic. *Kieler Meeresforsch., 27,* 129-134.

Zsolnay, A. (1973). Determination of aromatic hydrocarbons in submicrogram quantities in aqueous systems by means of high performance liquid chromatography. *Chemosphere, 6,* 253-260.

11

Transfer of Petroleum and Biogenic Hydrocarbons in the Sargassum Community

B. F. Morris, J. Cadwallader, J. Geiselman and *J. N. Butler*

Introduction

Between 10^{12} and 10^{13} g of petroleum are annually introduced into the marine environment (NAS, 1975) and occur as floating oil slicks and tar lumps, as particles (10 to 300 μm sized oil droplets, tar specks, or hydrocarbons adsorbed on other kinds of particles) suspended in the water column, and as dissolved or colloidal material dispersed throughout the water column.

Studies in the Sargasso Sea have indicated the concentrations of some of these fractions throughout the water column. At the sea surface, tar lumps average about 10 mg/m^2 (Morris, 1971; Butler *et al.*, 1973) while the dispersed fraction in the surface microlayer contains about 10 μg/m^2 (Keizer and Gordon, 1973; Wade and Quinn, 1975). Within the water column, hydrocarbon concentrations (total particulate and dispersed) range from 10 μg/l near the surface to less than 1 μg/l at depths greater than 100 m (NAS, 1975; Gordon and Keizer, 1974). About 0.5 μg/l is in the form of tar particles (Morris *et al.*, 1976), and the remainder is in the form of colloidal micelles or adsorbed onto seston. Only the lightweight and highly polar hydrocarbons are truly dissolved (McAuliffe, 1974).

The incorporation of hydrocarbons from the environment into marine organisms has been shown to occur via adsorption of dissolved compounds from the water (Lee *et al.*, 1972a, b; Stegeman and Teal, 1973) as well as by ingestion (Conover, 1971; Parker, 1971). Whether these compounds are then metabolized by the organisms, directly released, or subsequently transferred along the food web is known in very few cases. There is some evidence that food web magnification does *not* occur (Burns and Teal, 1973; Lee *et al.*, 1972b; Corner, 1975; Corner *et al.*, 1975). Marine organisms can synthesize hydrocarbons (Blumer *et al.*, 1964; Clark, 1966; Blumer *et al.*, 1971) but it is often difficult to distinguish petroleum pollutants from the natural (biogenic) hydrocarbons produced by the organisms themselves or transferred along the food chain.

A number of criteria for distinguishing biogenic from petroleum hydrocarbons have been developed (NAS, 1975). We have employed those based on gas chromatography, which depend on measuring the relative amounts of the most abundant nonpolar compounds (usually normal paraffins, isoprenoids, or olefins). Specifically:

1. Distinctive, relatively large peaks (often olefinic) are almost certainly biogenic.

235

2. A series of normal alkanes from C_{16} to C_{33} where the adjacent peaks are of essentially equal size usually indicates petroleum. However, some special biogenic sources, such as bacterial cultures (Davis, 1968) or mixtures of phytoplankton (Clark and Blumer, 1967) also show this pattern.

3. A series of normal alkanes where the odd-numbered compounds are about 1.7 times the concentration of even-numbered compounds is typical of biogenic sources, particularly phytoplankton (Clark, 1966; Clark and Blumer, 1967).

4. The presence of pristane (C_{19} isoprenoid) but not phytane (C_{20} isoprenoid) indicates a biogenic source (Blumer et al., 1964).

5. The presence of both pristane and phytane in a ratio of 1.5 to 2.5 is typical of petroleum (Blumer and Sass, 1972). Petroleum also may contain a C_{18} isoprenoid which is not biogenic.

6. A large envelope of unresolved compounds (alicyclic, branched aliphatic, etc.) is typical of petroleum, but its size relative to the alkane peaks depends on the column resolution. Hydrocarbons extracted from unpolluted organisms rarely, if ever, have so many compounds present in such small amounts. (Note that the extraction procedure and column chromatography on silica/alumina prior to gas chromatography removes aromatics, fatty acids, and other complex mixtures which might produce such a pattern.)

The pelagic *Sargassum* community contains a variety of organisms in a relatively simple ecological framework (Morris and Mogelberg, 1973). These organisms live in association with petroleum pollution, either in the form of tar lumps floating at the surface (Butler et al., 1973) or as dispersed hydrocarbons in the water column or in the surface microlayer (Wade and Quinn, 1975). Clark (1966) found a range of hydrocarbons in *Sargassum;* Burns and Teal (1973) reported a significant level of petroleum hydrocarbon contamination of both the algae and the fauna.

This work was undertaken to educidate the transfer processes of hydrocarbons in the *Sargassum* community, in hopes that it would help us understand the larger and more inaccessible open ocean ecosystem. In particular, the questions asked were (1) What are the amounts of petroleum hydrocarbons in *Sargassum* and in the organisms of its associated community? (2) What is the ratio of biogenic to petroleum hydrocarbons in the various members of the community? (3) How are these hydrocarbons transferred through the food web? (4) Are hydrocarbons absorbed directly from the environment? (5) Does the association of a particular *Sargassum* community with pelagic tar increase the amount of petroleum hydrocarbons in its members? (6) Does association with pelagic tar alter the ecology of the community?

Not all these questions have yet been answered. The results presented in this paper represent partial answers to the first three questions, for *Sargassum* and the larger animals associated with it. For the motile and sessile microfauna, or the epiphytic algae, the only information available concerns their position in the food web, or in some cases their abundance, but much work remains to be done.

Materials and Methods

The procedures used in the sampling, extraction, and gas-chromatographic analysis of the samples will be reported elsewhere in detail (Morris, Cadwallader and Butler, 1976). Essentially, these procedures follow the recommendations of Grice et al. (1972) for avoiding sample contamination and of Burns and Teal (1973) and Zafiriou et al. (1972), with some modifications suggested by Clark and Finley (1974) for extraction.

The samples were dip-netted near Bermuda, taken to the laboratory and sorted the same day, and stored frozen until analysis. Extraction was by the Soxhlet procedure. All reagents were redistilled and purity checked, and chemicals, handling equipment etc. were cleaned in twice-distilled pentane before use. The nonpolar hydrocarbon fraction resulting from silica-alumina column chromatography was analyzed by gas-liquid chromatography on a nonpolar column (Supelco 2100). The temperature was programmed from 50 to $300°C$, and n-alkanes were quantitatively determined in the C_{14} - C_{40} range, by comparison of peak height ratios with known standards. A standard and a blank were run with each day's sample for calibration. The unresolved envelope was measured by planimetry and quantified by comparison to the area of the n-alkane peak occurring at maximum envelope height. Typical chromatograms are shown in Figures 11-1 — 11-3.

Alkanes were identified by their retention indices. The quantities of the various hydrocarbons in each sample were determined from the calibration and the known amount of the total sample injected. These hydrocarbons were classified into four groups: "biogenic," "petroleum n-alkanes," "unresolved envelope," and "others." Biogenic peaks (normally near the retention index for a normal alkane) were distinguished by an abundance of greater than twice the adjacent n-alkane peaks. Petroleum n-alkanes were categorized as those whose abundance was similar along the series from C_{14} up to C_{40} (odd:even carbon ratio = 1). The unresolved envelope was measured between the base of the hydrocarbon peaks and the column bleed baseline as determined daily. Other measurable peaks with retention indices differing from the normal-alkanes occurred, usually in small quantities, and were probably either branched petroleum alkanes or biogenic alkenes (Blumer, Guillard and Chase, 1971; Clark and Blumer, 1967).

Sargassum samples were collected on 18 different occasions between 1972 and 1975, all from the vicinity of Bermuda. The individual species comprising the community were sorted and when enough material of one species was collected, it was extracted and analyzed for hydrocarbon content. To date, only

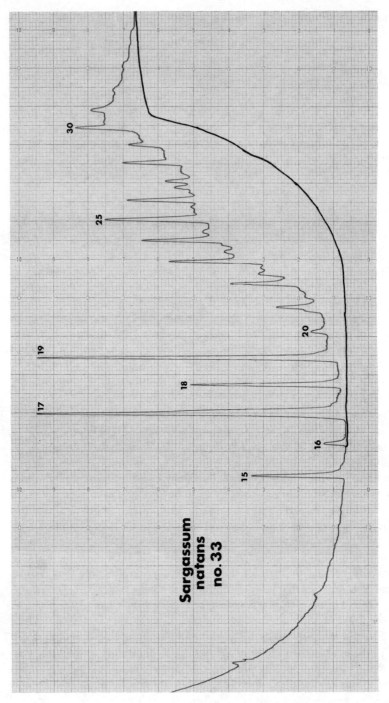

Figure 11-1. Typical gas chromatogram of extract from *Sargassum natans*.

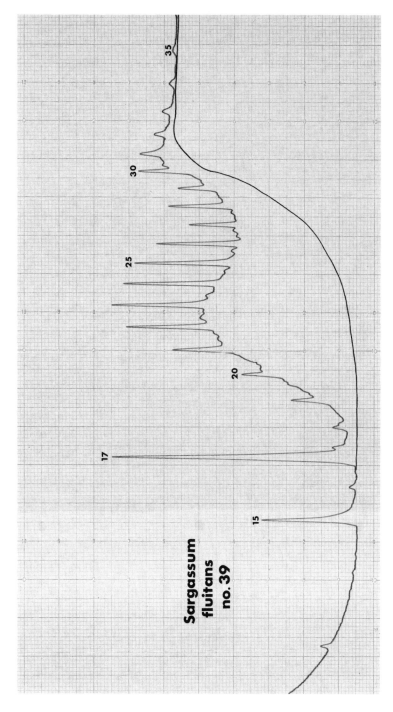

Figure 11-2. Typical gas chromatogram of extract from *Sargassum fluitans*.

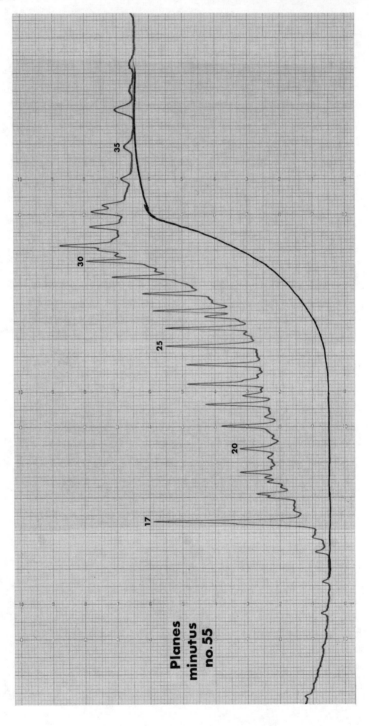

Figure 11-3. Typical gas chromatogram of extract from crab *Planes minutus*.

the larger organisms have been studied. Except for a few mixed species collections, the smaller organisms could not be obtained in sufficient quantity for a satisfactory analysis.

The pentane washings prior to the extraction procedure were often examined by GLC. These washings contained petroleum hydrocarbons, but the biogenic hydrocarbons were generally not abundant. This indicated that the surface of the *Sargassum* and organisms is often contaminated by tar particles and adsorbed hydrocarbons. But, since the amounts of hydrocarbons found in the washings averaged only about 2-3% (0.48 µg/l) of the total extractable hydrocarbons, we are confident that most of the hydrocarbons extracted were definitely contained within the sample tissues and that surface contamination did not affect our results. Burns and Teal (1973) also found low hydrocarbon levels in their pentane washings of the *Sargassum* algae.

Studies of the food web of the *Sargassum* community were performed both by gut analyses of the species and by direct observation of feeding behavior in the laboratory (Morris and Geiselman, 1976).

Results

Food Web

A preliminary summary diagram of the *Sargassum* community food web is given in Figures 11-4 and 11-5 (Morris and Geiselman, 1976). The community is supported by four major food sources: (1) the plankton and particulate organic matter in the surrounding water; (2) the *Sargassum* alga itself; (3) the epiphytes, including blue-green algae and diatoms, attached to Sargassum; and (4) the detritus produced within the community. At the first consumer level are the planktivores (filter feeding bryozoans, serpulid polychaetes, barnacles, hydroids), omnivores feeding on *Sargassum* and its epiphytes (amphipods, polychaetes, and other organisms with major alternative food sources), and detritus feeders (harpacticoid copepods, amphipods, and flatworms). Trophic pathways overlap even at the first trophic level, and above this level the distinction between levels is totally blurred.

For the higher level carnivores there are alternative food sources, and it is apparent that the limits of food availability within the *Sargassum* community have imposed a necessity for the largest organisms (crabs, shrimps, and fish) to have the greatest number of alternative food sources available to them. As an example, the common sargassum crab *Planes minutus,* seems to be an opportunistic feeder whose choice of food is determined primarily by its availability, success of capture, and the scarcity of preferred food sources. *Planes* has been observed to eat *Sargassum* and even tar lumps in the absence of other food sources.

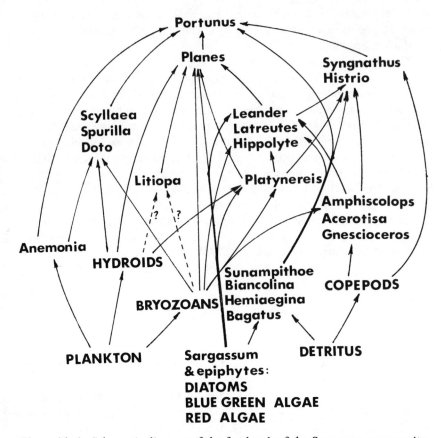

Figure 11-4. Schematic diagram of the food web of the Sargassum community.

These studies still require quantitative evaluation of the types and amounts of food consumed and of food selectivity. Further refinement of the food web is necessary, but a general indication of the complexity and hierarchy of the structure is apparent, and this is sufficient for a preliminary interpretation of the hydrocarbon analyses.

Hydrocarbon Levels

Analyses of various samples of ten species from the *Sargassum* community are summarized in Table 11-1. Only the means are reported in the table. The varia-- bility of the samples was quite large, and is indicated on some of the figures. Full details will be presented elsewhere (Morris, Cadwallader and Butler, 1976).

At the base of the *Sargassum* food web are the two algae with hydrocarbon

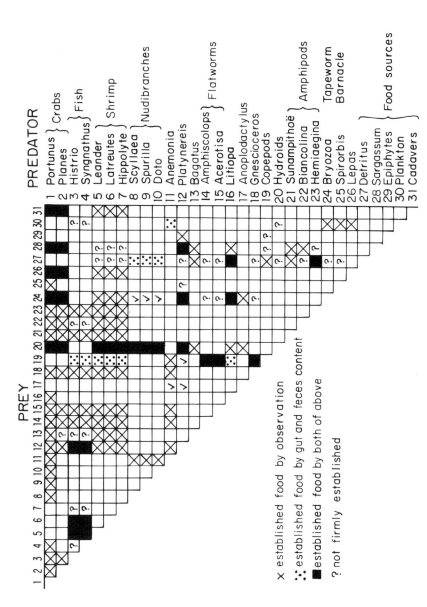

Figure 11-5. Summary diagram of the food sources of the Sargassum community fauna.

Table 11-1

Average Concentrations ($\mu g/g$ – Dry Weight) of Nonpolar Hydrocarbons in
Members of the Pelagic Sargassum Community

Organism (number of samples)	n-Alkanes Biogenic	Petroleum	Unresolved Envelope	Total	Petroleum/ Biogenic
Sargassum natans (37)	1.4[a]	5.2	17.6	23.1	16.3
Sargassum fluitans (10)	0.8[b]	3.3	7.8	12.3	13.9
Litiopa melanostoma (4)	0.3[c]	4.4	3.2	8.1	25.3
Scyllaea pelagica (1)	40.5	420	104	619	12.9
Latreutes fucorum (4)	0.2[d]	50.3	82.0	135	132.5
Leander tenuicornis (2)	7.2[d]	18.1	51.2	77.7	44.6
Planes minutus (5)	0.6[d]	2.5	13.1	16.5	26.0
Portunus sayi	0.8[d]	109	381	500	612.1
Syngnathus pelagicus	34.9[d]	1466	1764	3411	92.6
Carangidae ("Jacks")	3.5	21.2	53.9	81.0	21.2
Mixed Fauna	2.3	45.9	31.3	83.1	33.6

[a]C_{15} 0.2, C_{17} 1.0, C_{19} 0.2.

[b]C_{15} 0.3, C_{17} 0.5.

[c]C_{15} 0.1, C_{17} 0.2.

[d]Principally C_{17}.

contents between about 10 to 20 $\mu g/l$ (Figures 11-1, 11-2, 11-6, 11-7). Although biogenic hydrocarbons were noticeable, most of the hydrocarbons appeared to be petroleum: a series of n-alkanes between C_{20} to C_{36} with no odd/
even preference and a C_{18}/phytane ratio near unity when these two peaks were
resolved.

Biogenic hydrocarbon peaks for *Sargassum natans* (Figures 11-1 and 11-6)
occurred at retention indices corresponding to normal alkanes of C_{15}, C_{17}, and
C_{19}. The C_{17} peak was consistently the most abundant, and probably was com-
prised of several compounds which were not resolved by our column. Burns and
Teal (1973) identified six compounds in the multiplet, including n-heptadecane,
and four unidentified peaks with retention indices on Apiezon L of 1662, 1674,
1684, and 1696. These peaks could not be resolved on our column. A doublet
was also found at C_{15} which consisted of n-pentadecane and pentadecene. We
did not find the C_{21}:6 or C_{24} and higher biogenic compounds they also reported
and assume that they were associated with attached organisms rather than with

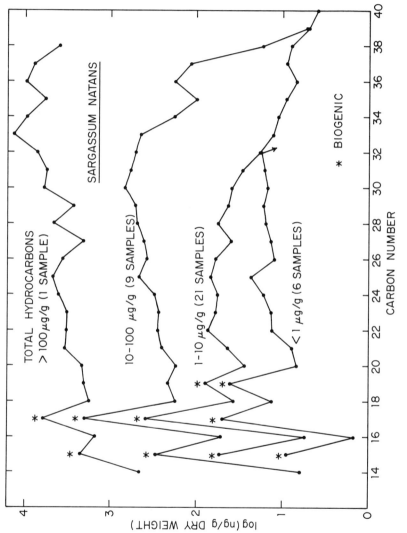

Figure 11-6. Hydrocarbon content of *Sargassum natans*.

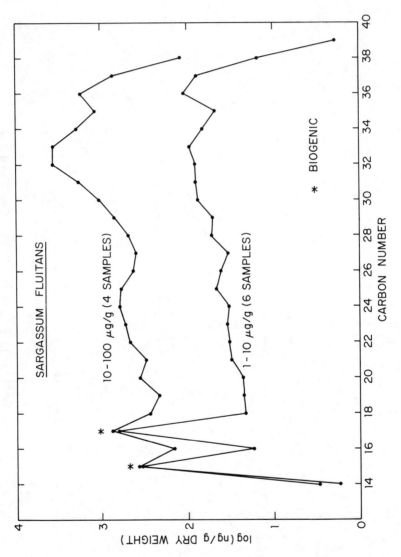

Figure 11-7. Hydrocarbon content of *Sargassum fluitans*.

Sargassum itself. *N*-nonadecane (C_{19}) was obvious as a biogenic hydrocarbon only in our samples which showed the lowest levels of petroleum contamination. At higher levels C_{19} abundance relative to adjacent *n*-alkanes was near unity; the biogenic component was masked by the C_{19} petroleum component.

The *Sargassum* community fauna all showed high levels of petroleum contamination. Biogenic hydrocarbons were at low levels and consisted primarily of C_{17} and in some cases C_{15}. The ratio of petroleum to biogenic compounds ranged from 13 to over 600. The average ratio for the mixed fauna samples was 34:1. In almost all cases a large unresolved envelope was present and phytane was often detectable (e.g., the C_{18} doublet in Figure 11-3).

The lowest contamination ratio occurred in the nudibranch *Scyllaea* (Figure 11-8) which feeds on anemones and hydroids. A distinct difference was found between the ratios for the two crabs. *Planes minutus* (Figure 11-9) feeds on a variety of foods, including *Sargassum,* and showed relatively low contamination (total 16 μg/g; ratio = 26). *Portunus sayi* (Figure 11-10) is the top predator in the community and had high hydrocarbon content (500 μg/g) and a petroleum biogenic hydrocarbon ratio of 612. The pipefish *Syngnathus* (Figure 11-11) had the highest hydrocarbon content (3300 μg/g). Hydrocarbon distribution for the snail *Litiopa* (Figure 11-12) and the shrimps *Leander* and *Latreutes* (Figures 11-13 and 11-14) are similar to the other animals.

We could not find any correlation between the hydrocarbon content of the various species and the amount of tar associated with the samples at the time of collection. This certainly indicated that our preparation procedures removed external tar, and this source of sample contamination was not a problem. But more importantly, it indicates that although the association between tar and *Sargassum* quantities fluctuates, the overall resulting petroleum load to the community is in response to the long-term general level of pollution, not to local concentration of tar.

Discussion

The only other study of hydrocarbons in the *Sargassum* community is that of Burns and Teal (1973). Table 11-2 compares their results with ours. This required us to correct their data to a dry weight basis, and this was done using wet/dry ratios determined in our laboratory. The agreement is within the range of variability encountered between individual samples for most species. The somewhat higher petroleum hydrocarbon contents we report probably reflect the higher level of pollution in the waters surrounding Bermuda than in the more northern Sargasso Sea where their samples were collected. Only the pipefish *Syngnathus* had very much higher petroleum levels in our samples. They may be due to our working with ungutted specimens, but we have to confirm whether this fish ingests tar. *Planes minutus* has been observed in the laboratory

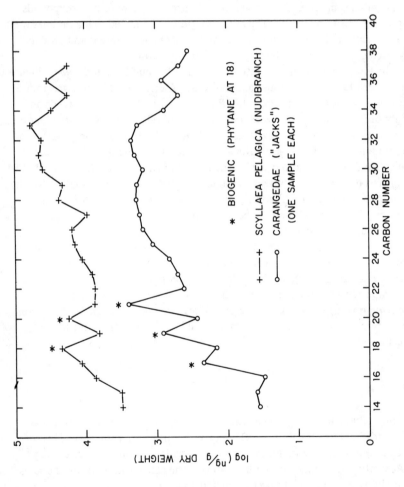

Figure 11-8. Hydrocarbon Content of *Scyllaea pelagica* and *Carangedae.*

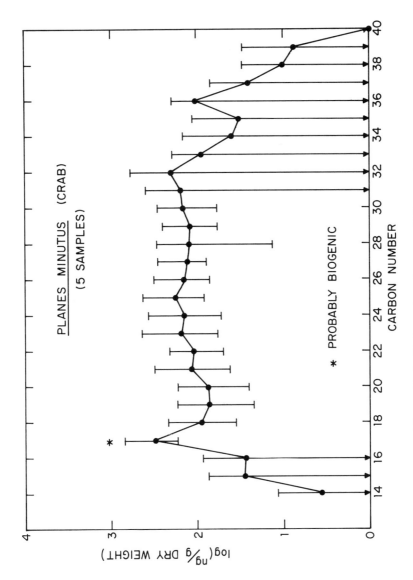

Figure 11-9. Hydrocarbon Content of *Planes minutus.*

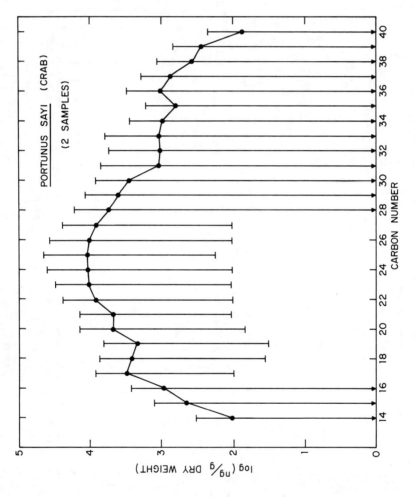

Figure 11-10. Hydrocarbon Content of *Porthunus sayi*.

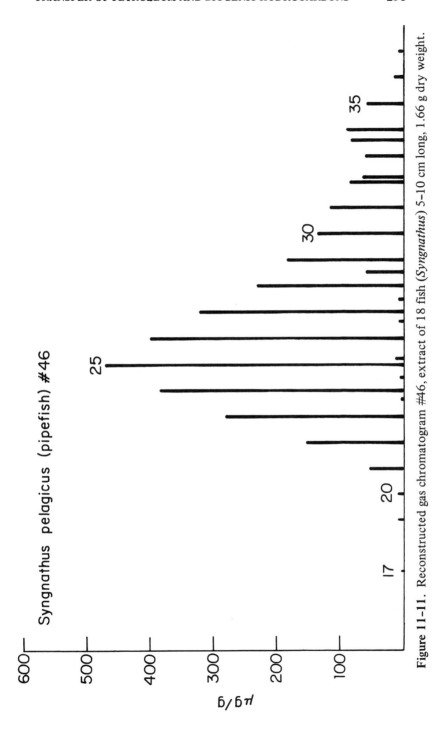

Figure 11-11. Reconstructed gas chromatogram #46, extract of 18 fish (*Syngnathus*) 5–10 cm long, 1.66 g dry weight.

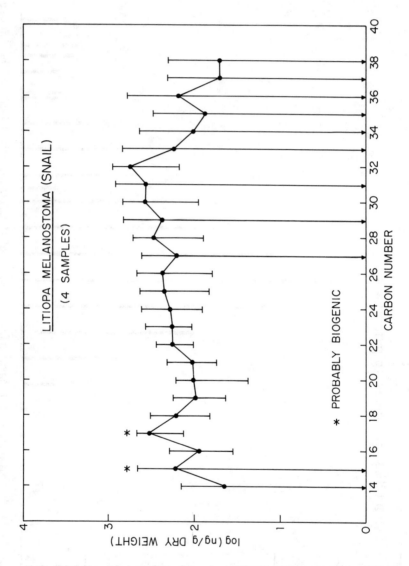

Figure 11-12. Hydrocarbon Content of *Litiopa melanostoma*.

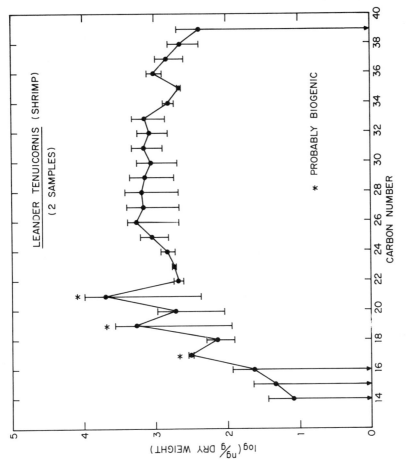

Figure 11-13. Hydrocarbon Content of *Leander tenuicornis.*

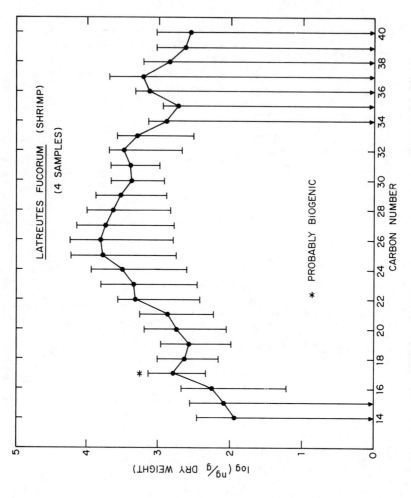

Figure 11-14. Hydrocarbon Content of *Latreutes fucorum.*

Table 11-2

Hydrocarbon Levels (μg/g Dry Weight) in Various Members of the Pelagic Sargassum Community

Organism	Dry/Wet Weight Ratio	Number of Samples	Biogenic	Petroleum
Sargassum natans algae	0.23	37(12)	1.4(7.7)	22.8(10.6)	16(1.4)
Sargassum fluitans algae	0.17	10(1)	0.8(69.4)	11.1(13.2)	14(0.2)
Planes minutus crab	0.22	5(1)	0.6(0.0)	15.6(49.5)	26(∞)
Portunus sayi crab	0.19	3(1)	0.8(0.0)	490(177)	613(∞)
Snygnathus pelagicus pipefish	0.27	2(1)	35(1)	3230(32.6)	93(31)
Leander tenuicornis shrimp	0.14	2(1)	7.2(1.1)	69.3(20.4)	45(18)
Histrio histrio fish	0.25	–(1)	–(0.2)	–(0.5)	–(31)
Canthidermis sp. fish	0.25	–(1)	–(0.3)	–(1.0)	–(21)
Litiopa melanostoma[a] snail	0.04	4(–)	7.5(–)	190(–)	25(–)
Scyllaea pelagica nudibranch	0.03	1(–)	40.5(–)	524(–)	13(–)
Latreutes fucorum shrimp	0.11	4(–)	0.2(–)	132(–)	133(–)
Carangidae sp. fish	0.18	1(–)	3.5(–)	75(–)	21(–)
Mixed Fauna	0.16	3(–)	2.3(–)	77(–)	34(–)

NOTE: The present data is compared to the results reported by Burns and Teal (1973) given in parentheses.

[a]Shell-less weight.

to ingest tar, and tar lumps have been found in fecal pellets of *Litiopa* (Figure 11–15). However, although our samples of *Planes* were not gutted, they did not show particularly high hydrocarbon levels.

The amounts of biogenic hydrocarbons in the two *Sargassum* species did not follow any clear pattern, except that the oldest portions of the plants were found to contain the highest abundance of the natural *n*-alkanes and new growth the least. Only one sample out of ten of *Sargassum fluitans* showed the high biogenic hydrocarbon levels reported by Burns and Teal (1973).

Our observations confirm the earlier reports that the *Sargassum* community

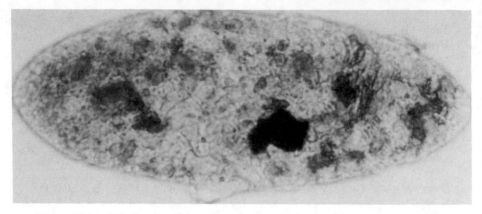

Figure 11-15. Fecal pellet from *Litiopa melanostoma* showing black tar lump inclusion.

is measurably contaminated by petroleum. Although the hydrocarbons may be accumulated in part through feeding (some top predators certainly had high hydrocarbon content), there does not appear to be any consistent magnification along the food web (Figure 11-16). No correlation was found between natural

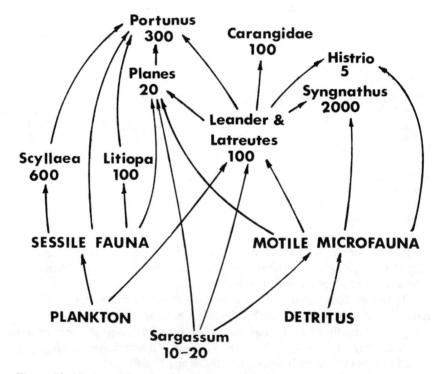

Figure 11-16. Simplified food web with approximate total hydrocarbon concentrations noted.

and biogenic hydrocarbon content, which might be expected if both were accumulated through the food web.

It seems likely that Burns and Teal are correct in suggesting that petroleum hydrocarbons are accumulated primarily across respiratory surfaces. Accumulation via the gut also seems likely for those species which ingest tar lumps. The levels of hydrocarbons in the different species must then be explained by differing body structure, age, and especially the physiological ability to store compounds found in petroleum.

Acknowledgments

This work was supported by the National Science Foundation, International Decade of Ocean Exploration, under Grant No. IDO 72-06411, and a grant from the Zemurray Foundation. The assistance of the Captain and crew of the R/V Panulirus, and various volunteers and students, particularly Barbara M. Baum, A. Barr Dolan, and Thomas D. Sleeter, is gratefully acknowledged.

References

Blumer, M., M. M. Mullin and D. W. Thomas (1964). Pristane in the marine environment. *Helgol. Wiss. Meeresunters, 10,* 187-201.

Blumer, M., R. R. L. Guillard and T. Chase (1971). Hydrocarbons of marine phytoplankton. *Marine Biology, 8,* 183-189.

Blumer, M. and J. Sass (1972). The West Falmouth Oil Spill II. Chemistry, *Woods Hole Oceanographic Institution Tech. Rept.* 72-19.

Burns, K. A. and J. M. Teal (1973). Hydrocarbons in the pelagic *Sargassum* community. *Deep-Sea Research 20,* 207-211.

Butler, J. N., B. F. Morris and J. Sass (1973). Pelagic tar from Bermuda and the Sargasso Sea. Special Publication No. 10, Bermuda Biological Station for Research.

Clark, R. C. (1966). *Saturated Hydrocarbons in Marine Plants and Sediments.* M.S. Thesis, MIT, Cambridge, Mass.

Clark, R. C. and M. Blumer (1967). Distribution of *n*-paraffins in marine organisms and sediment. *Limnology and Oceanography, 12,* 79-87.

Clark, R. C. and J. S. Finley (1974). Analytical techniques for isolating and quantifying petroleum paraffin hydrocarbons in marine organisms. *Marine Pollution Monitoring (Petroleum),* NBS Special Publication No. 409, pp. 209-212.

Corner, E. D. S. (1975). Proc. Royal Soc. B 189: 391 ff.

Corner, E. D. S., R. P. Harris, K. J. Whittle and P. R. Mackie (1975). Seminar Society of Experimental Biology, Liverpool.

Conover, R. J. (1971). Some relations between zooplankton and Bunker C oil in Chedabucto Bay following the wreck of the tanker *Arrow*. *J. Fish. Res. Board Can., 28,* 1327-1330.

Davis, J. B. (1968). Paraffin hydrocarbons in the sulfate-reducing bacterium *Desulfovibrio desulfuricans. Chem. Geol. 3,* 155-160.

Gordon, D. C. and P. D. Keizer (1974). Hydrocarbon concentrations in seawater along the Halifax-Bermuda section: lessons learned regarding sampling and some results. *Marine Pollution Monitoring (Petroleum),* NBS Special Publication No. 409, pp. 113-115.

Grice, G. D., G. R. Harvey, V. T. Bowen, and R. H. Backus (1972). The collection and preservation of open marine organisms for pollutant analysis. *Bul. Env. Contam. Toxicol., 7,* 125-132.

Keizer, P. D. and D. C. Gordon (1973). Detection of trace amounts of oil in seawater by fluorescence spectroscopy. *J. Fish. Res. Board Can., 30,* 1039-1046.

Lee, R. F., R. Sauerheber and A. A. Benson (1972a). Petroleum hydrocarbons: uptake and discharge by the marine mussel, *Mytilus edulis. Science, 177,* 344-346.

Lee, R. F., R. Sauerheber and G. H. Dobbs (1972b). Uptake, metabolism and discharge of polycyclic aromatic hydrocarbons by marine fish. *Marine Biology, 17,* 201-208.

McAuliffe, C. D. (1974). Determination of C_1-C_{10} hydrocarbons in water. In *Marine Pollution Monitoring (Petroleum),* NBS Special Publication No. 409, pp. 121-125.

Morris, B. F. (1971). Petroleum: Tar quantities floating in the Northwestern Atlantic taken with a new quantitative newston net. *Science, 173,* 430-432.

Morris, B. F. and D. Mogelberg (1973). Identification manual to the pelagic *Sargassum* fauna. Special Publication No. 11, Bermuda Biological Station. 63 pp. [This publication contains taxonomic information on 101 indigenous *Sargassum* species and a bibliography of 131 references to literature on the *Sargassum* community.]

Morris, B. F., J. N. Butler, J. Cadwallader and T. D. Sleeter (1976). Transfer of particulate hydrocarbon material from the ocean surface to the water column. In this volume.

Morris, B. F., J. Cadwallader and J. N. Butler (1976). Hydrocarbons in the *Sargassum* community (in preparation).

Morris, B. F. and J. Geiselman (1976). The food web of the *Sargassum* community (in preparation).

NAS (National Academy of Sciences) (1975). Petroleum in the Marine Environment. Washington, D. C., 107 pp.

Parker, C. A. (1971). The effect of some chemical and biological factors on the degradation of crude oil at sea. In *Water Pollution by Oil* (P. Hepple, ed.), Institute of Petroleum, London.

Stegeman, J. J. and J. M. Teal (1973). Accumulation, release, and retention of petroleum hydrocarbons by the oyster *Crassostrea virginica. Marine Biology, 22,* 37-44.

Wade, T. L. and J. G. Quinn (1975). Hydrocarbons in the Sargasso Sea surface microlayer. *Marine Pollution Bulletin, 6,* 54-57.

Zafiriou, O., M. Blumer and J. Myers (1972). Correlation of oils and oil products by gas chromatography. Woods Hole Oceanographic Institution Report 72-55.

12

Transfer of Higher Molecular Weight Chlorinated Hydrocarbons to the Marine Environment
R. W. Risebrough, B. W. de Lappe and W. Walker II

Introduction

At the beginning of the International Decade of Ocean Exploration Program in the early 1970s, there were insufficient data to assess the significance of contamination of the global marine environment by synthetic organic chemicals. Excessively high concentrations of DDT and PCB compounds were being reported from several oceanic areas; species of marine birds that are entirely dependent on marine food webs were unable to reproduce successfully in the marine environments of Southern California, New England and the Baltic. Moreover, organochlorine compounds were being detected in organisms from the remotest of marine environments, including those of Antarctica and the Arctic. It was not known whether the levels at that time would progressively increase as the DDT and PCB compounds entered the marine environment from the atmosphere, rivers, wastewater outfalls and other input sources. Such increases in contamination levels might result in the appearance of new kinds of deleterious effects, as well as a higher incidence of those already documented. Moreover, it was impossible at that time to reject the hypothesis that heavy metals of anthropogenic origin in the marine environment were contributing to the reproductive failures of marine fish-eating birds. Largely as a result of research carried out under the auspices of the International Decade of Ocean Exploration Program, it is now possible to begin to formulate a global perspective of the significance of chlorinated hydrocarbon pollution of the marine environment.

The present paper describes results of research on the higher molecular weight organochlorine pollutants carried out at the University of California's Bodega Marine Laboratory with support from the IDOE program. The lower molecular weight halogenated hydrocarbons and other synthetic organic compounds, which require analytical procedures different from those currently used to determine the higher molecular weight chlorinated hydrocarbons, are not discussed in this report. Research with heavy metals in the marine environment carried out during the IDOE program is briefly reviewed only to document the failure to find evidence for elevated levels of these elements at higher levels of marine food webs.

Use of the terms "DDT" and "PCB" as names of diverse and frequently very different compounds is no longer acceptable. DDT compounds in the environment are usually present in ratios quite different from the ratio in the original insecticide; PCB entering the sea in waste waters may consist predominantly of

261

the tri- and tetrachlorobiphenyls, but the PCB in marine mammals and birds frequently consists of pentra- and hexachlorobiphenyls. Use of the term "PCB" can describe both groups of compounds may therefore be misleading. Reference can readily be made to individual DDT compounds such as p,p'-DDE and o,p'-DDT; however at present there is no satisfactory alternative to identifying individual PCB compounds. Laboratories without a mass spectrometer are not sufficiently well equipped to do this. Usually the PCB in an environmental sample approximates one of the North American Aroclor preparations 1242, 1248, 1254, or 1260. Tri-, tetra-, penta-, and hexachlorobiphenyls, respectively, predominate in these mixtures, yet each contains both lower and higher chlorinated compounds. In the present report PCB mixtures most closely resembling Aroclor 1242, 1248, 1254 or 1260 are referred to respectively as trichlorobiphenyls, tetrachlorobiphenyls, pentachlorobiphenyls and hexachlorobiphenyls. This is clearly not an entirely satisfactory description, since commercial PCB mixtures contain a broader range of chlorobiphenyls.

Considerably more data on the distribution of the DDT and PCB compounds in the marine environment have been accumulated than for other synthetic organic pollutants. An understanding of the fluxes of the DDT and PCB groups consequently serve as predictive models for other pollutants with similar chemical and physical properties that might be present in the future. Production figures, usage patterns, and the physical and chemical properties of the DDT and PCB compounds should therefore be related to their present distribution in the global marine environment. Such data should also provide a determination of how changes in production and usage patterns will affect future environmental levels. Moreoever, it is necessary to relate existing levels to those which would harm the local marine ecosystem by affecting the viability or reproduction of the most senstivie species or groups of species.

Identification of Higher-Molecular Weight Chlorinated Hydrocarbons in the Marine Environment

A wide variety of synthetic chlorinated hydrocarbons has been used as insecticides in agriculture, forestry and public health; polychlorinated biphenyls are only one group of chlorinated compounds which have had widespread application in industry. Electron capture gas chromatography facilitates the detection of chlorinated hydrocarbons in environmental samples; accurate molecular identification is, however, considerably more difficult. Figure 12–1 shows chromatograms of extracts of seawater obtained off western Mexico at 15° 19.0'N, 115° 26.0'W on March 21, 1975; passing the extract through an alkaline side column (Miller and Wells, 1969) changes the gas chromatographic profile by converting some of the compounds to derivatives. Collection and analytical techniques are described below. The expected position of p,p'-DDE on these chromatograms is given as a point of reference. None of the peaks on these chromatograms

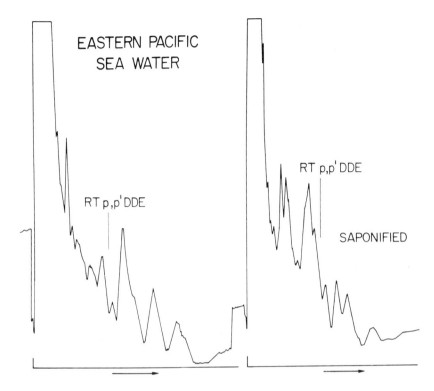

Figure 12-1. Gas chromatograms of an extract of seawater obtained off western Mexico at 15° 19.0' N, 115° 26.0' W on 21 March, 1975. Depth 3 meters, volume 137 liters. Chromatogram on the right is of the same extract after passage through a saponification side column. 3% OV-1, 100/120 Gas Chrom Q.

can be presently identified by the gas chromatographic techniques used for DDT and PCB analysis. Perhaps they are halogenated compounds of natural origin (Fenical, 1975). Their presence interferes with both the detection and measurement of polychlorinated biphenyl and DDT compounds, particularly when concentrations of the latter compounds are below 100 pg/l. In this sample concentrations of p,p'-DDT, p,p'-DDE and PCB were less than 5, 2, and 100 pg/l, respectively.

Figure 12-2 shows chromatograms of extracts of wastewater from a Southern California treatment plant. Wastewaters are major sources of pollutants to the Southern California marine environment (Schmidt *et al.,* 1971; D.R. Young, R. W. Risebrough, B. W. de Lappe, T. C. Heesen, and D. J. McDermott, unpublished data). None of the peaks on this chromatogram can be identified by the gas chromatographic techniques which are currently used routinely to determine

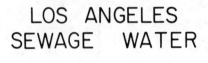

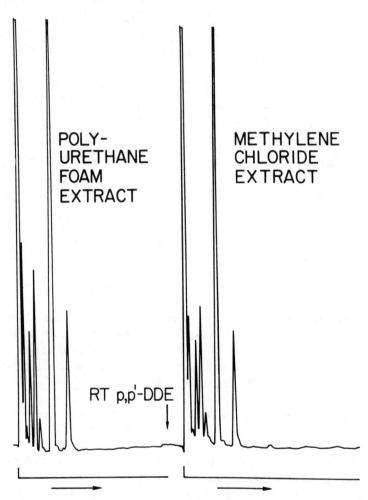

Figure 12-2. Gas chromatograms of replicate extracts of wastewater of Los Angeles County Sanitation Districts. Extraction procedures are described in the text. 6% florisil fraction. 2m, 2 mm ID glass column packed with OV-1, 100/120 mesh Gas Chrom Q. Some of these peaks have been identified as chlorinated anisoles (Burlingame *et al.*, 1976).

chlorinated hydrocarbons in environmental samples. More sophisticated techniques, such as gas chromatography/mass spectrometry would be required to determine unknown compounds. For example, analysis of these extracts by high resolution mass spectrometry showed the presence of trichloro-, tetrachloro-, and pentachloroanisoles (Burlingame *et al.,* in press, 1976). Perhaps these compounds derived from the methylation of the respective chlorinated phenols by bacteria. It is possible that they constitute some of the unidentified "early" peaks frequently seen in extracts of marine samples. In addition, dichloro-, trichloro-, tetrachloro-, and pentachlorobenzene were present in these extracts (Burlingame *et al.,* in press, 1976).

Polychlorinated biphenyl preparations have been found to contain chlorinated dibenzofurans as impurities. These compounds are highly toxic and lethal to embryos in submicrogram quantities (Vos and Koeman, 1970; Vos *et al.,* 1970; Bowes *et al.,* 1973). Their structural similarity to the chlorinated dibenzodioxins which are teratogenic at low concentrations (Sparschu *et al.,* 1971) suggested that they were possible causes of birth defects and other abnormalities observed in coastal marine birds (Hays and Risebrough, 1972). If present in the marine environment, these compounds would pose a hazard to both marine species and to human consumers. Zitko *et al.* (1972) examined marine samples from the Bay of Fundy for chlorinated dibenzofurans, but were unable to detect them. Detection limits were, however, substantially higher than the levels which would be expected in such tissues. The chlorinated dibenzofurans may be assumed to have entered the environment as a component of PCB, and might be present but as yet undetected in areas heavily contaminated with PCB compounds.

As a preliminary step in the search for chlorinated dibenzofurans in environmental samples, we measured the concentrations of these contaminants in various preparations of North American and European PCBs. Tetracholoro-, pentachloro-, and hexachlorodibenzofurans were detected in American PCB mixtures, with the exception of Aroclor 1016, in total concentrations from 0.8 to 2.0 $\mu g/g$. Higher concentrations were found in the French and German PCBs, Phenoclor DP=6 and Chlophen A-60 (Bowes *et al.,* 1975). Substantial quantities of these very toxic materials must then have been released to the environment over the past several decades with PCB. Little is known, however, of their environmental persistence or transport.

One of the Aroclor 1254 preparations, in which we measured the chlorinated dibenzofuran content, was incidentally used in a feeding experiment with Mallards (*Anas platyrynchos*). Subsequently, tissues of these ducks were pooled and the fat extracted. An aliquot of the lipid of known PCB content was analyzed for chlorinated dibenzofurans; a sample of corn oil with the same PCB content and with a known amount and composition of chlorinated dibenzofurans was simultaneously carried through the analytical procedures. Dibenzofurans could not be detected in the duck lipids, although they were recovered from the corn oil. It was concluded that the persistence of chlorinated dibenzofurans is low in ducks and probably also in other vertebrates (Norstrom *et al.,* in press). It therefore seemed unlikely that these compounds would persist in the marine environment and we have not continued our searches for them. The causes of the observed birth defects remain unknown, and the possibility that dibenzofurans may be formed *in vivo* from PCB compounds with structures favorable for their formation cannot be excluded.

Because levels of chlorinated hydrocarbons are frequently much higher in terminal members of marine food webs such as the birds and seals, compounds new to the marine environment have frequently been first detected in marine bird or mammal tissues. The majority of synthetic organic compounds so far identified in the marine environment have contained chlorine. Hexachlorobenzene (HCB), which has been detected in tissues of Great Cormorants (*Phalacrocorax carbo*), Sandwich Terns (*Thalasseus sandvicensis*) and Common Eiders (*Somateria mollissima*) from coastal areas of the Netherlands (Koeman and van Genderen, 1972; Koeman *et al.,* 1972b) has been considered a potentially hazardous marine pollutant (National Academy of Sciences, 1975); HCB is used as a fungicide but may enter the marine environment in significant quantities as a component of the tarry waste products from the manufacture of such chlorinated hydrocarbons as ethylene dichloride (U.S. Environmental Protection Agency, 1973).

Chlorinated styrenes have been identified by gas chromatography/mass spectrometry in tissues of Common Eiders, Sandwich Terns and Great Cormorants from the Netherlands (Ten Noever de Brauw and Koeman, 1972); their source apparently remains unknown. Chlorinated naphthalenes have also been identified in tissues of the Great Cormorant (Koeman *et al.,* 1973a).

Mirex has been reported in eggs of herons and of the White Ibis (*Eudocimus albus*) from estuaries of the U.S. Atlantic and Gulf coasts (Ohlendorf *et al.,* 1974) and has also been found in the blubber of a seal (*Phoca vitulina*) from the Netherlands (Ten Noever de Brauw *et al.,* 1973). In addition to its use as an insecticide, Mirex, under various trade names, is also used as a flame retardant.

Dieldrin has been found in eggs and tissues of several species of marine birds inhabiting coastal waters of Great Britain (Robinson *et al.,* 1967) and of New Zealand, and in addition has been found in pelagic species such as the Sooty Shearwater (*Puffinus griseus*) breeding in New Zealand subantarctic islands (Benington *et al.,* 1975). It is accumulated by Ospreys (*Pandion haliaetus*) feeding on coastal marine fish of the eastern United States (Wiemeyer *et al.,* 1975; Spitzer *et al.,* in press). Brown Pelicans (*Pelecanus occidentalis*), also feeding on coastal marine fish, have been found to contain endrin in Florida (Schreiber and Risebrough, 1972) and in the Gulf of California (Risebrough *et al.,* 1968b). In addition, heptachlor epoxide, toxaphene, and the several isomers of hexachlorocyclohexane (benzene hexachloride or BHC) are occasionally reported in local estuarine environments. Heptachlor epoxide and BHC isomers have been reported in antarctic birds breeding in the South Orkneys (Tatton and Ruzicka, 1967) but their identification has not been confirmed (Risebrough, in press, 1976).

Bidleman and Olney (1974a) reported the presence of both *cis*- and *trans*-chlordane in samples of marine air obtained on Bermuda and aboard ship during passage between Bermuda and Rhode Island. Concentrations occasionally exceeded those of *p,p'*-DDT. Chlordane has also been reported in rainwater of Hawaii (Bevenue *et al.*, 1972). The accumulation of chlordane compounds in food webs had, however, not been documented in these or other areas of the marine environment prior to 1974.

Eggs of the Common Tern (*Sterna hirundo*) from a marine population that breeds on Great Gull Island in Long Island Sound and winters in the Caribbean, and fish of several species from the vicinity of Great Gull Island were analyzed for chlordane compounds. Residues of alpha-(*cis*-)chlordane, gamma-(*trans*-)-chlordane (both of which may also include other chlordane compounds, and oxychlordane were detected. On a lipid basis the pooled sample of fish contained 0.22 ppm, 0.10 ppm and 0.05 ppm of alpha-(*cis*-)chlordane, gamma-(*trans*-)chlordane, and oxychlordane, respectively. Analysis of 27 individual eggs of the Common Terns yielded values of 0.86, 0.03 and 0.52, ppm lipid weight, of *cis*-chlordane, *trans*-chlordane, and oxychlordane, respectively. Two pooled samples of eggs, consisting of 10 unhatched eggs found intact and 109 eggs found broken, contained respectively 0.73 and 0.75 ppm of *cis*-chlordane, 0.03 and 0.02 ppm of *trans*-chlordane and 0.22 and 0.31 ppm of oxychlordane (P. Robinson and R. W. Risebrough, unpublished ms.). Chlordane compounds may therefore be more widespread in the marine environment than previously suspected. Several of the methodologies employed in routine analyses of chlorinated hydrocarbons do not detect the chlordane compounds; thus their presence has most likely been overlooked.

A dilemma is posed: a thorough examination of the fluxes of synthetic organic compounds in the marine environment requires detailed analytical studies of the very complex mixtures found in both seawater and in wastewaters which enter the sea. Yet such detailed chemical studies, which are both time-consuming and costly, may cause diversions from the overall assignment of understanding pollutant fluxes through the sea. This is particularly true if the compounds identified are of relatively minor importance in the marine environment. At the present time it is frequently difficult to assign priorities (Blumer, 1975).

Intercalibration

Intercalibration of different methods used to analyze marine samples and intercalibration among laboratories has become an integral and essential part of collaborative programs such as those undertaken under the auspices of the International Decade of Ocean Exploration program. Different laboratories frequently report widely differing results of analysis of the same sample (Holden, 1970; 1973). An initial intercalibration carried out among members of the IDOE program utilized a homogenate of shark liver. It contained approximately equivalent amounts of DDT and PCB compounds; substantially good agreement was obtained (Harvey *et al.*, 1974b). A subsequent intercalibration program co-ordinated by S. Pavlou and W. Hom of the University of Washington included IDOE investigators. A sample of marine sediment from Puget Sound was thoroughly mixed and then distributed among the participating laboratories.

The results indicated that surprisingly good agreement was obtained for the PCB determinations (coefficient of variation: 22); but agreement on the total DDT determinations was less satisfactory (coefficient of variation: 88). Pentachlorobiphenyls predominated in this sample, and analysis was less difficult than of samples containing not only PCB but a number of other organochlorines in equivalent concentrations. In addition, one laboratory reported the presence of chlordane compounds; these were not reported by other laboratories (Pavlou and Hom, in press).

Previous programs (Holden, 1973; Harvey *et al.*, 1974b) as well as the recent program undertaken by Pavlou and Hom (in press) designated participating laboratories by number or letter. However, identification of the laboratories seems desirable. As reported below, considerable discrepancy apparently exists among laboratories which utilize different methods in the determinations of chlorinated hydrocarbons in seawater. These differences should be identified and resolved through an intercalibration program as soon as possible.

Physical Chemistry of PCB and DDT Compounds

The behavior of PCB and DDT compounds in marine systems is dependent ultimately upon their physical and chemical characteristics; the available data are briefly reviewed here. The chemistry of PCB has been recently reviewed by Hutzinger *et al.* (1974). Selected solubility and vapor pressure data assembled by Mackay and Wolkoff (1973) and by Spencer and Cliath (1972) are presented in Table 12-1. Solubility of such compounds are very difficult to determine accurately as a result of the formation of aqueous emulsions, adsorption from emulsion by surfaces and loss into the atmosphere. Schoor (1975) has discussed in detail the problems associated with determining solubilities of Aroclor 1254. Using a technique in which organochlorine compounds were applied to florisil in acetone, evaporation of the acetone, and elution with distilled water, Weil *et al.* (1974) estimated the solubility of *p,p'*-DDE and *p,p'*-DDT to be 14 μg/l and 5.5 μg/l, respectively. The solubility of PCB isomers was determined to range from 0.016 μg/l to 4130 μg/l. The estimated solubility of *p,p'*-DDT by Weil *et al.* (1974) is therefore several times higher than the value presented in Table 12-1. Schoor (1975) estimated the solubility of Aroclor 1254 in water

Table 12-1
Solubility and Vapor Pressure Data for PCB and DDT Compounds

	Temperature (°C)	Solubility (mg/L)	Vapor Pressure (MM Hg)
p,p' – DDT	25	1.2×10^{-3}	1×10^{-7}
	30	–	7×10^{-7}
p,p' – DDE	30	–	65×10^{-7}
o,p' – DDT	30	–	55×10^{-7}
Trichlorobiphenyls[a]	25	0.24	4×10^{-4}
Tetrachlorobiphenyls[a]	25	5.4×10^{-2}	4.9×10^{-4}
Pentachlorobiphenyls[a]	25	1.2×10^{-2}	7.7×10^{-5}
Hexachlorobiphenyls[a]	25	2.7×10^{-3}	4.1×10^{-5}

Sources: Mackay and Wolkoff, 1973; Spencer and Cliath, 1972.

[a] Aroclors 1242, 1248, 1254 and 1260, respectively. The solubility and vapor pressure data apply therefore to complex mixtures of compounds.

containing 30 g/l NaCl to be approximately 40% of that in distilled water. Other data for solubility of these compounds in seawater are apparently not available. The mixture of pentachlorobiphenyls (Aroclor 1254) is approximately ten times as soluble in water as is p,p'-DDT (Table 12-1). On an individual compound basis, therefore, solubilities would appar to be approximately equivalent. Lower chlorinated PCBs are considerably more soluble than the penta- and hexachlorobiphenyls.

The order of magnitude difference in the vapor pressures of p,p'-DDT and p,p'-DDE would suggest significant differences in the atmospheric behavior of these two compounds. For example, 66% of the total DDT in the atmosphere over a field containing residual DDT consisted of p,p'-DDE (Cliath and Spencer, 1972). These authors concluded that most of the p,p'-DDT now present in well-aerated soils probably will be volatilized as p,p'-DDE. The vapor pressure of p,p'-DDE has been determined to be 6.5×10^{-6} mm Hg at 30°. The vapor pressure of Aroclor 1254 at 25°C has been estimated to be an order of magnitude higher. Yet this vapor pressure was determined for a mixture of compounds and therefore represents the sum of vapor pressures of a number of individual compounds. The vapor pressure of individual pentachlorobiphenyl isomers might therefore be equivalent to or only slightly higher than that of p,p'-DDE.

Mackay and Leinonen (1975) have estimated rates of the total evaporation process of low-solubility compounds from a water body to the atmosphere. The time required for concentrations to drop to half of the original value from a water depth of one meter was several hours for low molecular weight aromatic compounds, 74 hours for p,p'-DDT and 10.3 hours for Aroclor 1254. Several research groups (Bidleman and Olney, 1974a; Williams and Robertson, 1975) have reported that the concentrations in sea surface films were substantially higher than those in subsurface waters; the existence of the sea surface film may therefore impede the rate of evaporation. Nevertheless, the calculations of Mackay and Leinonen (1975) indicate that PCB and DDT compounds readily enter the atmosphere from water. In both marine and freshwater systems, however, DDT and PCB compounds also partition into the biota and inanimate particulates. The rate of loss into the atmosphere might therefore be significantly retarded.

Some of the difficulties in determination of solubilities and vapor pressure of individual compounds might be circumvented by measuring their concentrations in separate components of the marine environment including the atmosphere, subsurface water with particulates above a given size fraction removed, particulates, and selected members of the biota. Differences in partition coefficients pertaining to components would indicate differences in solubility and/or vapor pressure.

Patterns of Global Contamination of the Marine Environment by PCB and DDT Compounds

Although the data are few and sample sizes are frequently small, a reasonably complete picture of global marine contamination by DDT and PCB compounds can be presented using various species of marine birds as indicator organisms. Table 12-2 summarizes some of the available data on species of cormorants of the genus *Phalacrocorax*. All feed on fish and occupy similar ecological niches in marine food webs. With the exception of the residue values reported for the Double-Crested Cormorants in Southern California, which have been exposed to industrial contamination from the Los Angeles area (Gress *et al.*, 1973), the samples were obtained from areas reasonably remote from point sources of contamination. Samples obtained in estuaries and from inland waters are not therefore included. DDE residues in the Auckland Island Shags were somwhat lower than the majority of the DDE residues in cormorant eggs from Amchitka and Agattu in the Aleutians at the same equivalent latitude in the northern hemisphere. PCB values in the southern hemisphere birds were, however, one to two orders of magnitudes lower. Other data from biocoenetic equivalents in the two areas support the conclusion that at the fiftieth degree of latitude, DDE residues are equivalent or slightly lower in the southern hemisphere but PCB residues are one to two orders of magnitude lower. For example, DDE levels in an egg of a New Zealand Falcon (*Falco novaeseelandiae*) were equivalent to those in peregrines (*Falco peregrinus*) from Amchitka but PCB levels were an order of magnitude lower. Comparable differences were found between auklets (*Aethia pusilla* and *Cyclorrhynchus psittacula*) and the Tufted Puffin (*Lunda cirrhata*) of the Aleutians and the Diving Petrel (*Pelecanoides urinatrix*) from the Snares Islands

Table 12-2
PCB and DDE Residues in Cormorants, *Phalacrocorax spp.*
(*In ppm Lipid Weight*)

Locality, Date	Species	N	Tissue	% Lipid	DDE	PCB	PCB/DDE	Source
Amchitka, 1971	Red-faced cormorant	1	Yolk	20	3.8	19	5	1
Amchitka, 1974	Red-faced cormorant	1	Pectoral muscle	3.9	3.5	21	6	1
Agattu, 1974	Red-faced cormorant	1	Pectoral muscle	3.8	2.4	14	6	1
Amchitka, 1974	Pelagic cormorant	1	Pectoral muscle	4.0	0.8	8	10	1
Auckland Islands, 1972	Auckland Is. shag	4	Egg lipid	100	0.9	0.3	0.3	2
Iceland, 1973	Shag	10	Egg	5.0	3.8	23	6	3
	Great cormorant	13	Egg	4.8	3	10	3	3
Peru, 1969	Guanay	4	Egg lipid	100	12.2	15	1.2	4
Southern California, 1969	Double-crested cormorant	7	Egg lipid	100	754	87	0.1	5
Greenland, 1972	Great cormorant	3	Body fat	–	9.8	23	2.3	6

Sources: (1) White and Risebrough, in press; (2) Bennington *et al.*, 1975; (3) Sproul *et al.*, in press; (4) R.W. Risebrough, D.W. Anderson and J. McGahan, unpublished data; (5) Gress *et al.*, 1973; (6) Braestrup *et al.*, 1974.

of southern New Zealand as well as between seals and sea lions of the two island groups (White and Risebrough, in press; Bennington *et al.,* 1975).

Somewhat higher DDE levels in eggs of the Guanay (*Phalacrocorax bougain-villei*) suggest local sources of DDT contamination in Peru (R. W. Risebrough, D. W. Anderson and J. McGahan, unpublished data).

A comparison between the cormorant samples from the Aleutians and the eggs of two species, the Great Cormorant (*Phalacrocorax carbo*) and the Shag (*Phalacrocorax aristotelis*) from Iceland suggests that levels of DDE and PCB contamination in the two oceanic areas are similar. This conclusion is further-more supported by comparison of available residue data from fish. In five species of fish obtained from Amchitka in 1974, DDE residues on a wet weight basis ranged from 1 to 5 μg/kg; PCB residues ranged from 8 to 20 μg/kg, wet weight (White and Risebrough, in press). DDE residues in seven species of fish obtained from the coastal waters of Iceland in 1973 ranged from 1 to 9 μg/kg, wet weight (Sproul *et al.,* in press). On a lipid basis, PCB residues expressed as trichloro-, tetrachloro- or pentachlorobiphenyls ranged from 0.3 to 2 μg/g in the Amchitka fish and from 0.2 to 3 μg/g in the Icelandic fish.

Although sample sizes, particularly for the Amchitka samples, are small, a comparison of DDE and PCB residue levels in Black-Legged Kittiwakes (*Rissa tridactyla*), the Fulmar (*Fulmarus glacialis*), and the Thick-Billed Murre (*Uria lomvia*) suggest somewhat higher levels in the Icelandic birds although of the same order of magnitude and with comparable ratios (White and Risebrough, in press; Sproul *et al.,* in press). The differences probably reflect a higher level of contamination in those areas of the ocean where the Atlantic birds spent the winter months.

Earlier data (Risebrough *et al.,* 1968b) suggested that DDT compounds were more abundant than PCB in Pacific waters; the samples, however, were heavily biased towards coastal California waters where DDT contamination had been particularly severe.

Sproul *et al.* (in press) have presented a detailed comparison between residue levels in the breast muscles of Icelandic birds obtained in 1973 with those in birds obtained earlier by Bourne and Bogan (1972), principally in 1971 from areas north of Britain. Residue levels and PCB:DDE ratios were comparable.

In the northern Atlantic and Pacific, therefore, in areas comparatively remote from sources of contamination, PCB residues predominate over the DDT compounds. In the New Zealand area, however, including the subantarctic islands, DDT residues were found frequently to be present at higher concentra-tions than the sum of PCB (Bennington *et al.,* 1975). In the Antarctic, PCB compounds are also less abundant than those of the DDT group in resident species. Initially, PCB could not be detected in penguin eggs. Thus, maximum amounts of PCB in eggs of the Adelie Penguin (*Pygoscelis adeliae*) obtained from Cape Crozier near the big American base at McMurdo, in October 1967 were less than one eighteenth of the concentration of DDT compounds (Risebrough *et al.,*

1968b). Fourteen unhatched eggs of the Adelie Penguin obtained in January 1970 near Palmer Station, the American base on Anvers Island in the Antarctic Peninsula, were analyzed in 1972; PCB residues were not detected at that time because of the presence of interfering substances. It was subsequently found that these could be removed by rigorous saponification, permitting detection and measurement of PCB. PCB was also detected in eggs of the Adelie Penguin obtained on Anvers Island in December 1973 and in January 1975, as well as in eggs of *P. antarctica* and *P. papua* obtained elsewhere in the Antarctic Peninsula in 1975. The median DDE:PCB ratio was 3.0 (R. W. Risebrough, W. Walker II, T. T. Schmidt, B. W. de Lappe, C. W. Connors, unpublished data). Since it is likely that DDT use in the southern hemisphere has been greater than total PCB use, these ratios may be assumed to reflect these differences.

DDE residues in eggs of the Rockhopper Penguin (*Eudyptes crestatus*) from the Auckland Islands north of the Antarctic Convergence were only slightly but not significantly higher, 0.10-0.18 μg/g lipid weight, than those in penguin eggs from the antarctic continent (Risebrough, in press 1976) and were equivalent to residues in eggs of the three species cited above obtained in the Antarctic Peninsula in 1975. Similarly, DDE concentrations in eggs of the Wandering Albatross (*Diomedea exulans*) from the Auckland Islands ranged from 0.1 to 0.8 μg/g, lipid weight ($N = 4$), only slightly higher than the mean values ($N = 10$) in eggs and whole body homogenates of the Snow Petrel (*Pagodroma nivea*) from Hallett Station on the Antarctic Continent. DDE residues in four eggs of the Antarctic Tern (*Sterna vittata*) from the Auckland Islands ranged from 0.6 to 1.1 μg/g comparable to those in nine eggs from the Antarctic Peninsula with an average of 0.7 μg/g (Bennington *et al.,* 1975; Risebrough and Carmignani, 1972). From these limited data, it would appear that DDE levels are comparable over the Antarctic and Subantarctic regions and that no pronounced difference occurs in the area of the Antarctic Convergence.

Sproul *et al.* (in press) have examined available data for seabirds in the North Atlantic. PCB levels increased from west to east, between Iceland and Europe, with high levels reported in Gannets (*Morus bassanus*), Murres (*Uria aalge*) (Parslow and Jefferies, 1973; Parslow *et al.,* 1973), and Great Cormorants (Koeman *et al.,* 1973a). DDT residues were highest in the immediate vicinity of the Atlantic coast of the North American continent (Gilbertson and Reynolds, 1974; Zitko *et al.,* 1972) and dropped more rapidly than did PCB moving eastward into the Atlantic.

A population of the Sooty Tern (*Sterna fuscata*), a pelagic species, breeds on the Dry Tortugas in Florida and ranges widely over the tropical Atlantic. DDE and PCB levels in the breast muscle of 20 adults of known age were 2.5 and 7.8 ppm lipid weight, respectively (mean percent of lipid was 2.6%) (P. G. Connors, W. Robertson, S. A. Jacobs and R. W. Risebrough, unpublished data). These levels were equivalent to those in breast muscles of Arctic Terns (*Sterna paradisea*) that breed in the Shetlands but that migrate widely over the Atlantic and

winter in the Antarctic. They are also equivalent to those reported in Great
Shearwaters (*Puffinus gravis*), another species which ranges widely over the
North and South Atlantic, and in Manx Shearwaters (*Puffinus puffinus*) breeding
in Scotland (Bourne and Bogan, 1972). These levels and ratios appear therefore
to be typical of those in the mid-Atlantic at this level of the food web. Substantially higher values have been reported in Fulmars (*Fulmarus glacialis*) and
Black-Legged Kittiwakes that breed in the North Atlantic region (Bourne and
Bogan, 1972) and in Wilson's Petrels (*Oceanites oceanicus*) which breed in Antarctica but which spend the northern summer in the North Atlantic. The
PCB:DDE ratio of arithmetic mean concentrations in the Wilson's Petrels was
4.0; the average of individual ratios was, however, closer to 2.0 ($N = 9$) (Risebrough and Carmignani, 1972). Thus PCB:DDE ratios between 2 and 4 appear
to be characteristic of many seabirds of the Atlantic.

Degradation of DDT and PCB Compounds

The global contamination pattern by DDT and PCB compounds presented above
suggests a previously unsuspected degradative pathway of the DDT compounds.
Nisbet and Sarofim (1972) estimated that the total rate of loss to the environment of all PCBs in North America in 1970 was in the order of 1.5 to 2×10^3
tons/year into the atmosphere, 4 to 5×10^3 tons/year into fresh and coastal
waters and 1.8×10^4 tons/year into dumps and landfills. The major portion of
this PCB consisted of tri-, tetra- and pentachlorobiphenyls. In contrast to PCB,
the DDT compounds have been released directly into the environment. U.S.
production of DDT in the 1960s varied between 4.8×10^4 and 8.3×10^4 tons
per year. Global DDT production was estimated to be in the order of 1×10^5
tons per year at that time (National Academy of Sciences, 1971; Goldberg,
1975a). Approximately half of the DDT manufactured in the United States in
the 1960s was exported, principally to tropical countries. If the proportion of
pentachlorobiphenyls in the estimated total PCB input into the atmosphere and
into fresh and coastal waters of North America in the later 1960s is estimated
to be 2×10^3 tons/year, the environmental input of pentachlorobiphenyls was
substantially lower than that of *p,p'*-DDT.

Metabolism of *p,p'*-DDT to water soluble derivatives via *p,p'*-DDD has been
documented in a number of species, but the principal metabolite is usually
p,p'-DDE; moreover, this is the most abundant of the DDT compounds in the
great majority of environmental samples. Both DDE and PCB compounds are
metabolized to hydroxylated derivatives by marine mammals and birds (Jansson
et al., 1975). Nevertheless, the available data do not suggest a higher rate of
metabolic breakdown of *p,p'*-DDE than of pentachlorobiphenyls. In species
such as the Brown Pelican (*Pelecanus occidentalis*) ratios of DDE to pentachlorobiphenyls are equivalent in all individuals of a locality, are independent

of concentrations and are characteristic of a locality (Schreiber and Risebrough, 1972). Moreover, the same ratio is usually found in other vertebrates of the local ecosystem (Risebrough *et al.,* 1968b).

Under experimental conditions in the laboratory, DDE is degraded by ultraviolet light to a variety of derivatives (Kerner *et al.,* 1972). PCB compounds are also degraded, under experimental conditions, by ultraviolet light (Ruzo *et al.,* 1972; Safe and Hutzinger 1971; Herring *et al.,* 1972; Hutzinger *et al.,* 1972). The relative sensitivity of *p,p'*-DDE and the various chlorobiphenyls to ultraviolet light under environmental conditions, however, has apparently not yet been determined. A higher sensitivity of *p,p'*-DDE to degradation by ultraviolet light might explain why pentachlorobiphenyls are present in higher amounts than *p,p'*-DDE in the majority of marine ecosystems of the northern hemisphere.

The majority of polychlorinated biphenyls which have been produced have been tri- and tetrachlorobiphenyls. Frequently these cannot be detected in environmental samples which contain penta- and hexachlorobiphenyls. Wong and Kaiser (1975) have reported that bacteria present in lake water readily utilize Aroclor 1221 and Aroclor 1242 but not Aroclor 1254 as a sole carbon and energy source for growth. Metabolism of the lower chlorinated forms to polar derivatives in mammals and birds would also partially explain the lower persistence of these compounds in the environment.

Flux of Organochlorines Into the Sea from Rivers and Wastewater Outfalls

A committee of the National Academy of Sciences (1971) estimated that about 0.1% of the annual production of DDT reaches the oceans by surface runoff. Surface runoff of DDT into the sea from rivers in the United States during the 1960s was estimated to be in the order of 100 tons per year based on determinations of DDT compounds in principal rivers such as the Sacramento, the San Joaquin and the Mississippi. Input of PCB into the oceans from North American rivers has been estimated by Nisbet and Sarofim (1972) and the Panel on Hazardous Trace Substances (1972) to be on the order of 200 tons per year. For these calculations, one tenth of the total runoff from the continent (3×10^{15} liters/year) was assumed to contain a mean level of 500 ng/l, one third of the runoff was assumed to have a mean level of 50 ng/l, and the remainder was assumed to contain much lower amounts of PCB. Therefore, the total input of PCB into the oceans from North American sources other than the atmosphere, including losses from ships at sea was estimated to be less than 10^3 tons/ year. In the peak years of PCB use, 4 to 5×10^3 tons/year were estimated to have been discharged into all fresh and coastal waters. Unfortunately, relatively few measurements which include data on the kinds of DDT and PCB compounds are available for the peak years of use of both groups of compounds.

In coastal areas, wastewaters are frequently discharged directly into the sea rather than into a river; thus, in 1970 we began to measure chlorinated hydrocarbons in California wastewater effluents (Schmidt *et al.,* 1971). Four billion liters are currently discharged daily into the ocean from treatment plants in Southern California. In contrast, the total annual surface runoff from river, drainage channels, and urban storm sewers has averaged 400 billion liters per year between 1941 and 1970. On a daily basis, therefore, the amount of water discharged into the sea from the wastewater treatment plants is about three times greater than the total contributed by storm runoff (D. R. Young, R. W. Risebrough, B. W. de Lappe, T. C. Heesen and D. J. McDermott, unpublished ms.). Since the Southern California Bight had become highly contaminated with both DDT and PCB compounds (Risebrough *et al.,* 1968b), we expanded these studies in 1961 in collaboration with the Southern California Coastal Water Research Project to include determination of chlorinated hydrocarbons in all major wastewater outfalls in Southern California and in surface runoff from rivers and storm channels (Young *et al.,* unpublished ms.).

Measurements of DDT compounds and other chlorinated hydrocarbons in the waters entering the treatment plant of the Los Angeles County Sanitation Districts were begun in 1969 by personnel of the Sanitation Districts. One hundred forty-three separate analyses were carried out previous to April 1970. The mean value of these determinations, applied to a daily volume of 1.4 billion liters, indicated that approximately 290 kg of DDT compounds per day were entering the treatment plant from sources discharging into the sewage network of Los Angeles County. These sources included the world's largest manufacturer of DDT, the Montrose Chemical Corporation of California (Carry and Redner, 1970). Until April 1970, this factory passed its liquid wastes through a settling pond which was flushed with a minimum quantity of 1,900,000 liters of water per day. In April 1970, the company began to haul its DDT-containing liquors to a sanitary landfill (Sobelman, 1971), and amounts of DDT compounds entering the waste treatment plant from the sewage system dropped sharply thereafter (Carry and Redner, 1970). Between May and December of 1970, the DDT input into the water pollution control plant ranged between 50 and 100 kg per day (Carry and Redner, 1970). The organic deposits lining the sewer pipes below the DDT factory have proven to be a major storage site of DDT compounds, and thus they have been a continuing source of the DDT compounds entering the treatment plant since April 1970 (Redner and Payne, 1971).

Our program has included an extensive series of intercalibrations with the laboratory of the Los Angeles County Sanitation Districts, laboratories of the other sanitation districts of Southern California, and the laboratory of the Southern California Coastal Water Research Project (Young *et al., op. cit.,* unpublished ms.).

Input of DDT compounds into the Southern California Bight from the outfall of the County Sanitation Districts of Los Angeles County has continued to decline. In 1970 the total input was 21,600 kg; input was reduced to 1,440 kg in 1974 (Young *et al., op. cit.,* unpublished ms.). During this period DDT residues in Northern Anchovies (*Engraulis mordax*) and Brown Pelicans inhabiting the Southern California Bight have declined significantly (Anderson *et al.,* 1975). Some of the principal results of the program are summarized in Table 12-3. Most of the DDT, dieldrin and PCB in surface runoff enters the sea during the periods of storms; close correlations were observed between the chlorinated hydrocarbon concentrations and the particulate content of these waters. Dry weather flow during the remainder of the year contained substantially lower amounts of chlorinated hydrocarbons. Input of chlorinated hydrocarbons from all surface runoff was found to be substantially lower than input from municipal wastewaters. In some cases estimates of PCB are minimum values since all wastewaters were not sampled. The estimate for total PCB input in 1972, 20,000 kg or 20 tons, may be compared to the estimate by Nisbet and Sarofim (1972) of an input into fresh and coastal waters of North American of 5000 tons per year around 1970. As anticipated, trichlorobiphenyls predominated over the pentachlorobiphenyls in wastewaters entering the sea.

A chromatogram of an extract of seawater obtained from the Whites Point outfall area in May, 1973 is shown in Figure 12-3. The sample was obtained with the *in situ* system described below, at a depth of 25 meters. The DDT compounds p,p'-DDE and p,p'-DDD predominate; trichlorobiphenyls are also present. As discussed below, the wastewater outfalls of southern California have had a dominant effect on contamination levels in the Southern California Bight.

Aerial Transport of Chlorinated Hydrocarbons

The reported detection of DDT compounds in resident antarctic wildlife (Sladen *et al.,* 1966; George and Frear, 1966; Risebrough *et al.,* 1968b; Brewerton, 1969) indicated global contamination by synthetic organic compounds; the atmosphere appeared to be the most plausible medium of transport. The possibility that these DDT residues may have derived from human activities in Antarctica have been examined by Risebrough (in press) and were considered unlikely. Initially, PCB compounds could not be detected in a number of samples, including the eggs of the Adelie Penguin obtained near McMurdo and Hallett stations on the Antarctic Continent; DDT compounds were, however, present (Risebrough *et al.,* 1968b; Risebrough and Carmignani, 1972). A considerable amount of refuse has been burned at the bases of all nations participating in the antarctic program; discarded waste was likely to consist of material containing PCBs. (Contamination by PCB is more likely than contamination by the DDT compounds.)

Brewerton (1969), on the contrary, concluded that DDT contamination in the Antarctic most likely derived from scientific bases from a comparison of

Table 12-3
Comparison of Chlorinated Hydrocarbon Inputs to the Southern California Bight via Surface Runoff, Municipal Wastewater, and Other Routes
(In Kilograms/Year)

Route, Year	Total DDT	Dieldrin	Trichloro-biphenyls	Pentachloro-biphenyls	Total PCB	Source
Surface Runoff, 1971-72	106-108	25	96-201	96-119	192-320	1
Surface Runoff, 1972-73	320-325	66	0-552	248-282	248-834	1
Municipal Wastewater, 1972	6490[a]	100	≥19,200	≥ 260	≥ 19,500	1
Municipal Wastewater, 1973	3920	≤ 275	≥ 1,930	1,510	≥ 3,440	1
Municipal Wastewater, 1974	1560	84	5,480	1,020	6,500	1
Industrial Discharges[b], 1973	20	5	–	50	50	2
Antifouling Paints, 1973	<1	–	< 1	< 1	< 1	3
Aerial Fallout, 1973-74	1300	–	–	1,500	1,500	4

Sources: (1) D.R. Young, R.W. Risebrough, B.W. de Lappe, T.C. Heesen and D.J. McDermott, unpublished ms. (2) Young and Heesen, 1974; (3) Young et al., 1974; (4) SCCWRP, 1975.
[a] 1971 for total DDT: 21,600 kg.
[b] San Pedro Harbor only.

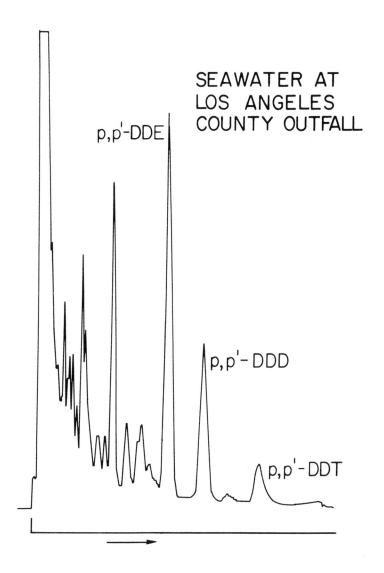

Figure 12-3. Chromatogram of seawater receiving input from the outfall of the
Los Angeles County Sanitation Districts, obtained at a depth of
25 meters in the vicinity of the diffusers. Volume: 15 liters.
3% OV-1 column. Outfall II cruise of *R./V. Thomas Thompson*,
May 1973.

residues in eggs of the Adelie Penguin from the vicinity of the large American base at McMurdo with those from the vicinity of a relatively small base at Cape Hallett. Inclusion of other data, however, showed that no significant differences in DDT contamination existed between the two areas (Risebrough, in press). Moreover, as discussed above, DDE residues do not appear to differ across the Antarctic Convergence. Since waters south of the Convergence derive from the melting of the antarctic icecap or the upwelling of deep waters, a barrier to transport by ocean currents exist. The DDT residues in resident antarctic wild-life therefore most likely derive from the atmosphere.

Peel (1975) was unable to detect PCB compounds in samples of antarctic snow obtained in December 1969 inland from the British Antarctic Station at Halley Bay (75° 31'S, 26° 42'W). The samples represented the previous 5-10 years snow accumulation. Total concentrations of p,p'-DDT and $p.p'$-DDE ranged from 0.1 to 2.0 ng/kg. On this basis, the argument was advanced that on a global scale the atmosphere is not the predominant mode of transport for PCBs. This conslusion appears inconsistent with that of Bidleman and Oleny (1974a; 1974b) and Harvey and Steinhauer (1974) who were able to detect PCB in samples of marine air in the North Atlantic at concenttrations considerably in excess of those of the DDT compounds.

Earlier studies in this laboratory (Risebrough *et al.*, 1968a) failed to detect PCB in samples of marine atmospheric particulates, collected on nylon mesh screens on Barbados which sampled the Northeast trades, and in La Jolla, California. Studies on Barbados were continued by Seba and Prospero (1971) and by Prospero and Seba (1972). No correlations were found between the air concentrations of the pesticides and that of airborne dust; pesticides were believed to have originated from the higher latitudes, either Europe or North America.

The failure to find polychlorinated biphenyls on the particulates trapped on nylon mesh screens may be explained from the results of Bidleman and Olney (1974a; 1974b). The majority of the organochlorine residues, particularly of PCB, were not trapped by a glass fiber filter placed before the polyurethane foam column through which air was passed; however, residues were recovered in the foam. Although some evaporation from particulates entrapped on such a filter might be anticipated, it would appear that the majority of PCB in the atmosphere is present in the vapor phase. The greater affinity for particulate matter of p,p'-DDT might be predicted from its low vapor pressure (Table 12-1).

Marine air trapped by Bidleman and Olney (1974a; 1974b) contained principally tri- and tetrachlorobiphenyls; PCB in seawater at these sites consisted predominantly of penta- and hexachlorobiphenyls. Such a partitioning between water and air might be expected from the vapor pressure data (Table 12-1), but the calculations of Mackay and Leinonen (1975) indicate equivalent residence times in water of mixtures of tri-, tetra-, penta-, and hexachlorobiphenyls. More data are clearly needed on the partitioning or organochlorines between water and the atmosphere.

In order to obtain additional data on the distribution of chlorinated hydro-carbons, particularly of PCB, in Antarctica, samples of antarctic snow were ex-tracted *in situ* on Doumer Island on the Antarctic Peninsula (64° 51.3'S, 63° 35.5'W) in January and February 1975; eggs of three species of antarctic pen-guins (*Pygoscelis adeliae, P. papua, P. antarctica*) were obtained to look for PCB in the food webs and to compare residues with those in eggs obtained five years previously. Eleven 99-liter samples of melted snow were extracted by tech-niques described elsewhere (R. W. Risebrough, W. Walker II, T. T. Schmidt, B. W. de Lappe, and C. W. Connors, unpublished ms.). PCB concentrations expressed as pentachlorobiphenyls ranged from 30 to 1200 picograms per liter; median concentration was 300 pg/l. The median total DDT:PCB ratio in these sam-ples was 6. Samples were obtained between the surface and a depth of 6 meters; lowest concentrations were recorded in four samples obtained between the sur-face and a depth of 1 meter; melting and evaporating of snow may account for the higher concentrations at greater depths. A gas chromatogram of an extract of the antarctic snow is shown in Figure 12-4a. The dominant component is *p,p'*-DDT; *o,p'*-DDT and *p,p'*-DDE were also present. In contrast, Peel (1975) recorded higher concentrations of *p,p'*-DDE than of *p,p'*-DDT but was unable to detect polychlorinated biphenyls. The chromatogram of Figure 12-4a shows no polychlorinated biphenyl compounds, but concentration and injection through the saponification side column permits both detection and quantifica-tion of pentachlorobiphenyls (Figure 12-4b).

The median DDE:PCB ratio in penguin eggs obtained between 1970 and 1975 in the Antarctic Peninsula was 3.0 (R. W. Risebrough, W. Walker II, T. T. Schmidt, B. W. de Lappe, and C. W. Connors, unpublished data). Total DDT: PCB ratios recorded in several of the subantarctic samples from the Auckland Islands ranged between 2 and 4 (Bennington *et al.*, 1975), comparable to those recorded in the antarctic penguin eggs. These findings further support the conclusion previously advanced that fallout patterns appear to be similar in areas north and south of the convergence, indicating the importance of atmospheric rather than oceanic transport of the pollutants.

The relative contributions of precipitation and dry fallout in the flux of chlorinated hydrocarbons into the sea from the atmosphere remains unknown; moreover, the interchange at the air-sea interface is poorly understood. Below we present results that indicate that "dry" fallout to the sea surface may be important and that precipitation is not the only route of entry of organochlorine compounds to the sea. This conclusion was advanced previously (Risebrough *et al.*, 1968a) with reference to fallout of organochlorine pesticides on atmos-pheric particulates into the tropical Atlantic.

Bidleman and Olney (1974a; 1974b) have shown that the majority of or-ganochlorine compounds in the atmosphere are present in the vapor phase. As equilibrium conditions between the atmosphere and the sea surface would pre-vail the lower chlorinated PCB compounds, which predominate in wastewaters

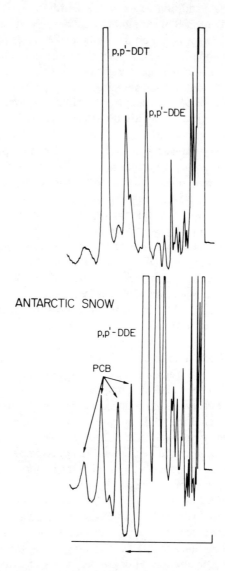

Figure 12-4. Chromatograms of antarctic snow, extracted *in situ* on Doumer
Island, Antarctic Peninsula, in January 1975. Figure 12-4a
(above) shows no discernible PCB peaks, surface snow, volume 99
liters. Figure 12-4b (below) shows the chromatogram of an extract that has been concentrated and passed through a saponification side column. 3% OV-1 column, 100/120 Gas Chrom Q.
Depth 4–5 meters, volume 99 liters.

entering the Southern California Bight (Table 12-3), would be expected to dissipate into the atmosphere. Ultraviolet degradation may therefore be a more significant degradative pathway than microbial breakdown.

Experimental studies of the volatility and flux into the atmosphere are more extensive for the DDT compounds than for the PCBs. Lloyd-Jones (1971) has estimated that about half of the DDT applied to field crops may enter the atmosphere. Spencer and Cliath (1972) and Cliath and Spencer (1972) have shown that the vapor pressure of p,p'-DDE is about nine times that of p,p'-DDT; much of the DDT present in the soil may therefore volatilize as p,p'-DDE. p,p'-DDT has been shown to volatilize readily from surfaces (Quraishi, 1970). Aroclor 1260, p,p'-DDT, and other organochlorine insecticides were found to escape readily into the atmosphere from waters obtained from two rivers and from a subtidal zone (Oloffs et al., 1972).

In 1970, a committee of the National Academy of Sciences (1971) speculated that up to 25% of the DDT applied to the land would eventually enter the sea. Determinations of fluxes across the air-sea interface, of fluxes through the water column, and fluxes between the water column and the sediments are needed to complete the global mass balance equation of the DDT compounds, as well as of the various kinds of polychlorinated biphenyls. Such models will be useful in predicting behavior of other synthetic organic pollutants. Needed, however, are estimates of past fluxes from the atmosphere to the sea.

Estimates of Integrated Fluxes from the Atmosphere to the Sea

Analysis of dated anaerobic sediments in the Santa Barbara Basin has permitted determinations of changes in the fluxes from the water column to the sediments (Hom et al., 1974). Analysis of varved anaerobic sediments in freshwater lakes and reservoirs will also be useful in determining changes in past fluxes. Permanent snowfields constitute other records which will permit determination of changes in the fluxes of organochlorines from the atmosphere over the time periods these compounds have been in use. Because local meterological conditions may significantly affect results, areas remote from input sources with a relatively uniform climate would be more suitable for total flux determinations.

The snowfield on Snow Dome on Mount Olympus, Washington, was sampled in September 1975 to a depth of 22.5 meters. Volumes analyzed ranged from 90 liters to 177 liters. A chromatogram of an extract is shown in Figure 12-5. As in the antarctic samples, p,p'-DDT was the predominant residues. The median concentration of p,p'-DDT was 10 ng/liter: concentrations of PCB were lower by two orders of magnitude (R. W. Risebrough and W. Walker II, unpublished results). In view of the general predominance of PCB compounds over DDT in the North Pacific (White and Risebrough, in press), it would appear that

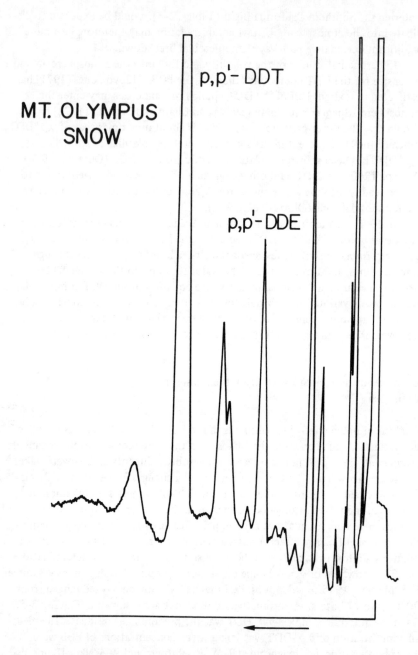

Figure 12-5. Chromatogram of snow from Mount Olympus, Washington, extracted *in situ,* September 1975. 3% OV-1 column, Gas Chrom Q.

precipitation contains only a small fraction of the PCB. Alternatively, high rates of evaporation on Mount Olympus, coupled with the higher vapor pressure of the PCB compounds, might result in a greater release of PCB into the atmosphere than the DDT compounds. Dating of these samples by lead-210 indicates that all samples were deposited within the last decade, in agreement with stratigraphy data (R. W. Risebrough, W. Walker and V. C. Anderlini, unpublished results). Snowfields with lower rates of precipitation, which would facilitate obtaining older samples, with more uniform deposition, and in colder climates, such that evaporation rates might be lower, would be more suitable for estimates of past changes in fluxes.

Determination of Chlorinated Hydrocarbons in Seawater: Development of an *In Situ* Sampling System

Accurate determinations of the chlorinated hydrocarbon content in seawater is preliminary to determinations of the fluxes between water and other components of the marine environment, particularly air, sediments, particulates and the biota. Concentrations are expectedly low and large volumes may be necessary to obtain sufficient quantities for accurate measurement.

Cox (1971) measured DDT residues in seawater obtained from a shipboard seawater system, using a continuous-flow, liquid-liquid extraction, along transects off Oregon and Washington and off the California coast. Concentrations of individual DDT compounds were not presented but total residues reported ranged from 2.3 ng/l off Oregon and Washington to 5.6 ng/l off Southern California. In 1970, when the samples were obtained, DDT contamination in the Southern California Bight, particularly in the area near Los Angeles, was exceptionally severe and a greater difference between Southern California and Oregon-Washington might be expected. Moreover, no differences were found between the outer regions of the Bight and the area immediately off Los Angeles. The volumes extracted ranged from 1.6 liters to 4.3 liters.

Williams and Robertson (1975) have reported PCB, *p,p*'-DDT and *p,p*'-DDE values in subsurface water from the outer California current and at two stations in the North Central Pacific Gyre, both in the vicinity of $31°N$, $155°W$. Water was collected in 2.5-liter glass bottles, 10–15 cm below the surface. Reported PCB values in the subsurface water were 2.5 ng/l in the California current and 4.5 ng/l at one of the North Central Pacific Gyre stations.

Scura and McClure (1975) utilized a column containing 5% activated carbon powder, 10% MgO and 85% refined diatomaceous earth for the determination of chlorinated hydrocarbons in 1-liter samples of seawater collected off the Southern California coast. Reported PCB concentrations in surface seawater at five stations in the Southern California Bight ranged from 3 ng/l to 9.6 ng/l. Concentrations reported in five samples obtained at depths from 500 to 1500 meters ranged from 2.3 to 10.3 ng/l. In the majority of samples, both p,p'-DDE and p,p'-DDT were detected; DDE concentrations ranged from <0.1 ng/l to 1.0 ng/l at the surface and from 0.3 to 1.2 ng/l at depths from 500 to 1500 meters. Concentrations of p,p'-DDT were somewhat lower.

Bidleman and Olney (1974a) have reported PCB and DDT concentrations at eight stations in the Sargasso Sea in 1973. PCB concentrations in subsurface water were lower than detectability (<0.9 ng/l) at four of the stations and ranged from 1.0 to 3.6 ng/l at the others. Concentrations of p,p'-DDT were 0.5 ng/l at one station but were below detectability (<0.15 ng/l) at the others. Samples were collected in 3.8-liter glass jugs at depths approximately 30 cm below the surface. Substantially higher concentrations were measured in the surface microlayer. These concentrations were calculated by reference to Aroclor 1260. Concentrations calculated by reference to Aroclor 1254 may have been up to two times higher (Risebrough and de Lappe, 1972).

Because of the difficulties in extracting large volumes of seawater with organic solvents, Harvey et al. (1973) developed an extraction system using Amberlite XAD-2 resin. Seawater, in volumes ranging from 19 and 80 liters, was obtained in a large volume sampler, passed through the resin, and analyzed on board ship. Reported PCB concentrations, calculated relative to Aroclor 1254, averaged 35 ng/l at the surface and 10 ng/l at 200 meters. At depths of 100, 1500 and 3000 meters reported concentrations were greater than one ng/l. Samples were obtained in the summer of 1972 (Harvey et al., 1973).

In 1973 and 1974, substantially lower values were recorded and it was concluded that PCB concentrations in North Atlantic surface waters had declined 40-fold over the two-year period as a result of restrictions on PCB use (Harvey et al., 1974a). At two stations in 1973 (09°N, 40°W and 32°N, 70°W) samples were obtained at various depths of the water column. PCB concentrations ranged from 4.3 ng/l at 10 meters to 1 ng/l at 3000 meters at the former site and from 0.4 to 1.9 ng/l at depths between 10 meters and 5100 meters at the latter.

Harvey et al. (1974a) estimated that the decline of PCB concentrations would require removal of 2×10^4 tons of PCB from the upper 200 meters over the two-year period, assuming a mean concentration in 1972 of 20 ng/l in the upper 200 meters of the North Atlantic between 26°N and 63°N; the volume was assumed to be 10^{18} liters. The estimate of the volume over this area at a depth of 200 meters appears, however, to be low, perhaps by an order of magnitude since the corresponding area would be 5×10^6 km^2. The area of the Atlantic, including the North and South Atlantic, is 82×10^6 km^2 (Sverdrup, Johnson and Fleming, 1942). Thus, since the PCB reported consisted predominantly of pentachlorobiphenyls, on the order of 10^5 metric tons of pentachlorobiphenyls would have to be removed from the upper 200 meters. From 1957 through 1974, a total of only 123,723 thousand pounds, equivalent to 5.6×10^4 metric tons of pentachlorobiphenyls (Aroclor 1254), was sold in the United States (Monsanto Chemical Company, 1975). Sales in 1970, the peak year of

U. S. production, amounted to 5.6 X 10^3 tons. The mean depth of the Atlantic
is 3926 meters (Sverdrup *et al.,* 1942) and since PCB was recorded by Harvey
et al. (1973) at depths to 3000 meters, the estimated pentachlorobiphenyl content of the North Atlantic would be correspondingly higher. With the assumptions of 1 ng/l concentration throughout the water column, 3900 meter average
depth and the area of the North Atlantic between the equator and 65°N in the
order of 47 X 10^6 km^2, the pentachlorobiphenyl content would be in the order
of 180,000 metric tons, higher than the total U.S. domestic use from 1957
through 1974.

Similarly, a mean PCB concentration of 1 ng/l in the North Pacific would
appear to be too high. If a mean value of 1 ng/l of pentachlorobiphenyls were
assumed for the upper 200 meters of the North Pacific and the area from the
equator to 65°N is 82 X 10^6 km^2, 16,000 tons of pentachlorobiphenyls would
be present in this volume. This represents almost one third of total U.S. use
from 1957 through 1974. Nisbet and Sarofim (1972) and the Panel on Hazardous Trace Substances (1972) estimated that 1.5 to 2.5 X 10^3 tons per year of
PCB, including tri- and tetrachlorobiphenyls, had been discharged into the
atmosphere from North America. If this estimate is correct, the atmospheric
input of pentachlorobiphenyls into the North Atlantic must have been substantially less, perhaps by an order of magnitude.

In an effort to reduce the very significant problems of contamination during
the collection and extraction of trace organics from seawater, we have worked
towards the development of an *in situ* sampling system. Such a system elimi-
nates the need for intermediary sampling and storage containers by providing
extraction at depth through the use of filter columns which have been precleaned
in the laboratory to known background levels. Columns may be transported to
and from the field in clean containers, with controlled exposure only at the time
of sampling. The essential component of such a sytem is a filter medium with a
high chemical affinity for chlorinated hydrocarbons in seawater. Such a substance must also retain particulates without clogging, provide suitable blanks
for sampling at pg/l concentrations and function efficiently at flow rates which
would permit samples of 100 liters or more to be extracted within a reasonable
period of time.

The use of polyurethane foam as an absorptive medium for the extraction
of a wide variety of metal derivatives and organic compounds from aqueous solutions was first reported by Bowen (1970). Following this initial report, Gesser
et al. (1971) demonstrated the ability of polyurethane foam to recover PCB
from distilled water spiked at 2 µg/l. Two foam plugs (22 mm diameter, 38 mm
long, density 0.027 gm/cm^3) with a total volume of 29 cm^3 produced recoveries
of 91-98% at flow rates of 250 ml/min. Subsequent studies by Uthe *et al.*
(1972) indicated that coating foam plugs with DC-200 enhanced the recovery of
ten chlorinated hydrocarbon insecticides from spiked distilled water solutions.
Recoveries averaged between 86-100% for DC-200 coated plugs and 45-84%

with untreated foam. An application of DC-200 coated foam plugs for monitoring a freshwater stream was described by Uthe *et al.* (1974).

Musty and Nickless (1974) have presented recovery data for PCB and thirteen chlorinated insecticides using four untreated foam plugs (22 mm diameter, 40 mm long, total volume 60.8 cm^3) at flow rates of 100 ml/min. Under these conditions, the extraction efficiency of all thirteen insecticides tested was greater than 90%. Recoveries of Aroclor 1242, 1248 and 1254 ranged from 87–99%; recovery of Aroclor 1260 was only 40%. Using two foam plugs (34 mm diameter, 44 mm long, total volume 80 cm^3) Bedford (1974) reported recoveries of Aroclor 1254 varying between 91% from distilled water to 69% from unfiltered lake water at flow rates of 250–275 ml/min. Bedford hypothesized that PCB was adsorbed to the surface of fine particles in the lake water which were not effectively retained by the foam. Experiments with montmorillonite clay suspensions, however, failed to produce a comparable condition. Recoveries of 91% of Aroclor 1254 were obtained from distilled water solutions with clay concentrations ranging from 0 to 30 mg/l.

Initial studies in our laboratory evaluated the extraction of chlorinated hydrocarbons from seawater by a high density, high ether content polyurethane foam (81% polyether glycol, 19% toluene di-isocynate, density 0.036 gm/cm^3, United Foam Company, Los Angeles) which had previously been shown effective in the removal of phthalate esters from solution (Carmignani and Bennett, in press). After determining that satisfactory blanks and recoveries could be obtained, equipment was prepared for a seawater sampling program off Southern California aboard the *R./V. Thomas Thompson* in May 1973. Polyurethane foam plugs (66 mm diameter, 57 mm thick) were cut from sheets, using a heated tube with a sharpened end. Plugs were Soxhlet-extracted with acetone prior to being packed in stainless steel columns (60 mm I.D., 280 mm long). Five plugs were used in each column, providing a total volume of 975 cm^3. Columns were rinsed with 300 ml portions of acetone and hexane which were collected and reduced on a rotary evaporator until appropriate blanks were obtained (at 300:1 concentration, no peaks greater than 1 pg *p,p'*-DDE). Columns were wrapped in aluminum foil for transport to and from the field.

Water was collected in a 60-liter Bodman sampler (Benthos, Inc., North Falmouth, Mass.) which was drained by gravity through stainless steel tubing, through a tee, splitting the sample between two columns. In this manner, replicate samples were taken which established analytical reproducibility. Columns were fitted to the stainless steel transfer lines by means of a stainless steel female adaptor fitted with a viton *O* ring. A second adaptor, fitted at the exit end of the column, provided the attachment of a stainless steel needle valve which was used to regulate the flow rate at 500 ml/min. Column effluent was collected and volumes measured for each column.

Columns were extracted with successive 500 ml volumes of glass distilled acetone and hexane (Burdick and Jackson, Muskeegon, Mich.). Elutants were transferred to a two-liter separatory funnel and shaken vigorously with 700 ml distilled water which had previously been cleaned by passage through a foam column. A considerable amount of salt is normally present in the column elutant, thus adding additional NaCl was unnecessary. After standing a minimum of two hours to provide for complete separation, the hexane fraction was collected. Cleanup was accomplished by passage through a column containing celite, sulfuric acid and fuming sulfuric acid (Stanley and LeFavoure, 1965). Samples were analyzed on a Tracor MT-220 gas chromatograph equipped with Ni_{63} electron capture detectors. Inlet, column and detector temperatures were 225°, 190° and 300°C, respectively. Glass columns, 2 meters long, with 4 mm I.D., were fitted with saponification side columns as described by Miller and Wells (1969). Two types of column packings were employed; 3% OV-1 and 1.5% OV-1, both on 100/120 Gas Chrom Q. The results of the 1973 Outfall II Cruise are presented in Table 12–4.

In the immediate vicinity of the outfall, concentrations of p,p'-DDE in May 1973 were 26 ng/l. DDE concentrations of 8–12 ng/l were recorded elsewhere in the Los Angeles area and dropped to 1 ng/l or less in waters further offshore (Table 12–4). Trichlorobiphenyls were present in concentrations of 27 ng/l in the immediate vicinity of the outfall and dropped to 2 to 3 ng/l in offshore waters in the coastal current. Because of the predominance of trichlorobiphenyls, pentachlorobiphenyls were not determined in the immediate vicinity of the Whites Point outfall; elsewhere concentrations were in the order of 1 to 3 ng/l. During the week of May 4–10, 1973, immediately prior to the offshore sampling, a composite of wastewaters entering the sea through the Whites Point outfall of the Los Angeles County Sanitation Districts treatment plant was analyzed for chlorinated hydrocarbons. Replicate analyses showed mean values of 4.1 μg/l of p,p'-DDE, 2.4 μg/l of p,p'-DDD, 0.29 μg/l of p,p'-DDT and 3.4 μg/l of trichlorobiphenyls (Aroclor 1242). Analytical procedures were identical to those described elsewhere (Young et $al.$, unpublished ms.). Thus, the organochlorine composition of the seawater in the vicinity of the outfall closely resembled that of the local effluent.

The concentrations of DDE and of PCB compounds recorded in San Francisco Bay in May 1973 were equivalent to those in offshore waters of the California Current (Table 12–4). The water sampled, however, may have been predominantly oceanic rather than estuarine.

In September 1973, a transect was run directly south of Point Conception, California, passing west of San Miguel Island. Replicate samples were taken at seven stations (Table 12–5). Seawater was collected in a stainless steel bucket from the bow of the ship, which was steaming slowly forward. The collected water was passed into two 15-liter aluminum pots. Water was then pumped from the pots through stainless steel columns prepared as previously described, using an impeller pump in-line behind the column. The volumes of all samples was 39 liters. "Santa Ana" conditions prevailed during the collections: winds

Table 12-4

Chlorinated Hydrocarbons in California Coastal Waters, May 1973, Outfall II Cruise, R.V. Thomas Thompson

(Concentrations in Nanograms per Liter)

Locality	Date	Depth (meters)	Volume (liters)	p,p'−DDE	p,p'−DDD	p,p'−DDT	Trichloro-biphenyls	Pentachloro-biphenyls
Whites Point Outfall	May 10, 1973	25	32.7	23.1	12.6	NM	24.9	NM
			19.2	29.3	14.0	7.7	28.3	NM
118° 16.7'W, 33° 40.8'N So. of Pt. Ferrin off San Pedro Bay	May 9, 1973	30	30.1	8.0	5.7	0.4	5.7	2.1
			19.0	9.0	6.8	0.6	NM	5.9
118° 23.2'W, 33° 41.6'N So. of Palos Verdes Peninsula	May 10, 1973	40	33.3	10.0	7.5	0.9	14.0	2.7
			22.8	12.0	8.4	0.8	14.0	3.1
118° 27.4'W, 33° 34.3'N San Pedro Channel	May 8, 1973	3	29.0	0.21	0.24	0.17	2.4	1.4
			24.3	0.23	0.23	0.24	3.2	1.2
	May 11, 1973	3	14.3	1.5	1.6	0.35	2.1	1.8
			30.0	1.1	1.0	0.13	3.2	1.9
122° 1.0'W, 35° 35.9'N	May 13, 1973	3	30.8	< 0.1	< 0.2	< 0.1	2.7	1.4
121° 55.0'W, 36° 40.5'N	May 13, 1973	3	24.3	0.13	0.17	0.30	NM	1.9
			34.3	0.15	0.20	0.25	NM	1.4
121° 49.2'W, 36° 47.9'N	May 14, 1973	3	33.6	0.09	0.13	0.09	2.8	1.4
			23.0	0.05	0.05	0.08	1.7	1.1
San Francisco Bay, 1 mile east of Golden Gate	May 15, 1973	3	24.0	0.11	0.30	< 0.1	NM	2.7

NM: Not measured.

Table 12-5
Chlorinated Hydrocarbons in Surface Seawater Along the Western Boundary of the Southern California Bight, September 1973, Obtained from the Vessel Paisano (*Concentrations in* ng/1)

Station	Locality	Column Number	p,p' – DDE	p,p' – DDT	Penta-chloro-biphenyls[a]
1 9-26-73, 0545	120° 35'W, 34° 26'N	1 2	.12 .10	.37 .32	 .27
2 9-26-73, 0900	120° 35'W, 34° 18'N	3 4	.09 .07	.35 .27	 .36
3 9-26-73, 1145	120° 35'W, 34° 10'N	5 6	.12 .13	.79 .82	 .49
4 9-26-73, 1500	120° 35'W, 34° 02'N	7 8	.10 .13	.51 .75	 .46
5 9-26-73, 1745	120° 35'W, 33° 54'N	9 10	.08 .09	.46 .53	 .44
6 9-26-73, 2300	120° 35'W, 33° 34'N	11 12	.13 .08	1.3 .79	 .49
7 9-27-73, 0200	120° 35'W, 33° 14'N	13 14	.06 .06	.47 .53	 .38

[a]Quantified from Aroclor 1254.

were from the land and temperatures were warmer than usual in the Southern California Bight. The predominant compound in these extracts was p,p'-DDT and the profiles were significantly different from those observed on other cruises when winds were onshore. We suggest that these results indicate that "dry" fallout to the sea surface may be important and that precipitation is not the only route of entry of organochlorine compounds to the sea. This conclusion was advanced previously (Risebrough *et al.*, 1968a) with reference to fallout of organochlorine pesticides into the tropical Atlantic.

Subsequent developments, utilizing a shipboard vacuum system capable of producing a 680 mm Hg vacuum under continuous operation, enabled columns to be lowered and pumped directly at depth (Figure 12-6). A stainless steel one-way flap valve with a viton O ring and 1/3-lb spring was fitted in-line between the column adaptor and the vacuum tubing to reduce the possibility of contamination from back-flow.

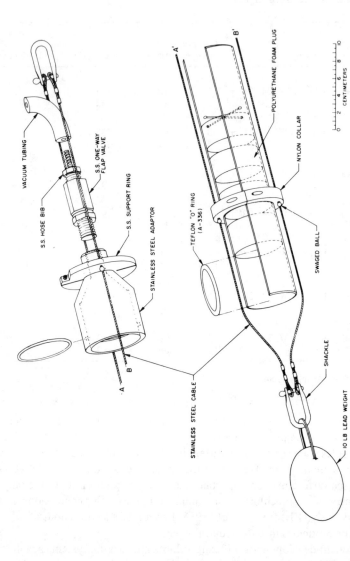

Figure 12-6. The "Slocum" (Captain Joshua Slocum 1844 – ?) selective organic *in situ* sampler for determining trace levels of organochlorine compounds in seawater. Seawater is pumped at depth through a stainless steel column packed with polyurethane foam plugs utilizing a shipboard vacuum system.

The results of an intercalibration study carried out in Los Angeles Harbor in 1974 are presented in Table 12-6. Solvent extractions were three stage in two-liter separatory funnels, using 400 ml of solvent at each stage. Water was sampled with foam columns in two manners. One set of columns was pumped *in situ* at a depth of 1 meter using a vacuum pump system. The second set received water from stainless steel bucket collections taken with efforts to avoid surface concentrations. This water was passed directly to a six-liter stainless steel reservoir fitted at the top end of a column held vertically in a ring stand assembly, with flow rates controlled by a valve connected by an adaptor to the exit end of the column. Reasonably consistent results were obtained from the four extraction procedures.

An additional intercalibration study was carried out during a wastewater sampling program at an experimental plant of the Los Angeles County Sanitation District in December 1974. Twenty-four liters of secondary effluent were passed by gravity flow through a polyurethane foam packed stainless steel column fitted with a six-liter stainless steel reservoir. A replicate sample was obtained by methylene chloride batchwise extraction, as described elsewhere (Burlingame *et al.*, 1976). Final extracts were reduced to a volume of 10 ml in hexane and fractionated on a florisil column eluting sequentially with 6% and 15% solutions of diethyl ether in hexane (Mills, P.A., 1961). Figure 12-2 presents the chromatograms of the 6% fractions of both extraction procedures. Under these conditions, both extracts were virtually identical in the composition of compounds producing a response with the electron capture detector.

The "Slocum" *in situ* sampler (Figure 12-6) was used to extract PCB and DDT compounds at depth from seawater during a cruise of *R./V. Cayuse* in the

Table 12-6

Intercalibration of Methodologies: Extraction of Organochlorines from Seawater, Los Angeles Harbor, May 1974

(Mean Concentrations with the Range (ng/1))

Method	N	Volume Extracted (liters)	$p, p' - DDE$	$p, p' - DDD$	PCB
Blank (column)	1	Assume 40	<0.02	<0.017	<0.14
Solvent extraction hexane	2	16.1	1.9 (1.8-1.9)	1.0 (0.9-1.0)	1.5 (1.4-1.6)
Solvent extraction methylene chloride	1	25.3	1.6	0.9	2.8
Foam column gravity	8	23-45	1.8 (1.1-2.3)	1.1 (0.7-1.3)	3.6 (1.7-7.4)
Foam column vacuum	3	26-79	1.8 (1.5-2.3)	1.1 (0.9-1.3)	3.0 (1.0-6.4)

Southern California Bight and from Mazatlan to San Diego in February-March 1975. Results of analysis of the samples obtained in Pacific offshore waters of Mexico are presented in Table 12-7. None of the samples obtained in the waters off western Mexico had detectable levels of any of the compounds looked for. Detection limits were imposed not only by background contamination of the system but by the presence of unknown compounds in the seawater (Figure 12-1).

In Table 12-8 are presented the results of analysis of seawater extracts obtained in the Southern California Bight in February 1975. Levels of p,p'-DDE were in the order of tens or hundreds of parts per quadrillion (pg/l) in the outer stations but approached 5-6 ng/l in the vicinity of the Whites Point outfall. Levels of pentachlorobiphenyls were below the limits of detectability, in the range of hundreds of picograms per liter at the outer stations and approached 1-2 ng/l in waters in the vicinity of the wastewater outfalls. The high levels of DDT compounds in the vicinity of the Whites Point outfall are assumed to have leached from the sewer pipes below the DDT manufacturing plant, which in the past was the principal source of DDT residues entering the Southern California Bight.

To date, we have no evidence of the channeling of water in the column systems used, which might result from the entrapment of air and compression of foam at greater depths. Another version of an *in situ* sampler with which we are currently experimenting prevents this possibility by requiring water to pass through concentric layers of polyurethane foam, Zitex (a pure Teflon filter material, Chemplast, Inc., Wayne, N.J.) and other absorbents. We are also experimenting with coating polyurethane foam with various silicone liquid phases which other investigators have shown to enhance the recovery of trace organics from solution (Uthe *et al.*, 1972; Aue *et al.*, 1972). Modifications have also been made to the "Slocum" *in situ* sampler which provide for a glass fiber filter to be fitted in-line prior to the polyurethane foam column. Such a unit offers accurate determinations of residue concentrations in both the particulate and dissolved fractions of seawater. Teflon columns have been found to provide a smoother connection with the stainless steel adaptors. These columns may also be adapted to form the body of a Soxhlet apparatus, simplifying extraction procedures and further reducing the possibilities of contamination.

The *in situ* sampling systems so far employed require that water be pumped to the ship and collected for volume determinations. Use of a submersible pump with a flow meter would eliminate this requirement. Such a system could also be coupled with standard physical oceanographic *in situ* probes (temperature, depth, conductivity, turbidity, etc.), providing a more detailed characterization of the water body sampled. The depth capabilities of such a fully submersible system would greatly exceed the limitations imposed by the shipboard vacuum pumping system.

Chlorinated Hydrocarbons in Plankton and Other Particulates

Because of the low water solubilities of the higher molecular weight chlorinated hydrocarbons, a substantial fraction of the chlorinated hydrocarbons in the water system can be expected to be associated with particulates, including plankton, and with other members of the biota. Satisfactory determinations have not yet been made, however, of the portion of DDT and PCB residues in a seawater system that are present in the "dissolved" state. The seawater analytical system employed by Harvey et al. (1973) included a glass fiber filter or a glass wool plug which removed much of the particulate matter before the water passed through the Amberlite resin column. Residues in the filter or glass wool plug were estimated to constitute a maximum of 10% of the PCB present in the filtered seawater samples. Thus, the residues reported would be in the "dissolved" phase. The analytical methods employed by other investigators have provided chlorinated hydrocarbon determinations of the total seawater sample, including plankton and other particulates.

Plankton samples obtained on R/V Atlantis II cruise 52 in 1969 with a #6 mesh plankton net were analyzed for chlorinated hydrocarbons. The data were compared with those subsequently obtained by G. R. Harvey and coworkers on R/V Atlantis cruises 59 and 60 (Risebrough et al., 1972). The chlorobiphenyls detected closely resembled the composition of Aroclor 1254, and PCB compounds were present in substantially higher concentrations than those of the DDT group. PCB in the zooplankton from the stations on the Continental Shelf and Slope ranged from 2.4 to 260 ppm, lipid weight, with a median value of approximately 40 ppm. Median percent lipid weight of dry weight was 3.8%. On a dry and wet weight basis, with the assumption that the dry weight constitutes 10% of wet weight, representative concentrations in zooplankton in the Shelf and Slope areas were estimated to be in the order of 1.5 ppm and 0.15 ppm, respectively.

Since these data were published we have examined the possibility that some if not all of the samples were contaminated by shipbottom antifouling paint. Such contamination would explain the close resemblance of the gas chromatographic profiles of the plankton extracts to that of Aroclor 1254 and would account for the exceptionally high PCB:DDT ratios; fish and other organisms higher in the food web were consistently found to have substantially lower ratios.

Shipbottom paint until recently has contained PCB (Young et al., 1974) and is therefore a possible source of such contamination. Thus, Jensen et al. (1972) found that the plankton samples obtained in the wake of a boat had significantly higher concentrations of PCB than did those obtained before the bow. Moreover, the gas chromatograph profile of plankton samples obtained in the wake closely resembled that of the commercial pentachlorobiphenyl preparation

Table 12-7

Maximum DDT and PCB Concentrations in Pacific Offshore Waters of Mexico, March 16-28, 1975

(*Concentrations in pg/l, Depth of 3 m*)

Longitude	Latitude	Volume (liters)	p,p' – DDE	p,p' – DDD	p,p' – DDT	Pentachloro- biphenyls
106° 34.0′ W	22° 59.0′N	96	< 2	< 3	< 5	< 30
107° 05.5′	22° 18.6′	96	< 10	< 10	< 10	< 70
108° 27.2′	20° 29.0′	114.5	< 1	< 10	< 2	< 70
109° 00.0′	19° 47.9′	96	< 3	< 5	< 10	< 50
110° 16.1′	18° 05′	114	< 1	< 1	< 1	<100
110° 43.0′	17° 27.0′	116	< 1	< 1	< 2	<100
112° 27.5′	15° 59.5′	115	< 10	< 10	< 20	<100
113° 06.0′	15° 24.1′	96	< 1	< 80	< 2	< 10
114° 44.5′	13° 42.0′	117	< 3	< 5	< 9	<300
115° 25.0′	13° 06.0′	95	< 2	< 3	< 6	<100
115° 26.0′	15° 19.0′	137	< 2	< 2	< 5	<100
115° 32.0′	16° 03.0′	114	< 2	< 3	< 6	<100
114° 48.0′	18° 09.5′	60	< 2	< 2	< 5	<100
114° 41.0′	18° 44.5′	91	< 2	< 3	< 5	< 30
114° 50.0′	20° 37.5′	104	< 1	< 1	< 3	<100
114° 39.0′	21° 28.5′	80.8	< 1	< 2	< 4	< 40
114° 54.0′	23° 33.5′	116	< 10	< 4	< 7	< 80

Longitude	Latitude	Volume (liters)	p,p' – DDE	p,p' – DDD	p,p' – DDT	Pentachloro-biphenyls
115° 02.0'	24° 22.8'	105	< 1	< 2	< 3	< 20
115° 10.2'	26° 12.8'	81.8	< 2	< 2	< 5	< 20
114° 47.5'	28° 19.8'	91.2	< 10	< 3	< 5	< 30
115° 28.2'	28° 47.3'	79.8	< 2	< 4	< 8	< 40
116° 46.7'	30° 25.3'	70.4	< 2	< 3	< 7	< 30
117° 02.0'	31° 04.3'	89.3	< 4	< 5	< 10	< 50
117° 20.5'[a]	32° 34.7'[a]	79.8	200[a]	80	< 2	500[a]

[a] In U.S. waters off San Diego

Table 12-8
DDT and PCB Residues in Waters of the Southern California Bight, February 21-27, 1975
(*Concentrations in* ng/l)

Longitude	Latitude	Depth (meters)	Volume (liters)	$p,p'-$DDE	$p,p'-$DDD	$p,p'-$DDT	Pentachloro-biphenyls
119° 13.1'W	34° 04.6'N	3	38	0.3	<0.04	<0.08	<0.4
119° 21.9'	33° 57.5'	3	38	0.2	<0.08	<0.09	<0.5
119° 54.5'	33° 46.8'	3	38	0.09	<0.03	<0.05	<0.4
120° 23.6'	33° 35.2'	3	38	0.07	<0.03	<0.06	<0.3
119° 31.0'	33° 44.8'	3	38	<0.05	<0.07	<0.1	<0.7
118° 48.0'	33° 54.1'	3	38	0.1	<0.03	<0.06	<0.3
118° 31.1'	33° 54.8'	3	38	0.3	0.2	<0.1	<0.6
118° 31.4'	33° 48.7'	3	38	0.07	0.04	<0.06	<0.03
118° 27.6'	33° 36.2'	3	38	0.05	0.04	0.05	<0.1
118° 20.3'	33° 23.5'	3	38	0.06	<0.03	<0.06	0.1
118° 19.5'	33° 37.6'	3	38	2.1	0.7	0.3	0.7
4.5 mi SW Pt. Fermin							
118° 17.9'	33° 41.4'	3	38	4.5	1.5	<0.4	1.5
0.9 mi S Pt. Fermin							
118° 19.4'	33° 41.8'	3	39	3.5	1.0	<0.2	1.0
Over Wt Point outfall							
118° 22.2'	33° 35.4'	3	40	0.7	0.3	0.3	0.04
7 mi SW Wt Point outfall							

Longitude	Latitude	Depth (meters)	Volume (liters)	p,p'–DDE	p,p'–DDD	p,p'–DDT	Pentachloro-biphenyls
118° 20.9'	33° 42.3'	3	38	5.9	2.2	0.5	2.0
1.3 mi NW Wt Point outfall							
118° 19.9'	33° 40.6'	3	38	3.5	1.7	0.8	0.3
1.3 mi SW Wt Point outfall							
118° 24.6'	33° 43.5'	3	38	4.0	1.2	0.4	0.7
S of Pt. Vicente							
118° 19.4'	33° 41.8'	16	23	5.0	3.0	<0.4	1.5
Over Wt. Pt. outfall							
118° 19.0'	33° 41.7'	36	23	4.0	2.0	<0.07	1.5
0.4 mi S Wt. Pt. outfall							
118° 21.7'	33° 42.2'		19	2.0	2.0	0.7	1.5
NW of Wt. Pt. outfall							
118° 20.0'	33° 40.0'	3	38	0.06	0.09	<0.09	0.3
Entrance Los Angeles Harbor							
118° 20.0'	33° 30.0'	3	38	<0.06	<0.09	<0.2	0.2
14 mi SW Pt. Fermin							
117° 17.2'	32° 40.8'	65	57	0.2	0.08	<0.1	0.2
		3	38	0.1	0.1	<0.1	0.3
0.5 mi NE Pt. Loma							

Clophen A-50, which closely resembles the American preparation Aroclor 1254.
DDT compounds observable in chromatograms of extracts of plankton obtained
before the bow were not visible in chromatograms of extracts of plankton obtained in the wake. Because of the close similarity in gas chromatographic profiles of plankton extracts obtained in both our laboratory (Risebrough *et al.,*
1972) and at the Woods Hole Oceanographic Institution (Harvey *et al.,* 1974b)
with those of the pentachlorobiphenyl preparation Aroclor 1254 it is possible
that these samples were contaminated with antifouling shipbottom paint.

Ware and Addison (1973) have determined PCB residues in plankton from
the Gulf of St. Lawrence. Extensive precautions were taken to ensure that the
samples were not contaminated. Estimations of the PCB content of #6 mesh
plankton net samples were comparable to those we had earlier reported from the
Atlantic (Risebrough *et al.,* 1972). PCB concentrations were inversely related
to particle size. Total PCB residues associated with #20 mesh plankton net
samples ranged from 20 to 930 pg/l at eight stations in the Gulf of St. Lawrence
in the summer of 1972. Concentrations therefore approached 1 ng/l of the
water mass. Concentrations in particulate material smaller than 73 μm, which
would pass through the #20 mesh plankton net, were not determined. Particles
in the size range 73 μm to 202 μm, however, ranged up to 31 ppm wet weight,
apparently the highest PCB concentrations so far reported in natural plankton.
It is therefore possible that particles smaller than 73 μm might contain still
higher PCB concentrations.

Simultaneous determinations of organochlorine concentrations of the total
water system and of one or more particulate size classes would appear to be
needed to understand the partitioning among and fluxes between components
of these systems.

Sediments: Fluxes Through the Water Column

Organochlorines may be expected to be associated with fecal pellets and other
organic detrital matter, as well as inorganic particulates, that pass through the
water column and are deposited in the sediments. Although a substantial portion of these residues might then be incorporated by benthic fauna, desposition
in the sediments provides a pathway for the ultimate removal of these chlorinated hydrocarbons from marine food webs.

Varved anaerobic sediments of the Santa Barbara Basin off Southern California, dated by counting of the varves and by ^{210}Pb and ^{228}Th/^{232}Th radio-
metric age measurements, were analyzed for chlorinated hydrocarbons. Deposition of DDE in the sediments began about 1952; deposition of PCB began about
1945. Estimated deposition rates (in grams per square meter per year) in 1967
of DDE and PCB were 1.9×10^{-4} and 1.2×10^{-4}, respectively (Hom *et al.,* 1974).
The anaerobic nature of these sediments minimizes uptake by benthic organisms

and recirculation in the marine biosphere. Although a high rate of local input of both DDT and PCB compounds into the waters of the Southern California Bight precluded extrapolation to other oceanic areas, the data suggest that a substantial portion of chlorinated hydrocarbons entering the sea from diverse sources was being deposited in the sediments.

Clues to the relative flux of DDT and PCB compounds through the water column may be obtained from the analysis of benthic organisms. The bathyl-demersal fish, *Antimora rostrata,* is a permanent resident of the Lower Continental Slope; bottom photographs indicate that these fish swim slowly over the bottom, crop benthic epifauna and root in the ooze for infauna (Marshall, 1971; Meith-Avcin *et al.,* in press). Specimens collected on the upper continental rise southeast of Cape Hatteras in 1972 at a depth of 2500 meters were found to contain 3 to 7 ppm of total DDT compounds in the livers (Meith-Avcin *et al.,* in press). Two of the livers were subsequently analyzed for PCB compounds. Lipid content was high, 84 and 87% of total dry weight, respectively. Concentrations recorded on a lipid weight basis were as follows: *p,p'*-DDE, 5.9 and 19 μg/g; *p,p'*-DDT, 2.4 and 12 μg/g; *o,p'*-DDT, 0.6 and 3.1 μg/g; *p,p'*-DDD, 2.4 and 5.2 μg/g; pentachlorobiphenyls, 3.8 and 12.5 μg/g (R. T. Barber, S. M. Warlen, R. W. Risebrough and P. J. Whaling, unpublished data).

The substantial concentrations of undegraded *p,p'*-DDT and *o,p'*-DDT in these deep-sea fish indicate that these compounds may be transported unchanged through the water column. The predominance of DDT compounds over PCB in these fish is at variance with the PCB:DDT ratios recorded in various species of mesopelagic and surface fish from the North Atlantic, which almost invariably have had more PCB than DDT (Harvey *et al.,* 1974b). In fish obtained below 200 meters and down through 900 meters, the median concentration of DDT in the lipid was 1 ppm (N = 36; Harvey *et al.,* 1974b). If the two PCB values recorded in *Antimora* are typical, the benthic environment would appear to be enriched with PCB in relation to the water column. Concentrations of DDT compounds are higher, suggesting that transport of DDT compounds through the water column may be more rapid than transport of PCB.

Estimates of the transport of chlorinated hydrocarbons through the water column and of rates of deposition in the sediments are hampered by the same deficiencies present in all other studies of chlorinated hydrocarbon fluxes through the marine environment. Data are too few, too scattered and too widely separated in both time and space to permit little more than speculation on what might be occuring.

Partitioning of Chlorinated Hydrocarbons Between Seawater and the Biota

The solubility and vapor pressure data presented in Table 12-1 suggested that

the solubility of p,p'-DDE might be comparable to that of individual pentachlorobiphenyls; vapor pressure of individual pentachlorobiphenyls appeared
to be somewhat higher than that of p,p'-DDE. Measurements, however, are as
yet insufficient to establish whether suspected differences are real. A number
of studies indicate, however, that pentachlorobiphenyls and p,p'-DDE behave
similarly in aquatic environments and in organisms. Using a laboratory model
ecosystem, Metcalf *et al.* (1975) compared the degradation pathways and biomagnification of radio-label tri-, tetra-, and pentachlorobiphenyl and p,p'-DDE
in an alga, mosquito, snail and fish. The ecological magnification (μg/g in or-
ganism/μg/g in water) of 2, 4, 5, 4', 5'-pentachlorobiphenyl and of DDE were
5000 and 11,000, respectively, for the alga (*Oedogonium cardiacum*), 60,000
and 36,000 for the snail (*Physa*), 17,000 and 60,000 for the mosquito (*Culex
pipiens*), and 12,000 for both compounds in the fish *Gambusia affinis*. The
failure to find consistent differences in ecological magnification factors between
the two compounds suggests that the observed differences resulted from experimental variables. Both compounds showed similar low rates of degradation to
polar products (Metcalf *et al.*, 1975). In a comparable experiment using the
Green Sunfish (*Lepomis cyanellus*) only 1% of the radio-labeled DDE and 0.7%
of the radio-labeled pentachlorobiphenyl was degraded to polar compounds by
the fish. The bioconcentration factors of the pentachlorobiphenyl, p,p'-DDE
and p,p'-DDT were estimated to be 1500, 900 and 17,500, respectively, in this
system. The uptake curves, however, were similar for the three molecules and
it is not clear whether the differences in bioconcentration factors are real, especially in view of the considerable discrepancy between those of p,p'-DDT and
p,p'-DDE (Sanborn *et al.*, 1975). Biological magnification factors for accumulation of [36]Cl-labeled Aroclor 1254 have been determined for several species of
aquatic invertebrates (Sanders and Chandler, 1972). The lower chlorinated
isomers were accumulated to a greater degree than the higher chlorinated isomers suggesting that availability to the organisms was limited by water solubilities and that the higher chlorinated isomers were not truly dissolved in the
water medium (Schoor, 1975). Thus, it is extremely difficult to approximate
natural conditions while undertaking uptake studies in the laboratory.

To begin investigations of the partitioning between seawater and the biota,
we measured chlorinated hydrocarbon concentrations in seawater, including the
particulate fraction, obtained at sites where mussels (*Mytilus edulis* or *M. californianus*) were also obtained. In Table 12-9 are presented the bioaccumulation
factors of chlorinated hydrocarbons between *Mytilus sp.* and seawater from four
sites on the Mediterranean coast of France and two in California. Equivalent factors were obtained in the samples from Marseille for PCB, predominantly pentachlorobiphenyls, and p,p'-DDT; for both the bioaccumulation
factor was 690,000. Lower bioaccumulation factors, particularly for p,p'-DDT,
were obtained at Valras-Plage and Stes.-Maries. The dominant DDT compound
in the water was p,p'-DDT; o,p'-DDT was also present, suggesting a recent input

Table 12-9
Bioaccumulation Factors of Chlorinated Hydrocarbons Between *Mytilus sp.*
and Seawater

Location		Mussels (ng/g, Wet Weight)	Seawater (ng/l)	Bioaccumulation Factors
		Chlorinated Hydrocarbon Concentrations		
France				
Valras-Plage	PCB	250	0.94	270,000
9 July 1974	p,p' – DDT	51	0.54	90,000
Stes-Maries	PCB	450	0.95	470,000
10 July 1974	p,p' – DDT	13	0.35	40,000
Carro	PCB	200	0.68	300,000
11 July 1974	p,p' – DDT	52	0.25	210,000
Marseille	PCB	1,100	1.6	690,000
12 July 1974	p,p' – DDT	900	1.3	690,000
California				
Royal Palms,	PCB	230 (N=2)	1.2 (N=4)	190,000
Los Angeles	p,p' – DDE	1,600 (N=2)	5.1 (N=4)	310,000
19 September				
1972				
San Francisco	PCB	68±24 (N=47)	0.98±.47 (N=26)	69,000
Bay	p,p' – DDE	5±2 (N=47)	0.11±.06 (N=26)	45,000
Means± SD				
January-March				
1975				

Source: R.W. Risebrough, B.W. de Lappe and T.T. Schmidt, unpublished.

source from aerial fallout. Possibly equilibrium between the mussels and the water of p,p'-DDT had not yet been reached. Bioaccumulation factors in San Francisco Bay were an order of magnitude lower than in Marseille. Factors which are different in the two localities include the particulate content, the salinity, and the temperature. If the particulate content is high, a larger proportion of the residues in the seawater system might be associated with them, thereby diminishing the relative amount found in mussels.

Chromatograms of saponified extract of *Mytilus* from Carro, on the Mediterranean coast of France between Marseille and the mouth of the Rhone, and of seawater obtained at the same site are shown in Figure 12-7. The seawater extract contained a large number of lower molecular weight compounds which were not significantly bioaccumulated by the mussels. The relative proportions of PCB compounds are somewhat different in the two extracts. Such differences presumably are a result of slightly different solubilities of individual PCB compounds (R. W. Risebrough, B. W. de Lappe and T. T. Schmidt, unpublished ms.).

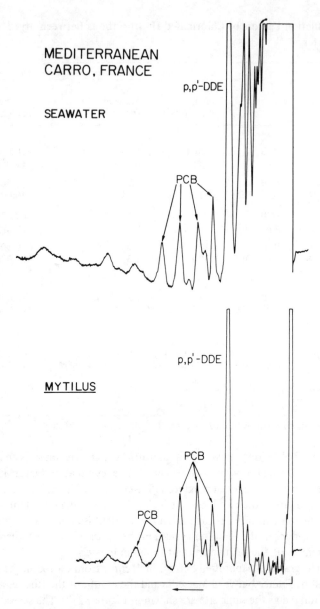

Figure 12-7. Chromatograms of extracts of seawater and *Mytilus* from Carro, on the Mediterranean coast of France, June 1974. Aliquots used for GC analysis were passed through a saponification side column, 3% OV-1 column, Gas Chrom Q (from R. W. Risebrough, B. W. de Lappe, and T. T. Schmidt, unpublished ms.).

Total DDT residues in *Mytilus californianus* at southern California sites in 1971 are shown in Figure 12-8. Concentrations were highest in the vicinity of the Whites Point outfall of the Los Angeles County Sanitation Districts. The PCB concentrations in mussels from these sites indicated a number of input sources. Concentrations of both DDT and PCB declined at some of these areas between 1971 and 1974. Thus, the mussels appear to be good indicators of local point sources of pollution (B. W. de Lappe, R. W. Risebrough, P. Millikin, and D. R. Young, unpublished ms.).

Other factors which may affect partitioning of chlorinated hydrocarbons between the seawater system and mussels or other invertebrates include size, sex, gonadal condition, time of year and proportion of time which the organisms spend exposed to the air at low tide. The effects of these factors on the chlorinated hydrocarbon concentrations in *Mytilus californianus* from the California coast were examined. No relationships with size were found; concentrations appeared to fluctuate randomly throughout the year and the samples examined showed no significant differences in the chlorinated hydrocarbon content between sexes. Mussels living in the mid-tidal range appeared to have higher concentrations than those living at the low or high tide levels, respectively (B. W. de Lappe, R. W. Risebrough, P. Millikin and D. R. Young, unpublished ms.).

The "Mussel Watch" program (Goldberg, 1975b) will be initiated in the United States in 1976 to examine the use of mussels as indicator organisms for a

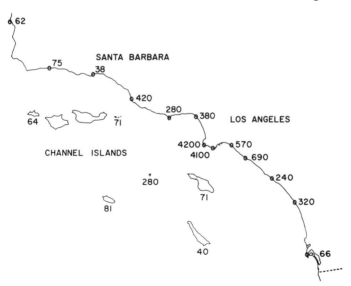

Figure 12-8. Total DDT residues in *Mytilus californianus* from Southern California, 1971. Concentrations in nanograms/gram of the wet weight. (Analytical procedures described in B. W. de Lappe, R. W. Risebrough, P. Millikin, and D. R. Young, unpublished ms.).

variety of marine pollutants. Determination of partition coefficients between concentrations in the mussels and in the ambient seawater system would significantly enhance the value of such studies. Moreover, all factors which affect the variance of the numbers obtained, including the precision of the analytical determinations, size, sex, particulate content of the water, salinity in the case of estuaries, lipid content, time of year, microhabitat, etc. should be investigated if the mussels are to be used to detect differences in residue concentrations in time or space.

Fluxes of Pollutants into Marine Birds and Mammals

Unlike the chlorinated hydrocarbons which have a background level of zero in the marine environment, heavy metals and trace elements are natural components of marine food webs. Frequently it is difficult or impossible to estimate anthropogenic input of a given element into a marine food web, since knowledge of background "natural" levels are lacking. The discovery of several abnormalities in some marine bird and mammal populations, discussed below, prompted investigations of levels of heavy metals and trace elements in those populations.

To determine whether a significant anthropogenic component might be present, levels of various heavy metals and trace elements were compared between two populations of the same species widely separated in breeding range and by comparing levels in species in the northern hemisphere with biocoenetic equivalents in the Antarctic. Concentrations of industrial chlorinated hydrocarbons such as the polychlorinated biphenyls in the antarctic biota have been one to several magnitudes lower than those in biocoenetic equivalents in the northern hemisphere. If levels of a particular metal recorded in the antarctic species were equivalent to those of a related species in the northern hemisphere, it would appear that both levels would imply anthropogenic input to the northern hemisphere populations.

Levels of silver, cobalt, lead, cadmium, chromium, nickel, mercury, copper and zinc were determined in breast muscle, liver and bone of four populations of petrels; the Snow Petrel (*Pagodroma nivea*) restricted to the pack ice zone of Antarctica, the Ashy Petrel (*Oceanodroma homochroa*) resident in coastal California waters and two populations of Wilson's Petrel (*Oceanites oceanicus*). One population of Wilson's Petrel breeds in the Antarctic Peninsula and winters in the North Atlantic during the nothern summer, the other breeds on the opposite side of the antarctic continent, and presumably winters in the Australian region. The distributions of copper and zinc showed no significant geographical or species variation. Silver, cobalt and lead were present in low concentrations and no marked variation in the populations examined was apparent. Differences in the levels of cadmium, chromium, nickel and mercury may have resulted from anthropogenic input. The differences, however, were relatively small and may have resulted from a combination of differences among species and geographical factors (Anderlini *et al.*, 1972).

Examination of levels of these metals in tissues of the Brown Pelicans inhabiting coastal waters of California and Florida showed no differences, with the exception of mercury which was several times higher in the Florida samples (Connors *et al.*, 1972).

Two populations of Common Terns were examined for heavy metal contamination (Ag, Cd, Co, Cr, Cu, Hg, Ni, Pb and Zn). One population is marine, on Long Island Sound; the other is a freshwater population breeding on Lake Ontario. Abnormalities had been noted in both populations, although different abnormalities were found in each colony. No differences were detected in the levels of these metals in bone, breast muscle, liver or kidney between the two populations (Connors *et al.*, 1975).

Breast muscle and livers of known age Sooty Terns (*Sterna fuscata*) from the tropical Atlantic were analyzed for Cu, Zn, As, Se, Hg, Fe, Mn, Sr, Cr, Pb, Ni and Cd by X-ray fluorescence or Zeeman-effect atomic adsorption. No increases in metal concentrations with age of the birds was detected after the age of three years, indicating that fluxes of these elements into the terns had reached equilibrium with metabolic processes that excrete them (P. G. Connors, W. Robertson, S. A. Jacobs and R. W. Risebrough, unpublished data). The mercury:selenium correlation found in marine mammals (Koeman *et al.*, 1973b; Martin *et al.*, in press) was not found in this species, suggesting that marine birds and mammals detoxify mercury and/or selenium in different ways.

In collaboration with other laboratories, tissues of aborting female sea lions and their premature pups (cf. below) and a series of female sea lions giving normal birth, and their pups, collected in 1972 on San Miguel Island, California, were analyzed for a series of heavy metals and trace elements, as well as for chlorinated hydrocarbons. No differences in metals were found between the two groups of animals with the exception of selenium, bromine and mercury which increased with age; they also increased in equimolar ratios, suggesting the existence of a molecular complex, containing selenium, mercury and bromine which functions either to detoxify the methyl mercury present in the fish upon which the sea lions feed, or to detoxify the selenium present with the mercury in fish. No evidence of anthropogenic input to the levels of metals and trace elements in the sea lion population of Southern California was found (Martin *et al.*, in press). Thus, a terminal carnivore dependent upon fish in an area of the sea known to receive high inputs of heavy metals from the atmosphere and from wastewater outfalls in the Los Angeles area does not appear to contain elevated levels of these metals. This observation reinforces the conclusions of other studies of the distribution of metals and trace elements in marine mammals carried out in collaboration with other laboratories that anthropogenic contributions to the burden of heavy metals and nonessential trace elements at this level of marine food webs are as yet small.

Within a marine ecosystem ratios of DDE to pentachlorobiphenyls tend to persist among different individuals of the population and among different species (Risebrough *et al.*, 1968b; White and Risebrough, in press). PCB was found to be highly correlated with DDE in tissues of known age Sooty Terns from the tropical Atlantic (Spearman's Rank Correlation Coefficients for DDE and PCB levels in whole body lipid were +0.84, N = 24, p <.001, in adults; 1.00, N = 8, p<.001, in juveniles) (P.G. Connors, W. Robertson, S.A. Jacobs and R.W. Risebrough, unpublished data). When differences in the DDE:PCB ratio have been observed for members of the same ecosystem, these differences may reflect inconsistencies in which polychlorinated biphenyl isomers were reported. Thus, tri- and tetrachlorobiphenyls recorded in the fish from the vicinity of Amchitka were not equivalent to the penta- and hexachlorobiphenyls recorded in birds and mammals (White and Risebrough, in press). Without consistency in reporting PCB, PCB:DDE ratios are meaningless.

The widely differing ratios of pentachlorobiphenyl : DDC between those observed in plankton (Ware and Addison, 1973; Harvey *et al.*, 1974b; Risebrough *et al.*, 1972) and in fish and other vertebrates remain largely unexplained. The available physico-chemical data suggest that DDE and pentachlorobiphenyls should show similar patterns of accumulation. Detailed studies in a few areas should resolve this apparent discrepancy.

A chromatogram showing a saponified extract of the Gentoo Penguin is shown in Figure 12-9. The profile is similar to that of the saponified extract of antarctic snow (Figure 12-4), except that proportions of PCB compounds are changed somewhat and the relative amount of DDE is reduced compared to that in the snow. The change in the ratio of total DDT:PCB from the snow to the penguin might result from (1) preferential degradation of p,p'-DDT to polar derivatives during passage through the food chain, or (2) direct input of PCB from the atmosphere to the sea with only a fraction of PCB scavenged from the atmosphere by precipitation.

Biological Effects of Marine Pollutants

Studies of pollutant transfer to the marine environment are "relevant" if levels found are related to those which damage a component of the marine ecosystem. Thus, input fluxes may be considered in relation to potentially harmful levels in the marine environment. As indicated above, there is yet no evidence for harmful effects of heavy metals and trace elements to marine birds and mammals. Apparently high mercury levels have been recorded in eggs of the White-Tailed Eagle (*Haliaetus albicilla*) in the Baltic (Henriksson *et al.*, 1966), but eagles in this area also contain excessively high levels of organochlorine compounds (Jensen *et al.*, 1972), probably accounting for the population decline of the species in this area. A decline in the reproductive success of the White-Tailed Eagles

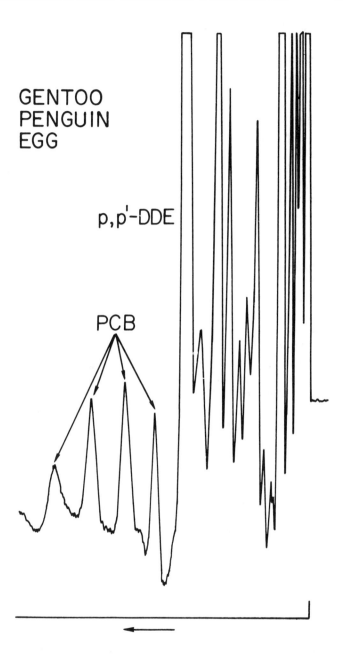

Figure 12-9. Chromatogram of saponified extract of a Gentoo Penguin egg,
Doumer Island, Antarctic Peninsula, January 1975. 3% OV-1
column, Gas Chrom Q.

breeding in Schleswig-Holstein has been attributed to DDE (Koeman *et al.*, 1972a).

Premature pupping of the California Sea Lions (*Zalophus californianus*) was first observed in the Channel Islands area of Southern California in the late 1960s (DeLong *et al.*, 1973) and correlations were observed between the premature births and the occurrence of high levels of DDT and PCB compounds. In 1972, ten females which had given birth prematurely, their dead pups, ten healthy females, and their pups were collected for pollutant residue determinations and for examination of the presence of disease organisms. As in the previous study, aborting animals had significantly higher DDE and PCB concentrations than did the nonaborting female sea lions; the ratios of DDE to PCB also differed between the two groups. It was postulated that the presence of these high levels of organochlorines lowered the resistance of the animals to a disease endemic in the population, producing the observed premature births (Gilmartin *et al.*, in press).

Extreme shell thinning of eggs of the Brown Pelican (*Pelecanus occidentalis*) and of the Double-Crested Cormorant (*Phalacrocorax auritus*), in the Southern California coastal area was discovered in 1969 (Risebrough *et al.*, 1971; Gress *et al.*, 1973). Extensive statistical studies of eggs of the Brown Pelican obtained before 1945 and now preserved in museums and of several hundred eggs obtained since 1969 from California, Mexico and Florida has shown that all of the thinning can be attributed to p,p'-DDE (R.W. Risebrough, W. Walker, D.W. Anderson, and R.W. Schreiber, unpublished data). With the decline of DDT input into the Southern California Bight following the change in waste disposal methods of the DDT company in Los Angeles, levels of p,p'-DDE in the Southern California Bight have decreased and the reproductive success of the pelicans has subsequently improved (Anderson *et al.*, 1975).

Dead chicks and unhatched eggs of the Bermuda Petrel (*Pterodroma cahow*) were examined to follow up a previous report that a decline in reproductive success of this species, a pelagic feeder in the Sargasso Sea adjacent to Bermuda, was attributed to DDT residues (Wurster and Wingate, 1968). Reproductive success has improved since 1968, but no change in DDE and PCB levels have been found over the period of the decline and recovery of reproductive success. Moreover, levels of both DDE and PCB were low in comparison to those of other petrels (D. Wingate and R.W. Risebrough, unpublished data).

Peregrine Falcons feeding on marine birds appear to be affected by shell thinning in all areas of their nearly global range so far studied and reproductive success is frequently low (Peakall *et al.*, 1975; Ohlendorf *et al.*, in press; Ratcliffe, 1972). Investigations of pollutant levels in the Peregrine Falcon in Amchitka, Greenland, and coastal Chile showed DDE levels sufficiently high to be associated with shell thinning (White and Risebrough, in press; Walker *et al.*, 1973a; Walker, *et al.*, 1973b). An unhatched egg of the related New Zealand Falcon (*Falco novaeseelandiae*) from the subantarctic Auckland Islands contained DDE residues sufficiently high to cause shell thinning in other species of falcons (Bennington *et al.*, 1975).

Birth defects in a colony of Common Terns (*Sterna hirundo*) and Roseate Terns (*S. dougallii*) in Long Island Sound were similar to those caused under experimental conditions by chlorinated dibenzodioxins (Hays and Risebrough, 1972). The possible existence of these and the closely related chlorinated dibenzofurans in marine food webs was postulated.

Mortality of Great Cormorants (*Phalacrocorax carbo*) along the Netherlands coast and debilitation of Glaucous Gulls feeding on the eggs of other seabirds in the Arctic has been attributed to PCB poisoning (Koeman *et al.*, 1973a; Bourne and Bogan, 1972). Moreover, high body burdens of organochlorines possibly weaken birds at a time of external stress such as bad weather and the low availability of food (Holdgate, 1971; Parslow and Jefferies, 1973; Parslow *et al.*, 1973; Lloyd *et al.*, 1974).

Other aspects of effects of organochlorine compounds and petroleum on marine birds have recently been reviewed by Ohlendorf, Risebrough and Vermeer (in press).

Potential effects of organochlorine compounds on marine phytoplankton were observed by Wurster (1968), who reported that DDT inhibited the photosynthesis of several species of marine phytoplankton. Subsequently, PCB compounds were found to have comparable effects (Mosser *et al.*, 1972; Fisher *et al.*, 1973). Organochlorines were subsequently shown to affect growth rather than photosynthesis *per se* (Fisher, 1975). The concentrations required to produce these effects have, however, been much higher than those which occur in marine environments. Ecological effects resulting from differential effects on phytoplankton growth appear therefore unlikely.

Research Needs

Research carried out over the past five years has generated results which in turn have opened up a bewildering variety of potential research projects. Some of these are strictly "scientific" in the more classical sense of the term; others have more immediate aims to determine whether levels of a given pollutant are increasing in a particular area and whether or not a given level is exerting harmful effects on the local marine environment. Among the many possibilities, we suggest the following priority. Since per capita production of synthetic organic compounds has continued to increase, it has become essential to determine which ones might be potentially significant marine pollutants (National Academy of Sciences, 1975). From the existing data on the global distribution of the DDT and PCB compounds, it is possible to present a preliminary model of the global mass balances. Many data, however, are required to complete the picture. Such a model would not only permit a definitive assessment of the effects of these pollutants on the marine environment but the model might be used to predict the global movements and effects of other relatively persistent organic pollutants which may appear in the future.

Acknowledgments

Research was supported by the International Decade of Ocean Exploration Program, grants number GX32885 and ID072 06412 A02, the Office of Polar Programs, National Science Foundation, OPP74-21254 and OPP75-23520, and by the Ecology and Systematics Division of the National Science Foundation, GB-36593.

References

Anderlini, V C., P. G. Connors, R. W. Risebrough and J. H. Martin (1972). Concentrations of heavy metals in some Antarctic and North American sea birds. In *Proceedings of the Colloquium, Conservation Problems in Antarctica* (B.C. Parker, ed.), Allen Press, Lawrence, Kansas.

Anderson, D. W., J. R. Jehl, Jr., R. W. Risebrough, L. A. Woods, Jr., L. R. Deweese and W. G. Edgecomb (1975). Brown Pelicans: improved reproduction off the southern California coast. *Science, 190,* 806–808.

Aue, W. A., K. Shubhender and C. R. Hastings (1972). The use of support-bonded silicones for the extraction of organochlorines of interest from water. *J. Chromatogr., 73,* 99–104.

Bedford, J. W. (1974). The use of polyurethane foam plugs for extraction of polychlorinated biphenyls (PCB's) from natural waters. *Bull. Environ. Contam. Toxicol., 12,* 622–625.

Bennington, S., P. G. Connors, C. Connors and R. W. Risebrough (1975). Patterns of chlorinated hydrocarbon contamination in New Zealand subantarctic and coastal marine birds. *Environ. Pollut., 8,* 135–147.

Bevenue, A., J. N. Ogata and J. W. Hylin (1972). Organochlorine pesticides in rainwater, Oahu, Hawaii 1971–72. *Bull. Environ. Contam. Toxicol., 8,* 238.

Bidleman, T. F., and C. E. Olney (1974a). Chlorinated hydrocarbons in the Sargasso Sea atmosphere and surface water. *Science, 183,* 516–518.

Bidleman, T. F. and C. E. Olney (1974b). High-volume collection of atmospheric polychlorinated biphenyls. *Bull. Environ. Contam. Toxicol., 11,* 442–450.

Blumer, M. (1975). Organic compounds in nature: limits of our knowledge. *Angewandte Chemie, 14,* 507–514.

Bourne, W. R. P. and J. A. Bogan (1972). Polychlorinated biphenyls in North Atlantic seabirds. *Marine Pollut. Bull., 3,* 171–175.

Bowen, H. J. M. (1970). Absorption by polyurethane foams; new method of separation. *J. Chem. Soc. (A)*, 1082-1085.

Bowes, G. W., B. R. Simoneit, A. L. Burlingame, B. W. de Lappe and R. W. Rise-brough (1972). The search for chlorinated dibenzofurans and chlorinated dibenzodioxins in wildlife populations showing elevated levels of embryonic death. *Environ. Health Perspectives: Experimental Issue No. 5*, 191-198.

Bowes, G. W., M. J. Mulvihill, B. R. Simoneit, A. L. Burlingame and R. W. Rise-brough (1975). Identification of chlorinated dibenzofurans in American polychlorinated biphenyls. *Nature, 256*, 305-307.

Braestrup, L., J. Clausen and O. Berg (1974). DDE, PCB and aldrin levels in arctic birds of Greenland. *Bull. Environ. Contam. Toxicol., 11*, 326-332.

Brewerton, H. V. (1969) DDT in fats of antarctic animals. *N. Z. Journal of Science, 12*, 194-199.

Burlingame, A. L., B. J. Kimble, E. S. Scott, J. W. De Leeuw, F. C. Walls, B. W. de Lappe and R. W. Risebrough. The molecular nature and extreme complexity of trace organic constituents in southern California municipal wastewater effluents. *Proceedings of the Symposium on the Identification and Analysis of Organic Pollutants in Water* (L.H. Keith, ed.), First Chemical Congress of the North American Continent, Mexico City, November-December 1975, Ann Arbor Science Publishers (in press).

Carmignani, G. M. and J. B. Bennett. Filter media for the removal of phthalate ester in water of closed aquaculture systems. *Aquaculture* (in press).

Carry, C. W. and J. A. Redner (1970). Pesticides and heavy metals: progress report. December 1970. County Sanitation Districts of Los Angeles County, 2020 Beverly Boulevard, Los Angeles, California 90057. 51 pp.

Cliath, M. M. and W. F. Spencer (1972). Dissipation of pesticides from soil by volatilization of degradation products. I. Lindane and DDT. *Environ. Sci. Technol., 6*, 910-914.

Connors, P. G., V. C. Anderlini, R. W. Risebrough, J. H. Martin, R. W. Schreiber and D. W. Anderson (1972). Heavy metal concentrations in brown pelicans from Florida and California. *Cal-Neva Wildlife 1972*, 56-64.

Connors, P. G., V. C. Anderlini, R. W. Risebrough, M. Gilbertson and H. Hays (1975). Investigations of heavy metals in Common Tern populations. *Canadian Field-Naturalist, 89*, 157-162.

Connors, P. G., W. Robertson, S. A. Jacobs and R. W. Risebrough. Accumulation of metals and chlorinated hydrocarbons by Sooty Terns (*Sterna fuscata*) of known age. Unpublished ms.

Cox, J. L. (1971). DDT residues in seawater and particulate matter in the California current system. *Fisheries Bull., 69*, 443-450.

DeLong, R. L., W. G. Gilmartin and J. G. Simpson (1973). Premature births in California sea lions: association with high organochlorine pollutant residue levels. *Science, 181*, 1168-1169.

Environmental Protection Agency (1973). Environmental contamination from hexachlorobenzene, July 20, 1973. Office of Toxic Substances, U. S. Environmental Protection Agency, Washington, D.C. 27 pp.

Fenical, W. (1975). Halogenation in the Rhodophyta — a review. *J. Phycology, 11,* 245–259.

Fisher, N. S. (1975). Chlorinated hydrocarbon pollutants and photosynthesis of marine phytoplankton: a reassessment. *Science, 189,* 463–464.

Fisher, N. S., L. B. Graham, E. J. Carpenter and C. F. Wurster (1973). Geographic differences in phytoplankton sensitivity to PCBs. *Nature, 241,* 548–549.

George, J. L . and D. E. H. Frear (1966). Pesticides in the Antarctic. *J. Appl. Ecol. 3* (Suppl.), 155–167.

Gesser, H. D., A. Chow, F. C. Davis, J. F. Uthe and J. Reinke (1971). The extraction and recovery of polychlorinated biphenyls (PCB) using porous polyurethane foam. *Analytical Letters, 4,* 883–886.

Gilbertson, M. and L. Reynolds (1974). DDE and PCB in Canadian birds, 1969 to 1972. Canadian Wildlife Service, Occasional Paper No. 19, 18 pp.

Gilmartin, W. G., R. L. DeLong, A. W. Smith, J. C. Sweeney, B. W. de Lappe, R. W. Risebrough, L. A. Griner, M. D. Dailey and D. B. Peakall (1967). Premature parturition in the California sea lion. *J. Wildl. Diseases, 12,* 104–115.

Goldberg, E. D. (1975a). Synthetic organohalides in the sea. *Proc. R. Soc. Lond. B., 189,* 277–289.

Goldberg, E. D. (1975b). The mussel watch — a first step in global marine monitoring. *Marine Pollut. Bull., 6,* 111.

Gress, F., R. W. Risebrough, D. W. Anderson, L. F. Kiff and J. R. Jehl (1973). Reproductive failures of double-crested cormorants in southern California and Baja California. *Wilson Bull., 85,* 197–208.

Harvey, G. R., W. G. Steinhauer and J. M. Teal (1973). Polychlorobiphenyls in North Atlantic Ocean water. *Science, 180,* 643–644.

Harvey, G. R. and W. G. Steinhauer (1974). Atmospheric transport of polychlorobiphenyls to the North Atlantic. *Atmospheric Environment, 8,* 777–782.

Harvey, G. R., W. G. Steinhauer and H. P. Miklas (1974a). Decline of PCB concentrations in North Atlantic surface water. *Nature, 252,* 387–388.

Harvey, G. R., H. P. Miklas, V. T. Bowen and W. G. Steinhauer (1974b). Observations on the distribution of chlorinated hydrocarbons in Atlantic Ocean organisms. *J. Mar. Research, 32,* 103–118.

Hays, H. and R. W. Risebrough (1972). Pollutant concentrations in abnormal young terns from Long Island Sound. *Auk, 89,* 19–35.

Henriksson, K., E. Karppanen and M. Heiminen (1966). High residue of mercury in Finnish white-tailed eagles. *Ornis Fennica, 43,* 38–45.

Herring, J. L., E. J. Hannan and D. D. Bills (1972). UV irradiation of Aroclor 1254 *Bull. Environ. Contam. Toxicol., 8,* 153-157.

Holdgate, M. W. (ed.) (1971). The sea bird wreck in the Irish Sea, Autumn 1969. Mimeographed, 76 pp., Natural Environmental Research Council, Alhambra House, 27-33 Charing Cross Road, London, W.C. 2.

Holden, A. V. (1970). International cooperative study of organochlorine pesticide residues in terrestrial and aquatic wildlife, 1967/1968. *Pesticides Monitoring Journal, 4,* 117-135.

Holden, A. V. (1973). International cooperative study of organochlorine and mercury residues in wildlife. *Pesticides Monitoring Journal, 7,* 37-52.

Hom, W., R. W. Risebrough, D. R. Young and A. Soutar (1974). Deposition of DDE and PCB in dated sediments of the Santa Barbara Basin. *Science, 184,* 1197-1199.

Hutzinger, O., S. Safe and V. Zitko (1972). Photochemical degradation of chlorobiphenyls (PCBs). *Environ. Health Perspectives, 1,* 15-20.

Hutzinger, O., S. Safe and V. Zitko (1974). The chemistry of PCB's. CRC Press, Cleveland, Ohio, 77 pp.

Jansson, B., S. Jensen, M. Olsson, L. Renberg, G. Sundstrom and R. Vaz (1975). Identification by GC-MS of phenolic metabolites of PCB and p,p'-DDE isolated from Baltic Guillemot and Seal. *Ambio, 4,* 93-97.

Jensen, S., A. G. Johnels, M. Olsson and T. Westermark (1972). The avifauna of Sweden as indicators of environmental contamination with mercury and chlorinated hydrocarbons. Fifteenth International Ornithological Congress, The Hague, 1970, 455-465.

Jensen, S., L. Renberg and M. Olsson (1972). PCB contamination from boat bottom paint and levels of PCB in plankton outside a polluted area. *Nature, 240,* 358-360.

Kerner, I., W. Klein and F. Korte (1972). Beiträge zur ökologischen chemie – XXXIII Photochemische reaktionen von 1, 1-dichlor-2 (p,p'-dichlorphenyl) äthylen (DDE). *Tetrahedron., 28,* 1575-1578.

Koeman, J. H. and H. Van Genderen (1972). Tissue levels in animals and effects caused by chlorinated hydrocarbon insecticides, chlorinated biphenyls and mercury in the marine environment along the Netherlands coast. *Marine Pollution and Sea Life,* pp. 1-8. Published in Great Britain by The Whitefriars Press Ltd., London and Tonbridge.

Koeman, J. H., R. H. Hadderingh and M. F. I. J. Bijleveld (1972a). Persistent pollutants in the white-tailed eagle in the Federal Republic of Germany. *Biol. Conserv., 4,* 373-377.

Koeman, J. H., T. Bothof, R. De Vries, H. Van Velzen-Blad and J. G. Vos (1972b). The impact of persistent pollutants on piscivorous and molluscivorous birds. *TNO-Nieuws, 27,* 561-569.

Koeman, J. H., H. C. W. Van Velzen-Blad, R. De Vries and J. G. Vos (1973a). Effects of PCB and DDE in cormorants and evaluation of PCB residues from an experimental study. *J. Reprod. Fert., Suppl. 19*, 353-364.

Koeman, J. H., W. H. M. Peeters, C. H. M. Koudspal-Hol, P. S. Tjioe, J. J. M. De Goeij (1973b). Mercury-selenium correlations in marine mammals. *Nature, 245*, 385-386.

de Lappe, B. W., R. W. Risebrough, P. Millikin and D. R. Young. Use of *Mytilus californianus* to monitor changes in chlorinated hydrocarbon contamination of California coastal waters. In preparation.

Lloyd, C., J. A. Bogan, W. R. P. Bourne, P. Dawson, J. L. S. Parslow and A. G. Stewart (1974). Seabird mortality in the north Irish Sea and Firth of Clyde early in 1974. *Marine Pollut. Bull., 5*, 136-140.

Lloyd-Jones, C. P. (1971). Evaporation of DDT. *Nature, 229*, 65-66.

McKay, D. and P.J. Leinonen (1975). Rate of evaporation of low-solubility contaminants from water bodies to atmosphere. *Environ. Sci. and Tech., 9*, 1178-1180.

MacKay, D. and A. W. Wolkoff (1973). Rate of evaporation of low-solubility contaminants from water bodies to atmosphere. *Environ. Sci. and Tech., 7*, 611-614.

Marshall, N. B. (1971). Explorations in the life of fishes. Harvard Univ. Press, Cambridge, Massachusetts.

Martin, J. H., V. C. Anderlini, D. Girvin, S. A. Jacobs, R. W. Risebrough, R. L. DeLong and W. G. Gilmartin. Mercury-selenium-bromine imbalance in premature parturient California sea lions. *Marine Biology*, in press.

Meith-Avcin, N., S. M. Warlen and R. T. Barber (1976). Organochlorine insecticide residues in a bathyl-demersal fish from 2,500 meters. *Environ. Letters*, in press.

Metcalf, R. L., J. R. Sanborn, P. Y. Lu and D. Nye (1975). Laboratory model ecosystem studies of the degradation and fate of radio-labeled tri-, tetra-, and pentachlorobiphenyl compared with DDE. *Arch. Environ. Contam. Toxicol., 3*, 151-165.

Miller, G. A. and C. E. Wells (1969). Alkaline pre-column for use in gas chromatographic pesticide residue analysis. *J.A.O.A.C., 52*, 548-553.

Mills, P. A. (1961). Collaborative study of certain chlorinated organic pesticides in dairy products. *J.A.O.A.C., 44*, 171.

Mosser, J. L., N. S. Fisher, T.-C. Teng and C. F. Wurster (1972). Polychlorinated biphenyls: toxicity to certain phytoplankters. *Science, 175*, 191-192.

Musty, P. R. and G. Nickless (1974). The extraction and recovery of chlorinated insecticides and polychlorinated biphenyls from water using porous polyurethane foams. *J. Chromatography, 100*, 83-93.

National Academy of Sciences (1971). Chlorinated hydrocarbons in the marine environment. National Academy of Sciences (U.S.), Washington, D. C.

National Academy of Sciences (1975). Assessing Potential Ocean Pollutants: A Report of The Study Panel on Assessing Potential Ocean Pollutants to the Ocean Affairs Commission on Natural Resources, National Research Council. National Academy of Sciences, Washington, D. C. 438 pp.

Nisbet, I. C. T. and A. F. Sarofim (1972). Rates and routes of transport of PCBs in the environment. *Environ. Health Perspectives, No. 1,* 21-38.

Norstrom, R. J., R. W. Risebrough and D. J. Cartwright. Elimination of chlorinated dibenzofurans (CDFs) associated with PCBs fed to mallards (*Anas platyrhynchos*). *Toxicology Applied Pharmacology*, in press.

Ohlendorf, H. M., E. E. Klaas and T. E. Kaiser (1974). Environmental pollution in relation to estuarine birds. Pages 53-81. In *Survival in Toxic Environments*, M. A. Q. Khan and J. P. Bederka, ed., Academic Press, New York, 553 pp.

Ohlendorf, H. M., R. W. Risebrough and K. Vermeer. Exposure of marine birds to environmental pollutants. Proc. Pacific Seabird Congress, in press.

Oloffs, P. C., L. J. Albright and S. Y. Szeto (1972). Fate and behavior of five chlorinated hydrocarbons in three natural waters. *Canadian J. Microbiol., 18,* 1393-1498.

Panel on Hazardous Trace Substances (1972). PCBs – Environmental Impact. *Environ. Research, 5,* 249-362.

Parslow, J. L. F. and D. J. Jefferies (1973). Relationship between organochlorine residues in livers and whole bodies of guillemots. *Environ. Pollut., 5,* 87-101.

Parslow, J. L. F., D. J. Jefferies and H. M. Hanson (1973). Gannet mortality incidents in 1972. *Mar. Pollut. Bull., 4,* 41-44.

Pavlou, S. P. and W. Hom. Interlaboratory calibration results from chlorinated hydrocarbon analyses in marine sediments. *Marine Chemistry*, in press.

Peakall, D. B., T. J. Cade, C. M. White and J. R. Haugh (1975). Organochlorine residues in Alaskan Peregrines. *Pesticides Monitoring J., 8,* 255-260.

Peel, D. A. (1975). Organochlorine residues in antarctic snow. *Nature, 254,* 324-325.

Prospero, J. M. and D. B. Seba (1972). Some additional measurements of pesticides in the lower atmosphere of the northern equatorial Atlantic Ocean. *Atmospheric Environment, 6,* 363-364.

Quraishi, M. S. (1970). Volatilization of DDT-^{14}C and Lindane -^{14}C from treated surfaces, and metabolism of DDT-^{14}C by house fly pupae. *Canadian Entomologist, 102,* 1189-1195.

Ratcliffe, D. A. (1972). The peregrine population of Great Britain in 1971. *Bird Study, 19,* 117-157.

Redner, J. A. and K. Payne (1971). Chlorinated hydrocarbons. Progress Report – December 1971. County Sanitation Districts of Los Angeles County, 2020 Beverly Boulevard, Los Angeles, California 90057.

Risebrough, R. W. (1976). Transfer of organochlorine pollutants to Antarctica. Adaptations Within Antarctic Ecosystems. Scientific Committee for Antarctic Research, G.A. Llano, ed., in press.

Risebrough, R. W., D. W. Anderson and J. McGahan. Chlorinated hydrocarbons in a Peruvian coastal ecosystem. Unpublished ms.

Risebrough, R. W., R. J. Huggett, J. J. Griffin and E. D. Goldberg (1968a). Pesticides: transatlantic movements in the northeast trades. *Science, 159,* 1233-1236.

Risebrough, R. W., P. Reiche, D. B. Peakall, S. G. Herman and M. N. Kirven (1968b). Polychlorinated biphenyls in the global ecosystem. *Nature, 220,* 1098-1102.

Risebrough, R. W. and B. W. de Lappe (1972). Accumulation of polychlorinated biphenyls in ecosystems. *Environ. Health Perspectives, 1,* 39-45.

Risebrough, R. W., F. C. Sibley and M. N. Kirven (1971). Reproductive failure of the Brown Pelican on Anacapa Island in 1969. *American Birds, 25,* 8-9.

Risebrough, R. W. and G. M. Carmignani (1972). Chlorinated hydrocarbons in Antarctic birds. Pages 63-78, in *Proceedings of the Colloquium, Conservation Problems in Antarctica,* B. C. Parker, ed., Allen Press, Lawrence, Kansas.

Risebrough, R. W., V. Vreeland, G. R. Harvey, H. P. Miklas and G. M. Carmignani (1972). PCB residues in Atlantic zooplankton. *Bull. Environ. Contam. Toxicol., 8,* 345-355.

Risebrough, R. W., B. W. de Lappe, and T. T. Schmidt. Bioaccumulation factors of chlorinated hydrocarbons between *Mytilus* sp. and seawater. Unpublished ms.

Robinson, J., A. Richardson, A. N. Crabtree, J. C. Coulson and G. R. Potts (1967). Organochlorine residues in marine organisms. *Nature, 214,* 1307-1311.

Robinson, P. and R. W. Risebrough. Chlordane compounds in aquatic food webs. Unpublished ms.

Ruzo, L. O., M. J. Zabik and R. D. Schuetz (1972). Polychlorinated biphenyls: photolysis of 3, 4, 3', 4'-tetrachlorobiphenyl and 4,4'-dichlorobiphenyl in solution. *Bull. Environ. Contam. Toxicol., 8,* 217-218.

Safe, S. and O. Hutzinger (1971). Polychlorinated biphenyls: photolysis of 2,4,6,2',4',6'-hexachlorobiphenyl. *Nature, 232,* 641-642.

Sanborn, J. R., W. F. Childers and R. L. Metcalf (1975). Uptake of three poly-chlorinated biphenyls, DDT, and DDE by the green sunfish, *Lepomis cyanellus* Raf. *Bull. Environ. Contam. Toxicol., 13,* 209-217.

Sanders, H. O. and J. H. Chandler (1972). Biological magnification of a poly-chlorinated biphenyl (Aroclor(®) 1254) from water by aquatic inverte-brates. *Bull. Environ. Contam. Toxicol., 7,* 257-263.

Schmidt, T. T., R. W. Risebrough and F. Gress (1971). Input of polychlori-nated biphenyls into California coastal waters from urban sewage outfalls. *Bull. Environ. Contam. Toxicol., 6,* 235-243.

Schoor, W. P. (1975). Problems associated with low-solubility compounds in aquatic toxicity tests: theoretical model and solubility characteristics of Aroclor 1254 in water. *Water Research, 9,* 937-944.

Schreiber, R. W. and R. W. Risebrough (1972). Studies of the Brown Pelican. *Wilson Bull., 84,* 119-135.

Scura, E. D. and V. E. McClure (1975). Chlorinated hydrocarbons in seawater: analytical method and levels in the Northeastern Pacific. *Marine Chemis-try, 3,* 337-346.

Seba, D. B. and J. M. Prospero (1971). Pesticides in the lower atmosphere of the northern equatorial Atlantic Ocean. *Atmos. Environ., 5,* 1043-1050.

Sladen, W. J. L., C. M. Menzie and W. L. Reichel (1966). DDT residues in adelie penguins and a crabeater seal from Antarctica. *Nature, 210,* 670-673.

Sobelman, M. (1971). Montrose Chemical Corporation, Torrance, California, *in litt.*

Southern California Coastal Water Research Project (1975). Coastal Water Research Project Annual Report. Southern California Coastal Water Re-search Project, El Segundo, California.

Sparschu, G. L., F. L. Dunn and V. K. Rowe (1971). Study of the teratogeni-city of 2,3,7,8-tetrachlorodibenzo-*p*-dioxin in the rat. *Food Cosmet. Toxi-col., 9,* 405-412.

Spencer, W. F. and M. M. Cliath (1972). Volatility of DDT and related com-pounds. *J. Agr. Food Chem., 20,* 645-649.

Spitzer, P. R., R. W. Risebrough, J. W. Grier and C. R. Sindelar. Eggshell-thick-ness–pollutant relationships among North American ospreys. Proc. North American Osprey Symposium, J. Ogden, ed., in press.

Sproul, J. A., Jr., R. L. Bradley, Jr. and J. J. Hickey. Polychlorinated biphenyls, DDE, and dieldrin in Icelandic seabirds. *Pesticides Monitoring J.,* in press.

Stanley, R. L. and H. T. Lefavoure (1965). Rapid digestion and cleanup of ani-mal tissues for pesticide residue analysis. *J. Ass. Off. Agric. Chem., 48,* 666-667.

Sverdrup, H. V., M. W. Johnson and R. H. Fleming (1942). The Oceans: Their Physics, Chemistry and General Biology. Prentice Hall, New York. 1087 pp.

Tatton, J. O'G. and J. H. A. Ruzicka (1967). Organochlorine pesticides in Antarctica. *Nature, 215,* 346-348.

Ten Noever de Brauw, M. C. and J. H. Koeman (1972). Identification of chlorinated styrenes in cormorant tissues by a computerized gas chromatography mass spectrometry system. *Sci. Total Environ., 1,* 427-432.

Ten Noever de Brauw, M. C., C. Van Ingen and J. H. Koeman (1973). Mirex in seals. *Sci. Total Environ., 2,* 196-198.

Uthe, J. F., J. Reinke and H. Gesser (1972). Extraction of organochlorine pesticides from water by porous polyurethane coated with selective absorbents. *Environmental Letters, 3,* 117-135.

Uthe, J. F., J. Reinke and H. O'Brodovich (1974). Field studies on the use of coated polyurethane plugs as indwelling monitors of organochlorine pesticides and polychlorinated biphenyl content of streams. *Environmental Letters, 6,* 103-115.

Vos, J. G. and J. H. Koeman (1970). Comparative toxicologic study with polychlorinated biphenyls in chickens with special reference to porphyria, edema formation, liver necrosis and tissue residues. *Toxicol. Appl. Pharmacol., 17,* 656-668.

Vos, J. G., J. H. Koeman, H. L. Van Der Maas, M. C. Ten Noever de Brauw and R. H. De Vos (1970). Identification and toxicological evaluation of chlorinated dibenzofuran and chlorinated naphthalene in two commercial polychlorinated biphenyls. *Food Cosmet. Toxicol., 8,* 625-633.

Walker, W. II, R. W. Risebrough, J. T. Mendola and G. W. Bowes (1973a). South American studies of the peregrine, an indicator species for persistent pollutants. *Antarctic Journal of the United States, 8,* 29-31.

Walker, W. II, W. G. Mattox and R. W. Risebrough (1973b). Pollutant and shell thickness determinations of peregrine eggs from west Greenland. *Arctic, 26,* 256-258.

Ware, D. M. and R. F. Addison (1973). PCB residues in plankton from the Gulf of St. Lawrence. *Nature* (London), *246* (5434/Supp.), 519-521.

Weil, L., G. Dure and K.-E. Quentin (1974). Wasserloeslichkeit von insektiziden chlorirten Kohlenwasserstoffen und polychlorierten Biphenylen im Hihblich auf eine Gewaesserbelastung mit diesen Stoffen. *Z. Wasser- Abwasser-Forsch., 7,* 169-175.

White, C. M. and R. W. Risebrough (1976). Polychlorinated biphenyls in the ecosystems of Amchitka Island. In *The environment of Amchitka Island, Alaska,* M. L. Merritt and R. G. Fullers, eds., Oak Ridge, Tennessee, Technical Information Center, Atomic Energy Commission, in press.

Wiemeyer, S. N., P. R. Spitzer, W. C. Krantz, T. G. Lamont and E. Cromartie (1975). Effects of environmental pollutants on Connecticut and Maryland ospreys. *J. Wildl. Manage., 39,* 124-139.

Williams, P. M. and K. J. Robertson (1975). Chlorinated hydrocarbons in sea-surface films and subsurface waters at nearshore stations and in the North Central Pacific Gyre. *Fishery Bull., 73,* 445-447.

Wong, P. T. S. and K. L. E. Kaiser (1975). Bacterial degradation of polychlorinated biphenyls. II. Rate studies. *Bull. Environ. Contam. Toxicol., 13,* 249-255.

Wurster, C. F. (1968). DDT reduces photosynthesis by marine phytoplankton. *Science, 159,* 1474-1475.

Wurster, C. F., Jr. and D. B. Wingate (1968). DDT residues and declining reproduction in the Bermuda Petrel. *Science, 159,* 979-981.

Young, D. R. and T. C. Heesen (1974). Inputs and distributions of chlorinated hydrocarbons in three southern California harbors. Rept. TM214. Southern California Coastal Water Research Project, El Segundo, California.

Young, D. R., T. C. Heesen, D. J. McDermott and P. E. Smokler (1974). Marine inputs of polychlorinated biphenyls and copper from vessel antifouling paints. Rept. TM212. Southern California Coastal Water Research Project, El Segundo, California.

Young, D. R., R. W. Risebrough, B. W. de Lappe, T. C. Heesen and D. J. McDermott (1976). Input of chlorinated hydrocarbons into the Southern California Bight from surface runoff and municipal wastewaters. Unpublished ms.

Zitko, V., O. Hutzinger and P. M. K. Choi (1972). Contamination of the Bay of Fundy – Gulf of Maine area with polychlorinated biphenyls, polychlorinated terphenyls, chlorinated dibenzodioxins, and dibenzofurans. *Environ. Health Perspective, 1,* 47-50.

13

High Molecular Weight Chlorinated Hydrocarbons in the Air and Sea: Rates and Mechanisms of Air/Sea Transfer

T. F. Bidleman, C. P. Rice and *C. E. Olney*

Introduction

Chlorinated pesticides and PCB are ubiquitously present in the environment, from arctic mammals to the antarctic snowfields. How are these pollutants translocated from their sites of use to these remote areas? In 1971, the National Academy of Sciences published "Chlorinated Hydrocarbons in the Marine Environment," a report which assessed the ecological impact and possible transport routes of these compounds. In the same year Woodwell *et al.* (1971) formulated a mass balance for DDT in the biosphere which considered transfer from the land to the atmosphere to the ocean and predicted concentrations in these reservoirs through the end of this century. A similar model was forwarded by Cramer (1973). These studies emphasized the importance of thinking on a global scale when considering transport and concluded that the atmosphere was the most likely source of these compounds in remote areas.

The above groups had very little data upon which to base their conclusions. At that time there were only two reports of pesticides in marine air and none in seawater, although chlorinated hydrocarbon contamination of marine biota was well recognized. Since then we have accumulated more information on concentrations of CHC in the air and ocean, and it seems safe to conclude that DDT concentrations in ocean mixed layer are nearly two orders of magnitude lower than predicted by both global circulation models. Concentrations of DDT in the troposphere are at least 100 times lower than Woodwell's estimate, but perhaps less than an order of magnitude from those predicted by Cramer. Based on what we have learned since the early 1970s, new estimates of air/sea CHC fluxes can be made which are useful for three reasons: they allow us to assess whether aerial input can potentially account for the CHC levels presently observed in the ocean; they suggest which atmosphere removal processes are likely to result in significant deposition into the oceans; and they provide a stimulus for the further research which will be needed to improve our knowledge of CHC transfer through the physical environment.

Production and Use of CHC

The U. S. production of PCB and chlorinated pesticides is summarized in Table 13-1, and for PCB and DDT the estimated world output or use is also given.

Table 13-1
Production of Chlorinated Hydrocarbons

Compound	Year	U.S. Production	U.S. Exports	World Production/Use	Reference
			(10⁶ kg/yr)		
Aldrin	1972	5.9	0.14		von Rumker et al., 1974
Chlordane	1972	9.1	2.3		von Rumker et al., 1974
Dieldrin	1970	0.27			von Rumker et al., 1974
Heptachlor	1970	3.6			von Rumker et al., 1974
Toxaphene	1972	35	8.2		von Rumker et al., 1974
DDT	1971-81			105, pred. use	Goldberg, 1975
	1970	27	31		von Rumker et al., 1974
	1969	56	37		von Rumker et al., 1974
	1975-3rd quarter	10	2.0		Papageorge, 1975
PCB	1972-74	18-19	2.3-3.6		Papageorge, 1975
	1971	16		50	Papageorge, 1975 Anon., 1971, 1973
	1970	39	6.2		Anon., 1971
HCB		>0.9			N.A.S., 1975

Most of the information on chlorinated pesticides is from a 1973 survey of U.S. usage patterns carried out by Midwest Research Institute and published by the E.P.A. (von Rumker et al., 1974). Dr. von Rumker, who headed the study, does not think these figures have been updated in any publicly released report. PCB production figures up to 1971 were published by Monsanto (Anonymous, 1971) and Dr. W. P. Papageorge supplied data through the third quarter of 1975.

Several regulatory decisions since 1970 have been responsible for a decline in CHC usage in the U.S. Monsanto restricted sales of PCB to closed-system applications (electrical capacitors and transformers) at the end of 1970 and several other countries have stopped production or limited sales as a result of a 1973 O.E.C.D. agreement (Anonymous, 1973). With the restriction in sales, presumably the flow of PCB into the environment would be halted, but industrial discharge of PCB in fresh waterways may still be a problem. Recently fish from several U.S. lakes and rivers were found to contain PCB residues exceeding the F.D.A. tolerance limit (Boyle, 1975).

All uses of DDT in the U.S. were cancelled as of December 1972, but the ban was temporarily lifted in the spring of 1974 when over 45,000 kg of the pesticide was sprayed over the northwestern pine forests to control a tussock moth outbreak (Orgill et al., 1974; 1976). Uses of DDT in South America, Africa, and Asia are predicted to remain stable or increase slightly through the end of the decade. Most of the 105 million kg/yr estimated requirement is for malaria mosquito and cotton insect control (Goldberg, 1975). Several other chlorinated hydrocarbon pesticides have recently come under restriction in the U.S. The manufacturer of aldrin and its persistent epoxide, dieldrin, was prohibited in October 1974, and chlordane and heptachlor use were suspended early in 1976 (Anonymous, 1975, 1976).

By far the most heavily used chlorinated insecticide in the U.S. today is toxaphene, a complex mixture of nearly 200 polychlorinated camphenes. About 23 million kg/yr are used annually on cotton in the south and southeast U.S. (von Rumker *et al.,* 1974; Sanders, 1975). Concern over the environmental effects of toxaphene have increased with recent reports that the pesticide can produce stunted growth and broken backs in fish exposed to nanogram-per-liter levels (Mehrle and Mayer, 1975a, b; 1976). So far, no restrictions on toxaphene use have been issued.

Hexachlorobenzene (HCB) is a persistent, biologically accumulated CHC which has been identified in several species of fish and in human tissues (N.A.S., 1975). Compared to other pesticides, only a small quantity of HCB is intentionally released into the environment (6×10^3 kg/yr, as a grain fungicide), but accidental release as a contaminant of other pesticides may exceed this figure. The most important source of HCB seems to be the tarry wastes from manufacturing light CHC such as carbon tetrachloride and perchloroethylene. The HCB output in these tars has been estimated to exceed 900,000 kg/yr (N.A.S., 1975).

Distribution of CHC in Ocean Waters

Contamination of marine organisms with PCB and DDT residues is well documented, but there is a dearth of information on levels of these pollutants in seawater itself, especially in the open ocean. Doubtlessly this is because the concentrations are very low and contamination during sampling and analysis is a more serious problem than with animal tissues. Within the last few years progress in sampling has been achieved by using macroreticular resins and other polymeric adsorbents to isolate trace organics from large volumes of water, so the technology to carry out a worldwide sampling program is now available (Harvey *et al.,* 1973, 1976a; Musty *et al.,* 1974a, b; Bedford, 1974; Junk *et al.,* 1974; Richard *et al.,* 1974; Osterroht, 1974; Leoni *et al.,* 1975; Dawson *et al.,* 1976). Harvey *et al.* (1973) used an XAD-2 resin column to isolate PCB from 40-60 liters of seawater and reported mean levels of 27-41 ng/l for 1972 North Atlantic mixed layer samples. A year later they found that the concentrations had dropped to about 1 ng/l and showed little variation with depth down to 500 m (Harvey *et al.,* 1974a). Harvey attributed the decrease to scavenging by sinking particles accompanied by a declining input resulting from world sales restrictions. It seems unlikely that sales restrictions would have such an immediate impact on PCB input — equivalent to "turning off the tap" — and it is tempting to explain the apparent loss of PCB to sampling and analytical variations. However, measurements by our group (Table 13-2) from the University of Rhode Island in 1971–73 agreed with Harvey's values. Thus we have two independent sets of data suggesting that PCB levels in the North Atlantic dropped by an order of magnitude between 1971 and 1973, with no satisfactory explanation. The few available data indicate that PCB concentrations in the North Pacific are comparable to those presently observed in the North Atlantic (Table 13-2).

Since most of the PCB producing and using countries are in the nothern hemisphere, we would expect PCB levels in the southern oceans to be much lower, but Harvey et al. (1974b) found that plankton from the North and South Atlantic contained similar concentrations of PCB. Recent measurements of PCB in South Atlantic water were lower than those in the North Atlantic (Harvey and Steinhauer, 1976b), but the difference between the two oceans was less than one might expect from a consideration of PCB use geography.

Thus far no one has reliably measured DDT in the open ocean. Harvey et al. (1973) could not detect DDT in the North Atlantic below their limit of 1.0 ng/l in 1972, and although we found p,p'-DDT in surface films in the Sargasso and western North Atlantic (Olney and Wade, 1972; Bidleman and Olney, 1974 a,b) the concentration was less than 0.1 ng/l in the subsurface water. Three out of twelve South Atlantic samples taken by Harvey and Steinhauer (1976b) in 1975 contained 0.3 ng/l p,p'-DDT, but the rest were less than 0.1 ng/l. Williams and Robertson (1975) reported less than 0.03 ng/l total DDT in the North Central Pacific. Measurable concentrations of DDT have been found off the California coast, ranging from 2 to 6 ng/l in 1971 and from 0.2 to 1.8 ng/l in 1973–74 (Table 13-2).

Several studies have shown that CHC are enriched at the air/sea interface relative to their concentrations in the underlying water (Table 13-2). The enrichment mechanism is thought to be scavenging of these compounds by bubbles rising through the water column and/or atmospheric deposition on the sea surface. The volume of the "surface microlayer" (about 300 micrometers thickness, as collected by Garrett's (1965) screen technique) is too small to be considered a major reservoir for CHC, yet surface films are important by virtue of their position at the air/sea interface. Enrichment of pollutants in this layer may make them more available to marine life associated with the interface. Cheng and Bidleman (1976) have found ng/g levels of PCB and DDT in the surface-skimming water striders *Halobates* and *Rheumatobates* collected off Baha, California. Also the presence of films on the surface may influence the transfer of CHC across the air/sea interface. Hoffman and Duce (1974) found that organic carbon in Bermuda aerosols was enriched hundreds to thousands of times relative to seawater, so the potential for ejection of CHC on sea salt particles is clear. On the other hand, lipids at the sea surface may accelerate the vapor flux of these pollutants from the atmosphere to the ocean, as CHC have lipid solubilities greatly exceeding their solubilities in water. These qualitative arguments point out the need for further work on the role of surface films in interfacial transfer processes.

CHC are probably removed from the water column by adsorption onto sinking particles, as suggested by Harvey *et al.* (1974c, 1976a). In fresh water environments PCB are generally found associated with suspended material, and sediments have much higher concentrations than do the overlying waters (Crump-Wiesner, *et al.,* 1974). Partition coefficients of CHC between sediments and water (P = g/kg sediment $\div$ g/kg water) range from about 10^3 for clay minerals to 10^5 for organic particulates (Huang and Liao, 1970; Poirrier, *et al.,* 1972; Benvenue, *et al.,* 1972a). Values of 7×10^4 and 2×10^5 have been reported for the distribution of DDT and PCB between seawater and marine sediments (Pierce, *et al.,* 1974; Harvey and Steinhauer, 1976a).

Hom, *et al.* (1974) estimated that PCB and DDT residues have been accumulating in the sediments of the Santa Barbara Basin at rates of 1.2×10^{-4} and 1.9×10^{-4} g/m^2 yr, equivalent to about 5–8 metric tons/yr over an area 200×200 km. Sedimentation rates in the open ocean appear to be considerably lower. Harvey and Steinhauer (1974c, 1976a) identified ng/g levels of PCB in the top two cm of cores taken from the Hudson Canyon and the Hatteras Abyssal Plain and obtained a preliminary estimate of open-ocean PCB deposition (5×10^{-6} g/m^2-yr), using a sediment trap moored under 2000 m of water.

CHC in the Atmosphere

Vaporization

High molecular weight CHC are often considered "nonvolatile." However, significant quantities of these compounds can escape into the atmosphere, and within the last few years it has been realized that vaporization losses of pesticides from plant surfaces and soils can proceed more rapidly than chemical and biological degradation. Pesticide volatility is related to the vapor pressure, which ranges from 10^{-5} to 10^{-7} mm Hg for most chlorinated insecticides (Table 13-3). Saturation vapor densities calculated from the vapor pressure using the ideal gas law are in the microgram-milligram per cubic meter range, and are far higher than atmospheric concentrations found in the environment.

A large number of laboratory studies have been conducted to determine pesticide vaporization rates from soils. Spencer *et al.* (1973a, 1975) recently reviewed and discussed the many variables influencing pesticide vaporization. Predicting outdoor evaporation rates from laboratory data is a difficult task, yet most estimates of pesticide vapor losses have used this approach. Some workers have estimated that up to 50–90% of the DDT applied to the soil would be vaporized in 5–12 months, based on loss rates from metal planchets (Lloyd-Jones, 1971) and dishes of sand (Ware *et al.,* 1975). In their model of DDT

Table 13-2
Chlorinated Hydrocarbons in Ocean Water

Location	Sampling Dates	PCB (10^{-9} g/L)		DDT (10^{-9} g/L)		Reference
		SM^a	SS^b	SM	SS	
Coastal						
Biscayne Bay, Fla	1968			185–12,700	<1	Seba and Corcoran, 1969
Florida current	1968			80	<1	Seba and Corcoran, 1969
Narragansett Bay, R.I.	1971	450–4200	50–150			Duce et al., 1972
California current	1971				2–6	Cox, 1971
California current	1971	8	2.5	0.4	0.1	Williams and Robertson, 1975
California current	1973		0.3–0.5		0.3–1.3	Young, 1975
California coastal	1971	11–50		12–15		Williams and Robertson, 1975
Gulf of Santa Catalina, California	1974		2–10		0.2–1.8	Scura and McClure, 1975
California Harbors	1974		10–36		0.6–6.5	Scura and McClure, 1975
N.Y. Bight	1971	150–660	120–270			Olney and Wade, 1972
Chesapeake Bay	1973	5–20	<0.8	0.6–1.9	0.08–0.11	Bidleman and Olney, 1973, 1974a
Swedish fjords	1972	1.3×10^7		3.8×10^7		Larsson et al., 1974
Irish Sea	1974		0.5–1.0		0.1–0.2	Dawson et al., 1976
Open Ocean						
Iceland-Nova Scotia	1971	40–1750	5–40			Olney and Wade, 1972
Azores-Bermuda	1972	6–40	1–22			Bidleman and Olney, 1973

Location	Sampling Dates	PCB (10⁻⁹ g/L)		DDT (10⁻⁹ g/L)		Reference
		SM^a	SS^b	SM	SS	
Newfoundland-Portugal	1972		41		<1	Harvey et al., 1973; Harvey et al., 1974a
Portugal-Norwegian Sea	1972		36		<1	Harvey et al., 1973; Harvey et al., 1974a
U.S.-Bermuda	1972		27		<1	Harvey et al., 1973; Harvey et al., 1974a
Sargasso Sea	1973	4-42	<0.9-3.6	0.3-2.1	<0.15	Bidleman and Olney, 1974b
Bermuda-U.S.	1973	4-7	<0.9-2.4	0.2-0.5	<0.05	Bidleman and Olney, 1973, 1974a
Sargasso Sea-N.Y. Bight	1973		0.8			Harvey et al., 1974a
Azores-Barbados	1973		2.0			Harvey et al., 1974a
New England Cont. Shelf	1974		0.8			Harvey et al., 1974a
Northeast trades region	1975		4-9			Harvey and Steinhauer, 1976a
North Atlantic, U.K.-U.S.	1975		1-7			Harvey and Steinhauer, 1976b
South Atlantic	1975		0.3-4		<0.1-0.4	Harvey and Steinhauer, 1976b
North Central Pacific Gyre	1972	5-6	5.0	<0.02	<0.03	Williams and Robertson, 1975

[a]SM = surface microlayer, thickness sampled varies with investigator.

[b]SS = subsurface water.

Table 13-3
Volatility of Chlorinated Hydrocarbons

Compound	Temperature (°C)	Vapor Pressure (mm Hg)	Saturation vapor density (mg/m³)	Reference
p,p' – DDT	30	7.3×10^{-7}	0.013	Spencer and Cliath, 1972
p,p' – DDT	20	2.5×10^{-7}	0.0049	Atkins and Eggleton, 1971
o,p' – DDT	30	5.5×10^{-6}	0.104	Spencer and Cliath, 1972
p,p' – DDE	30	6.5×10^{-6}	0.109	Spencer and Cliath, 1972
p,p' – DDD	30	1.0×10^{-6}	0.017	Spencer and Cliath, 1972
dieldrin	20	2.9×10^{-6}	0.060	Atkins and Eggleton, 1971
dieldrin	20	2.8×10^{-6}	0.058	Spencer and Cliath, 1969
dieldrin	20	0.8×10^{-6}	0.017	Porter, 1964
chlordane	25	1.0×10^{-5} [a]	0.221	Spencer, 1973b
Aroclor 1242	25	4.0×10^{-4} [b]	5.5	MacKay and Wolkoff, 1973
Aroclor 1248	25	5.0×10^{-4} [b]	7.9	MacKay and Wolkoff, 1973
Aroclor 1254	25	8.0×10^{-5} [b]	1.4	MacKay and Wolkoff, 1973
Aroclor 1260	25	4.0×10^{-5} [b]	0.73	MacKay and Wolkoff, 1973

[a]Refined technical product.

[b]Extrapolated from Monsanto data.

circulation Woodwell *et al.* (1971) assumed that all the DDT produced during the year entered the atmosphere; an input which, when coupled with an estimated atmospheric residence time of three years, led to predictions of tropospheric DDT levels 100 times higher than those observed (Table 13-5).

Actual loss rates in the field may be an order of magnitude lower. Willis *et al.* (1972) reported that 18, 2, and 7% of the dieldrin applied to the soil surface at 22 kg/hectare evaporated from moist, dry, and flooded plots, respectively, in a five-month period. Caro *et al.* (1971a, b) measured dieldrin and heptachlor fluxes from a field during the growing season and found that while vaporization losses exceeded those from water runoff and eroded soils, only about 3-4% of the applied pesticides evaporated during the season. There is a great need to carry out more of these field studies if we are to obtain meaningful estimates of pesticide input into the atmosphere.

PCB enter the atmosphere by evaporation from products containing these compounds, chiefly electrical transformers, capacitors and plastics. Although PCB are no longer used as plasticizers, thousands of tons of PCB-containing scrap have accumulated in dumps and landfills, where it will continue to bleed PCB into the environment — slowly by diffusion or at an accelerated rate during open-dump burning. Nisbet and Sarofim (1972) formulated a mass balance for PCB in the environment based on sales and estimated useful lives of PCB-containing products and estimated that about 1.3-1.8 million kg/yr are discharged into the atmosphere each year from North America by vaporization and burning, about 7-9% of the peak U.S. production (Table 13-1). To the best of our knowledge, no one has yet attempted to validate this estimate by measuring PCB vapor emissions from dumps and industrial areas.

CHC in Continental Air

Measurement of airborne pesticides and PCB has been hampered by a lack of good collection methods. Greenburg-Smith impingers, used by the E.P.A. to monitor airborne pesticides, are limited to low flow rates and only about 40 m^3 of air can be sampled per day — too small a volume for the method to be useful in remote locations. Within the last few years several high-volume sampling methods have been devised using solid or liquid-coated adsorbents to trap airborne CHC. These systems can sample hundreds of cubic meters of air per day. A brief survey of these collection methods is given in Table 13-4 and a more detailed review was recently published by Seiber and Woodrow (1975).

In 1967-68 Stanley et al. (1971) measured airborne pesticides in four cities and five rural locations.[a] Highest concentrations were found in agricultural areas, especially during times of spraying, when toxaphene and DDT reached $\mu g/m^3$ levels (Table 13-5). Maximum concentrations in areas where no spraying was reported were 10-100 times lower, and repeated measurements of airborne DDT in Dothan, Alabama (an unsprayed rural area), were in the 1-7 ng/m^3 range. A three-year study in the Mississippi Delta, an intensive cotton growing area, showed that airborne DDT residues decreased sixfold between 1972-73 following the DDT ban (Arthur et al., 1976). Even as late as 1974, however, average total DDT levels were still as high as 12 ng/m^3, apparently caused by vaporization from contaminated soils (Table 13-5).

A large-scale survey of pesticides in the atmosphere of the U.S. was carried out by E.P.A. in 1970-71. Over 1800 samples from 28 states were collected and analyzed during the two-year period. About a third of sites were in agricultural areas, while the rest were in rural nonfarm, urban residential, and commercial sections. A selection of this data for nine states was reported by Yobs et al. (1972) (Table 13-5). The means of the "average monthly levels" for these states were only a few ng/m^3, and even these values may be biased toward the high side as it appears from the authors' report that samples below the detection limit were not included in the averages. During this same time period a group from Syracuse University Research Corporation evaluated a high-volume collection system at five stations in New York, one in Florida, and one in Texas (Compton et al., 1972). Although airborne pesticides reached $\mu g/m^3$ concentrations in a few instances, 80% of the 200 samples contained no residues above the detection limits of 0.03-0.3 ng/m^3.

---

[a]Baltimore, Maryland; Buffalo, N.Y.; Dothan, Alabama; Iowa City, Iowa; Orlando, Florida; Salt Lake City, Utah; Fresno, California; Riverside, California; and Stoneville, Mississippi.

Table 13-4
High Volume Collection Systems for Airborne PCB and Pesticides

Collection Medium	Flow Rate (m³/min)	Sample Volume (m³)	Compounds	Reported Collection Efficiency, %	Reference
Polyurethane foam	0.5-1	70-2000	PCB[a] DDT, chlordane, dieldrin, toxaphene	≥80, ≥90	Bidleman and Olney, 1974b,c, 1975a,b
Polyurethane foam	2-3	70-160	DDT, DDE		Orgill et al., 1974
Silicone oil on ceramic ships	0.7	790	PCB[b], DDT	70	Harvey and Steinhauer, 1974d
Silica gel	0.15-1.0	100	DDT, DDE, lindane, heptachlor epoxide, dieldrin, endrin, parathion, ronnel, diazinon	50-100	Seiber and Woodrow, 1975
Paraffin oil on Chromasorb A	0.15-1.0	100	DDT, DDE, dieldrin, heptachlor epoxide, endrin, parathion, ronnel, diazinon	50-100	Seiber and Woodrow, 1975
Chromasorb 102	0.15-1.0	100	DDT, DDE, dieldrin, heptachlor epoxide, parathion	50-100	Seiber and Woodrow, 1975; Thomas and Seiber, 1974
Cottonseed oil on glass beads	0.3	430	DDT, DDE, endrin, heptachlor epoxide	80-100	Compton et al., 1972
Cottonseed oil on glass beads	0.3	430	aldrin, malathion, heptachlor, parathion	67-90	Compton et al., 1972

[a] Aroclors 1248 and 1254.
[b] Aroclor 1254.

Orgill *et al.* (1974, 1976) followed DDT translocation from freshly sprayed pine forests during the 1974 tussock moth control program. Samples were taken from 760–2440 m from small aircraft. The maximum concentration of total DDT encountered was 19 ng/m^3. and the mean in this altitude range was 2.2 ng/m^3. Four flights over the forest three months after spraying showed total DDT levels below 0.02 ng/m^3.

In our program to establish baseline CHC levels over land we collected samples in Kingston, R.I.; Sapelo Island, Georgia; Hays, Kansas; Organ Pipe National Park, Arizona; and the Canadian Northwest Territories. None of the sites were in areas where chlorinated pesticide spraying had been reported. Concentrations of pesticides were below 1 ng/m^3 except for toxaphene, which reached low ng/m^3 concentrations at the southern stations (Table 13-5).

Outside the U.S. we have very little information on airborne pesticides but a few measurements in England and West Germany suggest that the levels are similar to those over the U.S. (Table 13-5). From the above reports we can generally say that while airborne pesticides may reach μg/m^3 concentrations in agricultural situations, background continental levels are in the low ng/m^3 range and are probably less than 1 ng/m^3 for DDT—substantially lower than Woodwell's (1971) predicted 72 ng/m^3 DDT in the troposphere and in line with Cramer's (1973) 0.11 ng/m^3 estimate.

Most reports of pesticides in the air have not mentioned PCB. The mean concentration of PCB on suspended particulates in four cities in the late 1960s was 50 μg/g (E.P.A., quoted in Nisbet and Sarofim, 1972). Assuming an average TSP load of 70 micrograms per cubic meter (Lee, 1972) leads to an estimate of 3.5 ng/m^3 PCB in urban air. Measurements near urban and residential areas within the last three years are of this magnitude or higher, while levels in remote areas are considerably lower (Table 13-5).

CHC in Marine Air

Direct evidence of pesticide transport over the oceans was provided by Risebrough *et al.* (1968), Seba and Prospero (1971), and Prospero and Seba (1972) who used nylon screens to collect particulate matter blown from Africa to Barbados by the trades. The dust contained ng/g levels of dieldrin and DDT residues, and the total pesticide concentration in the air averaged between 7.8 × 10^{-5} and 1.4 × 10^{-4} ng/m^3 (Table 13-6). PCB were not identified. The screens collected 50% of the particles larger than 1 micrometer, so CHC on submicron particles or in the vapor phase would have escaped.

Recent measurements of total CHC (particulate + gaseous) over the western North Atlantic indicate that the screen samplers have very likely underestimated the pesticide burden of marine air and that most of the CHC is transported in the vapor phase. During 1973–74 we measured CHC from a tower on the south

Table 13-5
Chlorinated Hydrocarbons in Continental Air
(10^{-9} g/m^3)

Location, Date	No. Samples	Reference		p,p'-DDT	o,p'-DDT	p,p'-DDE	dieldrin	chlordane	toxaphene	PCB
U.S. cities, 1967-68, unsprayed	437	1	maximum	8-24	1-5	2-11				
U.S. rural areas, 1967-68, unsprayed	241	1	maximum	3-177	2-88	4-13				
U.S. agricultural areas, 1967-68, sprayed	197	1	maximum	950-1560	250-500	47-131				
28 states, 1970-71	1800	2	maximum	602	244	28	34	206	1684	
9 states, 1970-71	>1000	2	mean of "monthly avg. levels"	5.5	2.1	2.0	2.3			
N.Y., Fla., Texas, 1970-71	200	3	13% of samples / 85% of samples	0.5-8 / <0.3	0.2-2 / <0.15					
Miss. agricultural region, 1972 1973 1974	156 (3-yr total)	10	mean of "monthly avg. levels"	99.5 (total DDT) 16.0 (total DDT) 11.9 (total DDT)					258 82 159	
N.W. pine forest after DDT spraying 760-2440 m, 1974	12	4	maximum / 90% of samples	13 / <0.01-2.6	2.6 / <0.01-1.3	3.2 / 0.02-2.0				
Kingston, R.I., 1973-75	6	5	range	0.03-0.6	0.015-0.18			0.04-0.4		
Sapelo Island, Georgia, 1975	6	5	range	0.016-0.073				0.12-0.28	1.7-5.2	
Organ Pipe National Park, Arizona, 1974	6	5	range	0.11-0.45	0.046-0.26	0.059-0.16	0.011-0.12	0.082-0.14	2.7-7.0	<0.02-0.41

Location, Date	No. Samples	Reference		p,p'-DDT	o,p'-DDT	o,p'-DDE	dieldrin	chlordane	toxaphene	PCB
Hays, Kansas, 1974	3	5	range	0.006–0.017	0.004–0.018	0.004–0.056	0.010–0.021	0.022–0.11	0.083–2.6	<0.03
Northwest Territories, Canada, 1974	3	5	range	<0.001–0.009	<0.001–0.005		0.002–0.006	<0.001–0.007	0.04–0.13	<0.002–<0.07
La Jolla, California, 1973	6	6	range							0.5–14
Vineyard Sound, Mass., 1973	2	7	range							4–5
W. Germany, 1970		8	range	0.2–0.4 (total DDT?)						
London, U.K., 1965		9	range	3.0			18–21			

References: (1) Stanley et al., 1971. (2) Yobs et al., 1972. (3) Compton et al., 1972. (4) Orgill et al., 1974, 1976. (5) This work. (6) McClure and LaGrange 1973. (7) Harvey and Steinhauer 1974d. (8) Junge 1975a. (9) Abbott, et al., 1966.

Table 13-6
Chlorinated Hydrocarbons in Marine Air

Location and Date	No. of Samples	10^{-9} g/m³					Reference
		p, p' – DDT	Dieldrin	Chlordane	Toxaphene	PCB	
N.E. tradewinds, Barbados, 1965–66[a]	15	0.000061 (total DDT)	0.000006				Risebrough et al., 1968
N.E. tradewinds, Barbados, 1968–70[a]	28	0.000066–0.0002 (total DDT)	0.000065				Seba and Prospero, 1971; Prospero and Seba, 1972
Bermuda, 1973	4	0.017–0.053 (total DDT)				0.15–0.5	Harvey and Steinhauer, 1974d
Bermuda, 1973	7	0.009–0.022		0.027–0.053	<0.02–1.1	0.19–0.66	Bidleman and Olney, 1974a,b, 1975a,b
Bermuda, 1974	25	<0.003–0.062	<0.003–0.077	0.01–0.14	0.11–1.9	0.08–0.48	Bidleman and Olney, 1974a,b, 1975a,b
Cruises, Bermuda-U.S., 1973–74	17	<0.001–0.058	0.003–0.048	<0.002–0.17	<0.04–1.6	<0.05–1.6	Bidleman and Olney, 1974a,b, 1975a,b
Chesapeake Bay, 1973	3	0.014–0.037	0.051	0.070–0.18	0.39–0.77	1.0–2.0	Bidleman and Olney, 1974a, 1975a,b
Grand Banks, 1973	5	<0.001 (total DDT)				0.05–0.16	Harvey and Steinhauer, 1974d

[a]Particulate (dust-borne) only.

shore of Bermuda (Figure 13-1) and on cruises between Bermuda and the U.S. coast (Figure 13-2), and have identified PCB, DDT, chlordane, dieldrin, and toxaphene (Bidleman and Olney, 1974b, c, 1975a, b) (Table 13-6). More than 95% of these CHC passed through a glass fiber filter, which collects particles $\geqslant 0.015$ micrometer radius with 98% efficiency, and was trapped on polyurethane foam plugs placed behind the filter. Concentrations of toxaphene were an order of magnitude higher than those of the other chlorinated pesticides, probably because of its higher volatility (25 times that of DDT) and large-scale use. Mean concentrations (ng/m^3) for the 20-month period were: PCB = 0.2, p,p'-DDT = 0.02, dieldrin = 0.02, chlordane = 0.03, and toxaphene = 0.6.

The Bermuda samples were collected only when the wind blew from the open ocean and not over the island (Duce *et al.* 1973), and counts of Aitken condensation nuclei, available for about two thirds of the samples, were $\leqslant 500$-600 particles/cm³. This, plus the fact that compatible data were obtained from onboard ship (Bidleman and Olney, 1974b, 1975a, b) assured us that our Bermuda values were representative of marine air and were not influenced by local contamination.

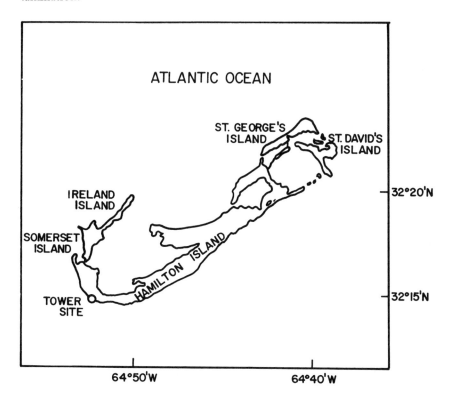

Figure 13-1. Site of atmospheric sampling tower in Bermuda.

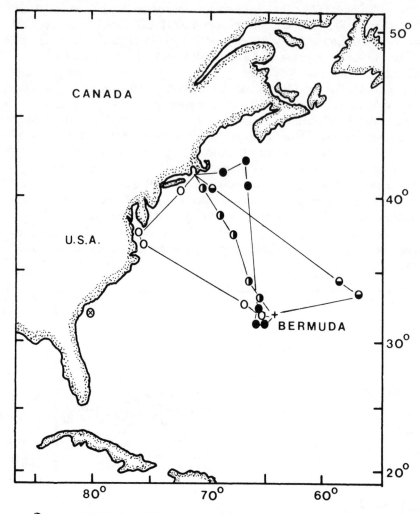

Figure 13-2. Atmospheric samples collected in the western North Atlantic.

For 1974 we were supplied with air mass trajectories by the National Oceanic and Atmospheric Administration (NOAA), which allowed us to follow air masses over Bermuda back in space and time. About half of the 1974 samples had trajectories similar to that in Figure 13-3. On these days air masses

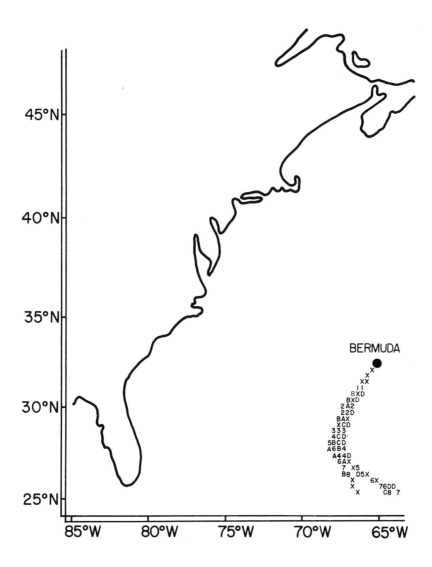

Figure 13-3. NOAA air mass trajectory, September 15, 1974. See Figure 13-4 for an explanation of symbols.

moved into Bermuda from the south-southeast open ocean and not from the North American continent. Some of this air may have been a component of the trade winds, and several of these samples had yellow or reddish brown dust on the filters suggestive of Sahara dust which is known to be transported from Africa by the trades. Other samples were taken on days when the air over Bermuda had previously passed over the continential U.S. An example of such a trajectory is shown in Figure 13-4.

On comparing the samples with no North American air source with those having trajectories passing over the continent, we found that the mean CHC levels of each set differed by no more than a factor of two (Figure 13-5). Similar results were reported by Harvey and Steinhauer (1974d) who reported PCB concentrations agreeing with ours over the western North Atlantic (Table 13-6) and found little variation in the levels with wind direction. While final conclusions await further measurements, these results imply that CHC have sufficiently long lifetimes to become well mixed in the North Atlantic atmosphere.

Aside from these few measurements in the North Atlantic we have no more direct information on worldwide atmospheric CHC transport. Serious thought should be given to the use of transcontinental airliners for atmospheric CHC monitoring, as had been done for particulates and trace gases (Perkins and Gustafsson, 1975). Considering the shift in DDT use from northern countries to the equatorial regions and the southern hemisphere (Goldberg, 1975), future work should be undertaken in these areas.

Removal from the Atmosphere and Atmosphere Lifetimes

Three atmospheric removal processes result in ocean deposition: rain, dry fallout, and vapor-phase deposition onto surfaces. The presence of CHC in rain and snow was one of the earliest indicators of aerial CHC translocation. Analyses of rain in Great Britain and the U.S. in the mid 1960s revealed μg-ng/1 levels of DDT, dieldrin, and BHC (Wheatley and Hardman, 1965; Abbott et al., 1965; Tarrant and Tatton, 1968; Cohen and Pinkerton, 1966). It was these figures that led Woodwell et al. (1971) to consider rainfall the most important process for removing DDT from the atmosphere and the National Academy of Sciences (1971) to suggest that up to 25% of the annual DDT production would be rained out into the world's oceans.

Rain samples over land are strongly influenced by local contamination, as was made clear by Bevenue et al. (1972b) in their study of CHC in Hawaiian rainwater. Dieldrin concentrations in rain collected in Honolulu, where the pesticide was used for termite control, were over five times higher than in samples taken only 25 km away. Unfortunately, we have no data on CHC in rain collected at sea.

Scavenging by falling particles may remove more CHC than rain, especially over land and near the coast. Sodergren (1972) and Bengston and Sodergren

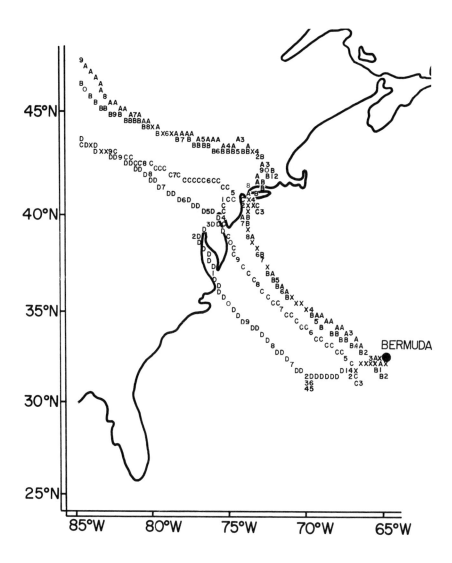

Figure 13-4. NOAA air mass trajectory, May 30, 1974. The letters A,B,C,D trace the history of air masses which were over Bermuda at 0000, 0600, 1200, and 1800 hrs Greenwich time. X indicates the crossing of two or more trajectories. The numbers refer to six-hour time intervals. Thus air mass C, which was over Bermuda at 1200 hrs, left the U.S. coast about 54-60 hrs previously.

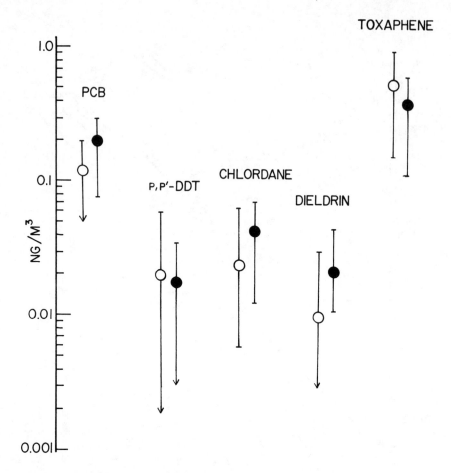

○ No North American Source, 11 samples
● ≥ 50% North American Source, 9 samples
↓ Indicates that the lower end of the range was the detection limit

Figure 13-5. Aerial CHC levels as related to air mass trajectory range and mean.

(1974) found that gram qualities of PCB and DDT per square km were deposited annually in Sweden and Iceland by dry fallout, and McClure and La Grange (1973) concluded that more CHC was deposited each year by dustfall than through washout. Young (1975) collected approximately 1000 fallout samples in California and found that aerial fallout contributed about as much DDT and one fourth the quantity of PCB to the Southern California Bight as did surface runoff and sewage outfalls. Particle fallout of DDT was four times higher within 50 km of the coast than in the 50–100 km region offshore. On the other hand, most of the DDT and about one third of the total PCB entered the Bight by ocean advection from the California Current. So it would seem that on a worldwide scale more CHC is deposited at sea and washed *into* shore than originates from runoff or aerial fallout at the continent-ocean boundaries. Similar conclusions were reached by the National Academy of Sciences (1971) who estimated that only about 0.1% of the annual DDT production enters the oceans by runoff.

A potentially important, though yet unexplored, mechanism of CHC input to the oceans may be vapor-phase transfer through the air/sea interface. Atkins and Eggleton (1971) reported that vapors of BHC, dieldrin, and DDT were deposited onto sections of turf in wind tunnel experiments with velocities of 0.04 cm/sec. Liss (1973) and Liss and Slater (1974) formulated a model for gas transfer across the air/sea interface assuming that the rate is limited by molecular diffusion across stagnant air and water films at the sea surface. The rate of exchange depends on the diffusion coefficients of the gas in each laminar layer, the film thickness, and the air/water partition coefficient (Henry's Law constant). For slightly soluble gases such as oxygen and nitrogen, the exchange rate is limited by transfer across the water film. In the case of highly soluble and/or rapidly hydrolyzed gases (such as SO_2), the exchange is gas-phase controlled. Which phase controls the transfer rate of high molecular weight CHC remains to be determined, but the chemical and physical properties of these compounds suggest that their exchange across the air/sea interface may also be limited by diffusion through the air laminar layer. Although CHC pesticides have very low water solubilities, they also have very low vapor pressures and the partition coefficients between air and water favor the water by about 10^3 (Atkins and Eggleton, 1971; Junge, 1975a). Moreover, it is likely that most of the CHC at the air/sea interface is associated with particulate matter and lipids, and that very little CHC exists in the dissolved state and able to participate in air/water partitioning. Thus the sea surface may act as a perfect sink for CHC vapors.

Assuming that CHC exchange is gas-phase controlled, the air-to-sea flux is given by:

$$F = k(C_b - C_s)$$

where C_b is the concentration in the bulk atmosphere, C_s is the concentration in the air at the interface, and k is the exchange constant (deposition velocity). If the surface microlayer is a perfect CHC absorber, the above equation becomes:

$$F_{max} = kC_b$$

and we can estimate upper limits to the aerial CHC input from the concentrations in the bulk atmosphere and the exchange constant.

Liss and Slater (1974) derived k for several trace gases from an experimental $k = 3000$ cm/hr for water molecules, assuming k varies inversely with the square root of the molecular weight of the gas. Thus k for $CCl_4 = 1030$ cm/hr. For DDT, $k = 680$ cm/hr = 0.19 cm/sec. This is five times higher than Atkins and Eggleton's (1971) experimental value and near the low end of the range of k for trace metal deposition (0.1-1 cm/sec) (Pierson, Cawse, Salmon, and Cambray,

1973). Using background levels of 0.1 ng/m^3 PCB and 0.01 ng/m^3 DDT over
the western North Atlantic and $k = 0.19$ cm/sec, we estimate maximum fluxes of
6×10^7 g/yr PCB and 6×10^6 g/yr DDT into the North American Basin, an area
of the western North Atlantic roughly 1×10^7 km^2 (about one third of the
North Atlantic) centered around the Bermuda Rise.

We can compare this vapor-phase transfer with estimates of particle deposition and rain input. From CHC vapor pressures and the number of particles/cm^3
in clean marine air, Junge (1975a,b) calculated that we might expect 10% of the
CHC over the ocean to be particulate. Actual data from the North Atlantic suggest the fraction of particulate CHC to be even lower. If we take 10% as an
upper limit and use $k = 1.0$ cm/sec for particle deposition, the corresponding
PCB and DDT particle fluxes are 3×10^7 g/yr and 3×10^6 g/yr.

Atkins and Eggleton (1971) reported washout ratios (g CHC/g rain $\div$ g
CHC/g air) in London air of 2-65 for BHC, dieldrin and DDT. Taking 65 as a
maximum value and applying it to background PCB and DDT levels over the
ocean, we estimate PCB and DDT in rain at sea to be 6.5 and 0.65 ng/l. If the
average rainfall over the western North Atlantic is about 80 cm/yr (Malkus,
1962), we get 5.2×10^7 and 5.2×10^6 g/yr PCB and DDT washout into the
North American Basin.

Thus the total estimated PCB and DDT input into the Basin by atmospheric
processes is 1.4×10^8 g/yr PCB and 1.4×10^7 g/yr DDT, and about 0.3% and
0.013% of the world's annual output of these chemicals. While these estimates
are only speculative and apt to be upper limits, they do indicate that all three
atmospheric removal processes are potentially important. Furthermore, they are
of the correct order of magnitude to account for the levels of PCB and DDT observed in the 100-m ocean mixed layer. The residence time of these compounds
in the mixed layer probably ranges from a year, as suggested by the 1972-73
decline in North Atlantic surface water PCB (Harvey *et al.*, 1974a) to as long as
10 years (about equal to the turnover rate of the water itself). The above atmospheric input would thus result in mixed layer concentrations ranging from 0.15-
1.5 ng/l PCB and 0.015-0.15 ng/l DDT, close to currently observed PCB concentrations in the North Atlantic and about equal to the present detection limit for
DDT in seawater (Table 13-2). Certainly, for DDT these estimates more nearly
reflect the true situation in the ocean than do either Woodwell's (1971) or
Cramer's (1973) predictions of 15 and 9 ng/l in the mixed layer.

Woodwell *et al.* (1971) estimated the atmospheric residence time of DDT as
3.3 yrs, assuming that all the DDT produced in a single year entered the atmosphere and was removed mainly by rainfall. Several lines of evidence suggest that
the residence time is more on the order of 0.05-0.3 yr.

1.   A deposition velocity of 0.2 cm/sec calculated for gaseous CHC air/sea
     transfer, would correspond to $\tau = 0.1$ yr.

2. If 10% of the CHC in the ocean atmosphere were adsorbed onto particles having a lifetime of seven days, τ for CHC would be 0.2 yr (Junge, 1975a,b).

3. Assuming the PCB burden of northern hemisphere air to be 0.1 ng/m^3 and dividing the total quantity held in the troposphere by twice Nisbet and Sarofim's (1972) estimated PCB input into the atmosphere (the world PCB production is roughly twice that of the U.S.) leads to $\tau = 0.05$ yr.

4. If CHC behave as other trace gases in the atmosphere, τ should be inversely related to the variation in atmospheric concentrations (Junge, 1974). From the coefficients of variation of our 1973-74 Bermuda CHC data (25-39 samples), $\tau = 0.16$–0.25 yr (Bidleman and Olney, 1975b).

Concluding Remarks

The PCB/DDT ratio in North Atlantic seawater is $\geqslant 10$. Harvey and Steinhauer (1976a) have argued that since the production of DDT has exceeded that of PCB, DDT must break down rapidly in the open ocean. We note that the PCB/DDT ratio in marine air is also about 10, and thus the ratio of these CHC in the ocean is not vastly different from what one might expect from aerial input. As an alternate explanation, we suggest that even though more DDT has been released into the environment, its translocation from the continents has been retarded relative to PCB by its much lower (100-fold) volatility.

A large number of high molecular weight CHC are being carried out to sea by the earth's wind systems; and aerial input, through a combination of rain, dry fallout, and vapor deposition may well represent the major source of PCB and DDT in the oceans. A critical area of research to be undertaken is the actual *measurement* of air/sea CHC fluxes at sea or from remote island stations. We hope that our discussions will encourage other investigators to pursue such investigations.

Acknowledgments

We have profited greatly from the cooperation and enthusiasm of Dr. Robert Duce and his research group. Many of the Bermuda air samples were taken by Mr. Paul DesLauriers, Ms. Barbara Ray, Dr. Ian Fletcher, and Dr. Eva Hoffman; and Mr. Randy Borys and Dr. Kenneth Rahn obtained continental samples for us from a mobile van. We would also like to thank the U. S. Naval Underwater Sound Laboratory, Tudor Hill, Bermuda, for the use of their land to construct the sampling tower. This work was supported by the National Science Foundation, Office of the International Decade of Ocean Exploration, under Grant GX-33777.

References

Anonymous (1971). *Chem. and Eng. News,* Dec. 6, p. 15.

Anonymous (1973). *Chem. and Eng. News,* Feb. 26, pp. 2-3.

Anonymous (1975). *Chem. and Eng. News,* Apr. 14, p. 12.

Anonymous (1976). *Chem. and Eng. News,* Jan. 5, p. 5.

Abbott, D. C., R. B. Harrison, J. O'G Tatton, and J. Thomson (1965). Organo-chlorine pesticides in the atmospheric environment. *Nature, 208,* 1317.

Abbott, D. C., R. B. Harrison, J. O'G. Tatton, and J. Thompson (1966). Or-ganochlorine pesticides in the atmosphere. *Nature, 211,* 259.

Arthur, R., J. Cain, and B. Barrentine (1976). Atmospheric levels of pesticides in the Mississippi Delta. *Bull. Environ. Contam. and Tox., 15,* 129-134.

Atkins, D. H. F. and A. E. J. Eggleton (1971). Studies of atmospheric washout and deposition of BHC, dieldrin, and p,p'-DDT using radiolabeled pesticides. Internat. Atomic Energy Agency, SM/142a/32, 521.

Bedford, J. (1974). The use of polyurethane foam plugs for extraction of PCB from natural waters. *Bull. Environ. Contam. and Tox., 12,* 622-625.

Bengtson, S. A. and A. Sodergren (1974). DDT and PCB residues in airborne fallout and animals in Iceland. *Ambio, 3,* 84-86.

Bevenue, A., J. W. Hylin, Y. Kawano and T. W. Kelly (1972a). Organochlorine pesticide residues in water, sediment, algae, and fish, Hawaii — 1970-71. *Pest. Monit. J., 6,* 56-64.

Bevenue, A., J. N. Ogata and J. W. Hylin (1972b). Organochlorine pesticides in rainwater in Hawaii. *Bull. Environ. Contam. and Tox., 8,* 238.

Bidleman, T. F. and C. E. Olney (1973). Reports to NSF-IDOE Grant No. GX-33777, April 30, Oct. 1.

Bidleman, T. F. and C. E. Olney (1974a). Report to NSF-IDOE Grant No. GX-33777, Oct. 1.

Bidleman, T. F. and C. E. Olney (1974b). Chlorinated hydrocarbons in the Sar-gasso Sea atmosphere and surface water. *Science, 183,* 516-518.

Bidleman, T. F. and C. E. Olney (1974c). High volume collection of atmospher-ic PCB. *Bull. Environ. Contam. and Tox., 11,* 442-450.

Bidleman, T. F. and C. E. Olney (1975a). Long range transport of toxaphene insecticide in the western North Atlantic atmosphere. *Nature, 257,* 475.

Bidleman, T. F. and C. E. Olney (1975b). Transport of PCB and chlorinated pesticides in the North Atlantic atmosphere. *XVI Symp. of the Internat. Union of Geodesy and Geophysics,* Grenoble, France, Aug. 25-Sept. 6.

R. H. Boyle (1975). Poisoned fish, troubled waters. *Sports Illustrated,* Sept. 1 14-17. Ibid (1975). The Spreading Menace of PCB. Dec. 1, 20-21.

Caro, J. H., A. W. Taylor and E. R. Lemon (1971a). Measurement of pesticide concentrations in air overlying a treated field. Internat. Symp. on Ident. and Meas. of Environ. Pollutants, Ottawa, Ontario, Canada, 72-77.

Caro, J. H. and A. W. Taylor (1971b). Pathways of loss of dieldrin from soils under field conditions. *J. Agr. Food Chem., 19,* 379.

Cheng, L. and T. F. Bidleman (1976). Chlorinated hydrocarbons in marine insects. *Estuarine and Coastal Mar. Science* (in press).

Cohen, J. and C. Pinkerton (1966). Widespread translocation of pesticides by air transport and rainout. In *Organic Pesticides in the Environment* (R. F. Gould, ed.), Amer. Chem. Soc., Adv. in Chem. Ser. Vol. 60, 163-176.

Compton, B., P. Bazydlo, and G. Zweig (1972). Field evaluation of methods of collection and analysis of airborne pesticides. Symp. on Pesticides in the Air, 163rd ACS Meeting, Boston, MA, April 11; Report to EPA contract CPA-70-145, May 1972.

Cox, J. L. (1971). DDT residues in seawater and particulate matter in the California current system. U. S. N.O.A.A. Fish. Bull., *69,* 443.

Cramer, J. (1973). Model of the circulation of DDT on earth. *Atmos. Environ., 7,* 241.

Crump-Wicsner, H. J., H. R. Feltz, and M. L. Yates (1974). A study of the distribution of PCB in the aquatic environment. *Pest. Monit. J., 8,* 157-161.

Dawson, R., J. P. Riley, and R. H. Tennant (1976). Two samplers for large-volume collection of chlorinated hydrocarbons. *Mar. Chem., 4,* 83-88.

Duce, R. A., J. G. Quinn, C. E. Olney, S. R. Piotrowicz, B. J. Ray and T. L. Wade (1972). Enrichment of heavy metals and organic compounds in the surface microlayer of Narragansett Bay, R. I. *Science, 176,* 161-163.

Duce, R. A., G. L. Hoffman, J. L. Fasching, and J. L. Moyers (1973). The collection and analysis of trace elements in atmospheric particulate matter over the ocean. Tech. Conf. on Observation and Measurement of Atmospheric Pollution, Helsinki, Finland.

Garrett, W. D. (1965). Collection of slick-forming materials from the sea surface. *Limnol. and Ocean., 10,* 602.

Goldberg, E. D. (1975). Synthetic organohalides in the sea. *Proc. R. Soc. Lond. B., 189,* 277-289.

Harvey, G. R., W. G. Steinhauer and J. M. Teal (1973). Polychlorobiphenyls in North Atlantic ocean water. *Science, 180,* 643.

Harvey, G. R., W. G. Steinhauer, and H. Miklas (1974a). Decline of PCB concentrations in North Atlantic surface water. *Nature, 252,* 387-388.

Harvey, G. R., H. P. Miklas, V. T. Bowen, and W. G. Steinhauer (1974b). Observations on the distribution of chlorinated hydrocarbons in Atlantic Ocean organisms. *J. Mar. Res., 32,* 103-118.

Harvey, G. R. (1974c). In *Pollutant transfer to the marine environment*. Deliberations and Recommendations of the National Science Foundation, Pollutant Transfer Workshop, Port Aransas, Texas, Jan. 11-12.

Harvey, G. R. and W. G. Steinhauer (1974d). Atmospheric Transport of Polychlorobiphenyls to the North Atlantic. *Atmos. Environ., 8,* 777-782.

Harvey, G. R. and W. G. Steinhauer (1976a). Biogeochemistry of PCB and DDT in the North Atlantic. In *Environmental Biogeochemistry* (J.O. Nriagu ed.), Ann Arbor Science Pub., Ann Arbor, Michigan.

Harvey, G. R. and W. G. Steinhauer (1976b). Woods Hole Oceanographic Institution, Woods Hole, Mass., personal communication.

Hoffman, E. J. and R. A. Duce (1974). The organic carbon content of marine aerosols collected on Bermuda. *J. Geophys. Res., 79,* 4474-4477.

Hom, W., R. Risebrough, A. Soutar, and D. R. Young (1974). Deposition of DDE and PCB in dated sediments of the Santa Barbara Basin. *Science, 184,* 1197-1199.

Huang, J. and C. Liao (1970). Adsorption of pesticides by clay minerals. *J. Sanit. Eng. Div., Proc. Amer. Soc. Civil Eng., 96* (SA5), 1057-1078.

Junge, C. E. (1974). Residence time and variability of tropospheric trace gases. Tellus, *26,* 477-488.

Junge, C. E. (1975a). Transport mechanisms for pesticides in the atmosphere. *Pure and App. Chem., 42,* 95-104.

Junge, C. E. (1975b). Basic considerations about trace constituents in the atmosphere as related to the fate of global pollutants. ACS Meeting, Philadelphia, Pa., April.

Junk, G. A., J. J. Richard, M. D. Greiser, D. Witiak, J. L. Witiak, M. D. Arguello, R. Vick, H. J. Svec, J. S. Fritz, and G. V. Calder (1974). Use of macroreticular resins in the analysis of water for trace organic contaminants. *J. Chromatog., 99,* 745-762.

Larsson, K., G. Odham, A. Sodergren (1974). On lipid surface films on the sea. I. A simple method for sampling and studies of composition. *Mar. Chem., 2,* 49-57.

Lee, R. E. (1972). The size of suspended particulate matter in air. Science, *178,* 567.

Leoni, V., G. Puccetti, and A. Grella (1975). Preliminary results on the use of tenax for the extraction of pesticides and polynuclear aromatic hydrocarbons from surface and drinking waters for analytical purposes. *J. Chromatog., 106,* 119.

Liss, P. A. (1973). Processes of gas exchange across an air-water interface. *Deep-Sea Res., 20,* 231.

Liss, P. S. and P. G. Slater (1974). Flux of gases across the air-sea interface. *Nature, 247,* 181-184.

Lloyd-Jones, C. P. (1971). Evaporation of DDT. *Nature, 229,* 65.

MacKay, D. and A. W. Wolkoff (1973). Rate of evaporation of low solubility contaminants from water bodies to the atmosphere. *Environ., Sci. and Tech., 7,* 611.

Malkus, J. S. (1962). In *The Sea,* Vol. I (M. N. Hill, ed.), Interscience Publishers, N. Y., pp. 130-131.

McClure, V. E. and J. LaGrange (1973). Transport of chlorinated hydrocarbons in the atmosphere. Scripps Institution of Oceanography, LaJolla, Calif., unpub. manuscript.

Mehrle, P. M. and F. L. Mayer, Jr. (1975). Toxaphene effects on the growth and bone composition of fathead minnows, *Pimephales promelas. J. Fish. Res. Bd. Can., 32,* 593; Toxaphene effects on the growth and development of brook trout, *Salvelinus fontinalis.* Ibid., *32,* 609.

Mehrle, P. M. and F. L. Mayer, Jr. (1976). Bone development and growth of fish as affected by toxaphene. In *Fate of Pollutants in Air and Water Environments.* (T. H. Suffet, ed.), Adv. in Environ. Sci. and Tech., Wiley-Interscience (N.Y.).

Musty, P. R., and G. Nickless (1974). The extraction and recovery of chlorinated insecticides and PCB from water using porous polyurethane foam. *J. Chromatog., 100,* 83-93.

Musty, P. R. and G. Nickless (1974). Use of amberlite XAD-4 resin for the extraction and recovery of chlorinated insecticides and PCB from water. *J. Chromatog., 89,* 185.

National Academy of Sciences (1971). Chlorinated hydrocarbons in the marine environment. Washington, D. C.

National Academy of Sciences (1975). *Assessing potential ocean pollutants,* Washington, D. C.

Nisbet, I. C. T., and A. F. Sarofim (1972). Rates and routes of transport of PCB in the environment. *Environ., Health Perspectives, 1,* 21-37.

Olney, C. E. and T. L. Wade (1972). Chlorinated hydrocarbons in the marine atmosphere and sea surface microlayer. IDOE Workshop on Baseline Measurements, Brookhaven, N. Y., May 24-26.

Orgill, M. M., M. R. Peterson and G. A. Sehmel (1974). Some initial measurements of DDT resuspension and translocation from Pacific Northwest forests. Symposium on the Atmosphere-Surface Exchange of Particulate and Gaseous Pollutants, Richland, Washington, Sept. 4-6. Battelle Pacific Northwest Laboratories Report BNWL-SA-5126.

Osterroht, C. (1974). Development of a method for the extraction and determination of nonpolar dissolved organic substances in seawater. *J. Chromatog., 101,* 289.

Papageorge, W. P. (1975). Monsanto Corp., personal communication.

Perkins, P. and U. R. C. Gustafsson (1975). An automated atmospheric sampling system operating on 747 airliners. Internat. Conf. on Environ. Sensing and Assessment, Las Vegas, Nevada, Sept. 14-19.

Pierce, R. H., C. E. Olney and G. T. Felbeck, Jr. (1974). *p,p'*-DDT adsorption to suspended particulate matter in seawater. *Geochim. Cosmochim. Acta, 38,* 1061-1073.

Pierson, D. H., P. A. Cawse, L. Salmon, and R. S. Cambray (1973). Trace elements in the atmospheric environment. *Nature, 241,* 252-256.

Poirrier, M. A., B. R. Bordelon and J. L. Laseter (1972). Adsorption and concentration of dissolved carbon-14 DDT by coloring colloids in surface waters. *Environ. Sci. and Tech., 6,* 1033-1035.

Porter, P. E. (1964). In *Analytical Methods for Pesticides, Plant Growth Regulators, and Food Additives,* Vol. II, Chap. 12, (G. Zweig, ed.), Academic Press, N. Y.

Prospero, J. M. and D. B. Seba (1972). Some Additional measurements of pesticides in the lower atmosphere of the northern equatorial Atlantic Ocean. *Atmos. Environ., 6,* 363-364.

Richard, J. and J. Fritz (1974). Adsorption of chlorinated pesticides from river water with XAD-2 Resin. *Talanta, 21,* 91.

Risebrough, R., R. J. Huggett, J. J. Griffin and E. D. Goldberg (1968). Pesticides: transatlantic movements in the northeast trades. *Science, 159,* 1233-1235.

Sanders, H. J. (1975). New weapons against insects. *Chem. and Eng. News,* July 28, pp. 18-21.

Scura, E. D. and V. E. McClure (1975). Chlorinated hydrocarbons in seawater: analytical method and levels in the northeastern pacific. *Mar. Chem., 3,* 337-346.

Seba, D. B. and E. F. Corcoran (1969). Surface slicks as concentrators of pesticides in the marine environment. Pest. Monit. J., *3,* 190-193.

Seba, D. B. and J. M. Prospero (1971). Pesticides in the lower atmosphere of the northern equatorial Atlantic Ocean. *Atmos. Environ., 5,* 1043-1050.

Seiber, J. N. and J. E. Woodrow (1975). Determination of pesticides and their transformation products in air. In *Environmental Dynamics of Pesticides.* (R. Hague and V. Freed, eds.), Plenum Pub. Co., N. Y.

Sodergren, A. (1972). Chlorinated hydrocarbon residues in airborne fallout. *Nature, 236,* 395-397.

Spencer, W. F. and M. M. Cliath (1969). Vapor density of dieldrin. *Environ. Sci. and Tech., 3,* 670.

Spencer, W. F. and M. M. Cliath (1972). Volatility of DDT and related compounds. *J. Agr. Food Chem., 20,* 645.

Spencer, W. F., W. J. Farmer and M. M. Cliath (1973a). Pesticide volatilization. *Res. Rev., 49,* 1-104.

Spencer, E. Y. (1973b). *Guide to the chemicals used in crop production,* 6th ed., Agriculture Canada, Ottawa, Ontario.

Spencer, W. F. and M. M. Cliath (1975). Vaporization of chemicals. In *Environmental Dynamics of Pesticides* (R. Hague and V. Freed, eds.), Plenum Pub. Co., N. Y. pp. 61-78.

Stanley, C. W., J. E. Barney, M. R. Helton and A. R. Yobs (1971). Measurement of atmospheric levels of pesticides. *Environ. Sci. and Tech., 5,* 430-435.

Tarrant, K. and J. O'G. Tatton (1968). Organochlorine pesticides in rainwater in the British Isles. *Nature, 219,* 725.

Thomas, T. C. and J. N. Seiber (1974). Chromasorb 102 an efficient medium for trapping pesticides from Air. *Bull. Environ. Contam. and Tox., 12,* 17.

von Rumker, R., E. W. Lawless and A. F. Meiners (1974). Production, distribution, use, and environmental impact potential of selected pesticides. U. S. Environ. Protection Agency, EPA/540/1-74-001.

Ware, G. W., W. P. Cahill and B. J. Estesen (1975). Volatilization of DDT and related materials from dry and irrigated soils. *Bull. Environ. Contam. and Tox., 14,* 88.

Wheatley, G. A. and J. A. Hardman (1965). Indication of the presence of organochlorine insecticides in rainwater in central England. *Nature, 207,* 386-388.

Williams, P. M. and K. J. Robertson (1975). Chlorinated hydrocarbons in sea-surface films and subsurface waters at nearshore stations and in the north central Pacific gyre. *U. S. N.O.A.A. Fish. Bull., 73,* 445-447.

Willis, G. H., J. R. Parr, R. I. Papendick and B. R. Carroll (1972). Volatilization of dieldrin from fallow soil as affected by different soil water regimes. *J. Environ. Qual., 1,* 193.

Woodwell, G. M., P. P. Craig, and H. A. Johnson (1971). DDT in the biosphere: Where does it go? *Science, 174,* 1101-1107.

Yobs, A. R., J. A. Hanan, B. L. Stevenson, J. J. Boland and H. F. Enos (1972). Levels of selected pesticides in the ambient air of the U.S. Symp. on Pesticides in the Air, 163rd ACS Meeting, Boston, Ma, April 11.

Young, D. R. (1975). A synoptic survey of chlorinated hydrocarbon inputs to the southern California Bight. Progress Report to E.P.A. Grant No. R801153, 31 January.

14 Environmental Concentrations and Fluxes of Some Halocarbons
C. Su and E. D. Goldberg

For the past several years we have been measuring the concentrations of some synthetic halocarbons in the environment. These substances qualify as potential oceanic and atmospheric pollutants on a number of counts. First, they are produced on a global basis in significant amounts, of the order of megatons or so per year. Secondly, they are markedly persistent in the atmosphere and in natural waters. Also, they can interfere with normal metabolic processes of living organisms. Finally, the chlorofluorocarbons are postulated to alter the ozone levels of the stratosphere through the release of chlorine atoms. The rates of production of many of the halocarbons are increasing at the present time and it appeared worthwhile to us to assess the factors that may govern their concentrations in various environmental reservoirs and to obtain a sense of their global distributions. Our presentation will initially consider their production and uses, as well as their chemical and physical properties, and then their environmental concentrations and fluxes.

Production and Use Data

Trichloroethylene

The primary use of trichloroethylene is as a vapor degreaser of metal parts in industrial metal fabrication plants. It accounts for 90 to 95% of production. Other uses include decafinating coffee and removing oleorescence from spices, as well as its employment as a cleaning solvent, diluent or carrier in paints and adhesives, solvent for removing oils and waxes from fibers and as a chemical intermediate. Peak production in the U. S. occurred in 1970 with a value of 2.77×10^{11} grams/yr, falling to 1.97×10^{11} grams/yr in 1974. This decrease has been attributed to its classification as a photochemically reactive smog contributant by Los Angeles County in the United States and to a consequential shift to other solvents for metal degreasing. The above information was derived from Seltzer (1975). Pearson and McConnell (1975) estimate a world production capacity for 1973 of 1.01×10^{12} grams which clearly is in accord with the above U.S. production value, for in general U.S. production of such chemicals is about one half to one quarter of the rest of the world.

Tetrachloroethylene

Commonly known as perchloroethylene or per, this halocarbon's primary use is in dry cleaning. In 1969, 65 - 70% was used for this purpose; 17% in vapor degreasing; 11% as chemical intermediates; and the remainder in miscellaneous applications. The first industrial production began in Germany and the United Kingdom shortly before World War I and in the United States in the 1920s where it was in demand as a dry cleaning chemical. Over the past 45 years production has increased annually at a rate of 9% in the U.S. and reached a value of 3.33×10^{11} grams/yr in 1972. The above information was taken from NAS (1975) which estimates a world production of 10^{12} grams/yr in agreement with that of Pearson and McConnell (1975), who submit a world production capacity for 1973 of 1.05×10^{12} grams/yr. NAS (1975) indicates U.S. per production will increase at a rate of 5 - 7% over the next few years.

1, 1, 1-Trichloroethane (TCE)

The major use of TCE is as a cold cleaning solvent, especially where there is a requirement to remove greases, oils, tars and similar substances from surfaces at ambient temperatures. It has a broad spectrum of industrial uses including those in aircraft, electronics, textile uses and molding industries. It has minor applications as a solvent for biocides and as a component in industries in the formulation of shoe polishes. In the U.S. commercial production began in 1951 and today there are four companies involved in its manufacture. There has also been large scale production in the United Kingdom and continental Europe. In 1973, the U.S. production was 2.49×10^{11} grams, which corresponds to an estimated world production capacity of 4.80×10^{11} grams/yr (Pearson and McConnell, 1975), if the U.S. is considered to account for about half of the world's chemical productivity. Much of the above information was derived from NAS (1975).

Chlorofluoromethanes

The two dominant members of this group in the environment are CCl_2F_2 and CCl_3F, both of which are primarily used as aerosol propellants at an estimated level of 230×10^9 grams per year in the U.S. out of a total chlorofluorocarbon production of 335×10^9 grams per year. Other members of industrial importance, $CHClF_2$, $CClF_2CFCl_2$ and $CClF_2CClF_2$ have an annual cumulative production of 68×10^9 grams in the United States. Other uses for these halocarbons are found in refrigeration, as solvents and as foaming agents. An integrated U.S. production of these materials up through 1973 is estimated to be 3.4 megatons. For CCl_3F, the estimated release from U.S. sources is given as 0.82

megatons. The ratio of CCl_2F_2/CCl_3F production is about 1.6 on a weight for weight basis. The above information was gathered from data in Howard and Hancett (1975).

Total world production is estimated to be slightly over one million metric tons per year (OECD, 1975). CCl_2F_2, CCl_3F and $CHClF_2$ constitute over 90% of this amount. Seventy-five percent of the CCl_3F and 50% of the CCl_2F_2 are used as propellants for aerosol products on a global basis, while 35% of the CCl_2F_2 is used as a refrigerant and 15% of the CCl_3F is employed as a foaming agent.

Worldwide production involves 43 companies in 23 countries (OECD, 1975). World production is expected to increase at a rate of 4% per annum. Since 1970, 7.6 metric megatons of fluorocarbons have been synthesized, of which about 3 million metric tons were CCl_3F and 4.6 million metric tons CCl_2F_2.

Naturally Occurring Halocarbons?

There is a near ubiquitous occurrence of three other halocarbons with the previously cited anthopogenic ones: methyl iodide, carbon tetrachloride and chloroform. Present evidence of natural origins is spotty yet it is difficult to assign an anthropogenic source for them.

Some guidance to sources may be found in the recent investigations of Burreson *et al.* (1975) on the occurrence of halocarbons in the essential oils of the red algae *Asparagopsis taxiformis* (*Rhodophyta*). The species $CHBr_2Cl$, $CHClBIr$, $CHBr_2I$ and CBr_4 were found. The authors suggest that they are the degradation products of 1, 1, 1-trihalomethyl ketones.

We have made similar analyses on another species of the red algae, *Asparagopsis Armata* and find by gas chromatographic assay CH_3I, $CHCl_3$, CCl_4, $CHBrCl_2$, CH_2ClI or $CBrCl_3$, $CHBr_2Cl$ and $CHBr_3$. Of interest is the observation that carbon tetrachloride and chloroform are in similar concentrations in the algae.

The quantitative importance of algae in providing haloforms to the environment is yet to be established, yet clearly they may provide an explanation for their widespread occurrences.

Lovelock *et al.* (1973) hypothesized a biological methylation of iodine species in seawater to methyl iodide and a subsequent transfer in part to the atmosphere. These investigators submit that the methyl iodide plays an important role in mobilizing iodine about the surface of the earth, from the oceans to the atmosphere and hence to land in washout or fallout. The remarkably high concentrations of iodine relative to chloride in the atmosphere is attributed to this flux of methyl iodide. On the other hand, Wilkniss *et al.* (1973) were unable to detect methyl iodide in the marine atmosphere, as did Lovelock *et al.* (1973). The former investigators claimed methyl iodide was a contaminant, associated with a glass syringe in the sampling equipment.

356   MARINE POLLUTANT TRANSFER

In the atmosphere carbon tetrachloride shows a uniform global distribution (Wilkniss *et al.*, 1973; Lovelock *et al.*, 1973). A few measurements made by the latter investigators in surface seawaters to a depth of 300 m indicate a decrease in CCl_4 concentrations with an e^{-1} depth of 50 m. They argued for an atmospheric source. Some confirmation of this has come from the work of Liss and Slater (1974) who on the basis of a two-layer model of the air-water interface derive a flux of carbon tetrachloride from the atmosphere to the oceans.

It would be surprising if the carbon tetrachloride in the environment were anthropogenic since the primary use of the chemical is in the production of the chlorofluorocarbons. There is reportedly a very small loss of carbon tetrachloride in this use as a chemical intermediate. Lovelock *et al.* (1973) suggest that CCl_4 is the product of a natural inorganic chemistry, perhaps involving atmospheric methane and chlorine. Such a mechanism might also serve for the production of $CHCl_3$, whose concentrations in seawater are greater than those of CCl_4 (Murray and Riley, 1973).

U.S. Production Data

The U. S. production data since 1955 is given for the halocarbons in Table 14-1. Most of these chemicals were produced in continuously increasing amounts up to 1973 (CCl_2F_2, CH_3Cl, CCl_3F, CH_3I, CH_3CCl_3, CCl_4). A few seemed to have their productions level off or even decrease ($CHCl_3$, $CHCl=CCl_2$, $CCl_2=CCl_2$). For comparison the world production data, estimated for 1973, is given.

Physico-Chemical Properties

The physico-chemical properties, boiling point and water solubility, of the halocarbons are given in Table 14-2. There are discrepancies concerning the water solubilities of 1, 1, 1-trichloroethane and tetrachloroethylene, the sources of which are unknown.

Analytical Procedures

The collection, isolation and analytical techniques employed for the halocarbons are described in Su (1976).

Atmospheric Concentrations

The concentrations of the halocarbons in air collected in the U.S. and in several areas in Europe and Asia are given in Table 14-3.

Table 14-1
U.S. Productions of Halocarbons (10^9 grams/year)

Year	CCl_2F_2[b]	CH_3Cl[a]	CCl_3F[b]	CH_3I[a]	$CHCl_3$[a]	CH_3CCl_3[a]	CCl_4[a]	$CHCl=CCl_2$[a]	$CCl_2=CCl_2$[a]
1955		17					130	143	81
56		19			18		137	157	84
57		21			21		144	153	89
58	59	20	23		26		142	134	85
59	71	30	27		21		167	163	92
60	75	38	33		32		169	160	95
61	78	48	41	0.011	35	~9	174	140	102
62	94	49	56	0.0091	45	~18	219	162	145[d]
63	98	52	63	0.0064	48		235	167	147
64	103	61	67		54		243	168	166
65	123	85	77		69		269	197	195
66	130	107	77	0.0050	81	110	294	218	210
67	141	125	83	0.0086	87	122[c]	324	222	242
68	148	138	93	0.010	82	136	346	235	289
69	167	183	108	0.0091	98	147	400	271	288
70	170	192	111	0.0041[c]	109	166	459	277	321
71	177	198	117	0.0082	105	170	458	234	320
72	199	206	136	0.0086	106	200	452	194	333
73	221	247	147		115	249	475	205	320
World Production[e]									
1975	400		300			480		1000	1000

[a]1955–1962 values were taken from *Kirk-Othmer Encyclopedia of Chemical Technology*, Interscience, N.Y., London, 1966, 2nd ed. 1963–1973 values were taken from U.S. Tariff Commission. Synthetic organic chemicals: U.S. Production and Sales (1963–1973).

[b]Howard, P.H. and A. Hancett, *Science* 189, 217 (1975).

[c]Sales amount.

[d]*Chemical and Engineering News*, June 4, 13 (1973).

[e]See text.

Table 14-2
Physical Chemical Properties of the Halocarbons

Substance	Boiling Point[a] (°C)	Water Solubility (in ppm w/w)
Dichlorodifluoromethane	− 29.8	280[a]
Trichlorofluormethane	23.8	110[a]
Methyl iodide	42.4	14,000[b1]
Chloroform	61.7	7950[b2]
1, 1, 1-trichloroethane	74.1	1300[d]; ~4000[b2]
Carbon tetrachloride	76.5	800[b2]
Trichloroethylene	87	1,100[b2]
Tetrachloroethylene	121	400[e]; 150[b2]

[a]*Handbook of Chemistry and Physics.* 55th edition. 1974-1975. CRC Press.

[b]*Kirk-Othmer Encyclopedia of Chemical Technology*, 2nd edition, Interscience Publishers.
[b1]Vol. 11, 1966, [b2]Vol. 5, 1964.

[d]C. Marsden and S. Mann. *Solvents Guide.* Cleaver-Hume Press Ltd., London (1963).

[e]A. Seidell. *Solubilities of Organic Compounds*, 3rd ed. Vol. II.D. Van Nostrand, New York (1941).

La Jolla, California

The atmosphere of La Jolla, California has been sampled over the past year and a half. The airs are both a mixture of continental and marine masses and probably reflect the variations in halocarbon concentrations that may be expected in northern Hemispheric samples. Variations for each of the halocarbons extend over about an order of magnitude (averages are given in parentheses):

CCl_2F_2 0.12 - 2.5 $\times 10^{-9}$ ml/ml of air   (0.73)
CCl_3F 0.10 - 0.56 $\times 10^{-9}$ (0.22)
CH_3CCl_3 0.13 - 1.1 $\times 10^{-9}$ (0.37)
$CHClCCl_2$ 0.09 - 5.4 $\times 10^{-9}$ (1.4)
$CCl_2 = CCl_2$ 0.08 - 3.1 $\times 10^{-9}$ (0.58)
$CHCl_3$ 0.17 - 2.8 $\times 10^{-9}$ (0.51)
CCl_4 0.06 - 0.22 $\times 10^{-9}$ (0.12)

Chloroform in these measurements is always more abundant than carbon tetrachloride for the presumably naturally produced halocarbons. Of the synthetic halocarbons, trichloroethylene is usually the most abundant, although this position is sometimes occupied by perchlorethylene, trichlorethane or dichlorodifluoromethane. The ratio of the fluorocarbon concentrations, CCl_2F_2/CCl_3F, is equal to 3.3, substantially higher than the reputed production ratio of 1.6. The reason for this anomaly is not yet understood.

High concentrations of CCl_2F_2, CCl_3F, CH_3CCl_3, $CHCl=CCl_2$ and $CCl_2=CCl_2$ are found in department stores and supermarkets where they leach from articles of commerce and/or refrigeration units.

Marine Air Samples

The chlorofluorocarbon concentrations in the marine air samples taken off the California coast are at the lower end of the range found in La Jolla, as are the concentrations of CCl_4 and CH_3CCl_3. The concentrations of $CHCl_3$, $CHCl=CCl_2$ and $CCl_2=CCl_2$ were closer to the average values, or even on the high side, of those found in the La Jolla area. Not too much significance can be placed upon such comparisons as the marine samples were collected within a period of 16 days and probably reflect a temporal condition rather than a general one.

Continental United States and Europe

The continental United States and Europe air masses have fluorocarbon concentrations substantially higher than the marine values and at the higher end of the spectrum of values observed in La Jolla (Table 14-3). The U.S. dichlorodifluoromethane concentrations averaged 1.2×10^{-9} ml/ml of air while their continental European counterparts were somewhat higher with an average of 3.9×10^{-9} ml/ml of air. For the other synthetic halocarbons, CH_3CCl_3, $CCl_2=CCl_2$ and $CHCl=CCl_2$, the ranges in continental areas were similar to those in La Jolla. The variations from site to site and time to time probably reflect specific entries of pollutants from the area of sampling. The highest value of $CCl_2=CCl_2$ was found in Washington, D. C., of $CHCl=CCl_2$ in Geneva and of CH_3CCl_3 in Grenoble, France.

The continental carbon tetrachloride atmospheric concentrations are similar to those of La Jolla and sometimes do attain values slightly higher than those found in the Southern California marine airs. This is probably a consequence of industrial usages and consequent leakages. Similarly, the chloroform concentrations in continental airs span the same range as that of the La Jolla samples. Again, the marine values are somewhat lower. There may be industrial inputs. Another possible source of these two halocarbons is from chlorination plants where they are produced (see subsequent section on halocarbons in waters).

Table 14-3
Atmospheric Halocarbon Concentrations
(In Units of 10^{-9} ml/ml of Air)

Location	Date	CCl_2F_2	CCl_3F	$CHCl_3$	CH_3CCl_3	CCl_4	$CHCl=CCl_2$	$CCl_2=CCl_2$
Scripps Pier La Jolla, Calif.	4-9-74 1500	0.22	0.10	0.36	0.17	0.11	0.41	1.1
	4-11-74 1200 1400	0.19 0.17	0.13 0.11	0.36 0.28	0.25 0.20	0.1 0.09	0.78 <0.10	0.19 0.12
	4-18-74 1200	0.15	0.1	0.36	0.14	0.07	2.0	2.3
	7-25-74 1600	0.15	0.12	0.19	0.13	0.12	0.41	<0.03
	11-11-74 1200	0.59	0.24	0.49	1.04	0.22	2.9	0.20
	1-14-75 1200 1300	1.3 0.88	0.24 0.31	<0.006 <0.06	<0.06 0.20	0.14 0.14	5.4 <0.10	0.34 0.51
	2-18-75 1200	1.6	0.32	0.2	0.92	0.12	1.8	0.46
	3-26-75 1200	2.5	0.56	1.3	0.60	0.08	3.9	1.1
	4-3-75 1300	0.33	0.12	1.1	0.34	0.06	1.6	0.32
	4-4-75 1100 1600	0.98	0.23 0.22	0.47 <0.06	0.16 0.16	0.07 0.06	4.0 0.99	3.1 0.35
	4-7-75 1000	2.5	0.45	0.26	0.41	0.10	2.3	0.32

Location	Date	CCl_2F_2	CCl_3F	$CHCl_3$	CH_3CCl_3	CCl_4	$CHCl=CCl_2$	$CCl_2=CCl_2$
	4-8-75 1000 (Raining)	1.0	0.38	1.0	1.1	0.10	0.99	0.71
	6-24-75 1030	1.0	0.30	2.8	0.36	0.13	1.2	0.41
	1200	0.12	0.17	0.33	0.29	0.13	0.15	0.078
	1500	0.16	0.14	0.24	0.28	0.13	<0.10	<0.05
	7-7-75	0.30	0.14	0.40	0.15	0.18	0.21	0.15
	10-2-75 1730	0.32	0.21	0.17	0.33	0.13	0.29	0.14
	10-3-75 1100	0.18	0.21	0.17	0.42	0.15	<0.10	0.14
	1-6-76 1900	0.31	0.19	<0.01	0.17	0.06	<0.023	0.1
	1630	0.35	0.20	<0.01	0.1	0.06	<0.023	0.1
Marine, off California Coast								
Osborn Bank basin	5-9-74	0.12	0.11	0.28	0.14	0.044	0.78	0.59
Santa Cruz basin	5-9-74	0.11	0.14	0.36	0.14	0.053	1.0	1.2
San Pedro basin	5-9-74	0.12	0.18	0.53	0.30	0.067	5.1	1.9
Santa Barbara basin	5-8-74	0.15	0.16	0.68	0.24	0.08	5.8	2.7
San Diego trough	5-24-74	0.15	0.13	0.36	0.14	0.09	0.78	<0.03

Table 14-3 continued

Location	Date	CCl_2F_2	CCl_3F	$CHCl_3$	CH_3CCl_3	CCl_4	$CHl=CHl_2$	$CCl_2=CCl_2$
Continental United States								
Washington, D.C.	3-29-74 0800	1.4	1.6	2.6	0.5	0.1	0.78	7.3
Los Angeles Chinatown	4-6-74 1130	1.0	0.78	0.36	0.34	0.35	2.0	0.78
Santa Monica residential area	4-6-74 1400	4.8	4.9	0.64	1.3	0.12	0.6	2.3
Orange County	4-16-74 0830	0.68	0.59	0.36	0.40	0.11	1.1	0.78
	1200	0.37	0.33	0.45	0.27	0.1	2.4	1.1
	1700	0.5	0.34	0.62	0.47	0.13	–	0.42
Chicago, downtown loop	4-19-74 0730	0.68	0.21	1.8	0.2	0.05	1.4	0.41
Chicago, airport	4-19-74	0.37	0.26	0.68	0.32	0.08	2.6	1.5
Mt. Cuyamaca, near San Diego, Calif. (6000 ft.)	3-15-75	0.33	0.16	1.1	0.41	0.08	1.2	0.22
El Cajon, San Diego	4-9-75 1200	1.9	0.40	0.79	0.72	0.1	2.3	0.31
Montgomery Pass, Nevada	4-12-75 1800	1.1	0.23	0.3	0.34	0.03	0.55	0.09
Lytton Lake, Calif. (7300 ft.)	12-29-75 1540	0.29	0.16	<0.01	0.07	<0.01	<0.023	0.1

Location	Date	CCl_2F_2	CCl_3F	$CHCl_3$	CH_3CCl_3	CCl_4	$CHCl=CCl_2$	$CCl_2=CCl_2$
Continental Europe								
Brussels, downtown	3-22-74 0830	3.9	3.3	3.0	1.1	0.1	1.1	3.2
Brussels, suburbs	3-22-74	0.74	0.26	1.1	0.55	0.03	5.2	0.95
Geneva	3-24-74	2.4	2.6	2.6	1.1	0.05	5.8	6.8
Moscow	11-16-74 0500	0.52	0.16	0.49	<0.04	0.02	2.9	0.20
Moscow	11-16-74 0730	0.74	0.20	0.90	0.12	0.088	2.6	0.04
Moscow	11-17-74 1800	0.41	0.16	0.49	0.12	0.10	5.3	0.03
Paris	8-28-75	2.5	>1.1	0.17	0.37	0.05	0.75	0.31
Grenoble, France	9-5-75	9.5 11.4 6.9	>1.1 >1.1 >1.1	0.83	overscale 0.11 0.4	>1.5 0.05 0.05	4.4 1.2 5.3	1.8 1.6 0.65
Brussels	10-10-75 0700	5.7	>1.1	0.49	1.3	0.081	5.8	1.8
Kyoto, Japan	11-22-75	3.0	0.55	0.21	0.19	0.022	0.94	1.4
Tokyo, Japan	11-26-75	3.0	0.58	0.21	0.05	<0.001	0.33	2.3

The halocarbon concentrations in marine and continental atmospheres have been gathered from the literature and collated in Table 14-4. For the chlorofluorocarbons CCl_2F_2 and CCl_3F, marine atmospheric concentrations of 0.1×10^{-9} ml/ml of air appear to be reasonable estimates on the basis of the limited number of analyses available. For the continental atmospheric concentrations there are more data and a wide range of values reported. Perhaps a reasonable average value for either of the chlorofluorocarbons would be 0.5×10^{-9} ml/ml of air, although we recognize that much of the reported differences in concentration may result from inadequacies in sampling or in analysis.

Only two investigations have considered CH_3CCl_3 concentrations in marine airs (Table 14-4) and their averages, 0.075×10^{-9} ml/ml for the Atlantic and 0.19×10^{-9} ml/ml for the Southern California samples lead to a reasonable estimate of about 0.1×10^{-9} ml/ml. The broad range of concentrations of this halocarbon in continental airs $(0.01 - 14.4 \times 10^{-9}$ ml/ml) does not allow for a ready estimation of an average value. We propose 0.5×10^{-9} ml/ml as a tentative value.

There appears to be two groups of results on the ethylenes, tetrachlor and trichlor, for both the marine and continental airs. One set has exceptionally low values compared to the others and includes the work of Lovelock (1974) and of Murray and Riley (1973), while the other investigators (Table 14-4) have found values at least one order, and in some cases up to three orders of magnitude higher. This situation causes difficulties in trying to estimate averages for the different air masses. However, one interesting observation does come out of the rather limited data: there are small differences between the marine and continental concentrations of trichlorethylene. This can be seen from the following data:

Source	Marine Air	Continental Air
Murray and Riley (1971)	0.001×10^{-9} ml/ml	0.00129×10^{-9} ml/ml
This paper	2.7×10^{-9} ml/ml	$1.5 - 3.5 \times 10^{-9}$ ml/ml

Because of the wide variations, we do not propose an average value at the present time for either continental or marine airs.

This discrepancy between high and low values by the English workers, Murray and Riley and Lovelock, and our laboratories continues for the chloroform analyses. The English data show more carbon tetrachloride than chloroform, in contrast to the U.S. results. Again, we can propose no average atmospheric values.

Halocarbon Concentrations in the Hydrosphere

The marine concentrations appear to reflect the atmospheric concentrations and the solubilities, while the atmospheric concentrations appear to be related to the production levels for the individual halocarbons (Table 14-5). Both of these observations are based upon the assumption that their residence times in the troposphere with respect to transfer to the stratosphere or with respect to degradation are similar. For example, CCl_3F and CH_3CCl_3 have similar production rates (0.3 and 0.48 megatons per year) and similar atmospheric concentrations (Table 14-3). The much higher concentrations of CH_3CCl_3 in marine waters and rains probably is in response to the higher solubility of the trichloroethane in water (Table 14-2). Similarly, $CHCl=CCl_2$ and $CCl_2=CCl_2$ have like production rates (both one megaton per year) and like atmospheric concentrations (Table 14-3), and the higher concentration of the trichloroethylene appears to be related to its higher solubility relative to the tetrachloroethylene.

The values of the halocarbons in hot springs (Table 14-5) are usually very low and provide perhaps a baseline from which the other measurement in natural waters can be considered. The two high values of trichloroethane and of chloroform in the hot spring sample from Southern California are not explicable at the present time.

Other patterns of halocarbon distribution emerge from Table 14-5. Carbon tetrachloride is always less abundant than chloroform. Within limits of variation, there is in general a uniform distribution of all of the halocarbons in natural waters excluding the hot springs. The snows usually have similar concentrations of the halocarbons, with the exception of an Alaskan snow which contained remarkably high values of all of the halocarbons except methyl iodide.

Environmental Sinks

Table 14-6 provides some indications of the present day sinks for the halocarbons CCl_2F_2, CCl_3F and CH_3CCl_3. It appears that for these chemicals the estimated world production remains primarily in the atmosphere with substantially smaller amounts in the surface ocean. On the basis of these approximate calculations, there is no evidence for any substantial destruction of any of these species.

Acknowledgments

Logistic support for snow collection provided by the Naval Arctic Research Laboratory, Barrow, Alaska. We thank Dr. W. Fenical and O. McConnell for giving us the pentane extraction of *Asparagopsis Armata.*

Table 14-4
Atmospheric Contents of Halocarbons by Other Investigators
(10^{-9} ml/ml of Air)

Location, Date	CCl_2F_2	CCl_3F	CH_3I	$CHCl_3$	CH_3CCl_3	CCl_4	$CHCl=CCl_2$	$CCl_2=CCl_2$	Reference
Marine Atmospheres									
Atlantic 60°N - 60°S 1971 - 1972		0.037-0.008 (0.05±0.007)	(0.0012±0.01)			0.044-0.078 (0.071±0.007)			A
Pacific 20°N - 78°S Nov.-Dec. 1972		0.038-0.15 (0.061±0.013)				0.038-0.097 (0.075±0.008)			B
North Atlantic Oct. 1973	(0.115±0.033)	(0.089±0.004)		(0.019±0.015)	(0.075±0.009)	(0.138±0.015)	(<0.005±0.006)	(0.021±0.003)	C
Pacific 20°N - 20°S March-April 1974		(0.080±0.006)							D
Arctic 80°N Winter 1974		(0.12)							D
Northeast Atlantic 1972				$0.13\text{-}0.88 \times 10^{-3}$ ($0.33\pm0.18 \times 10^{-3}$)		$0.58\text{-}2.0 \times 10^{-3}$ ($1.01\pm0.37 \times 10^{-3}$)	$0.37\text{-}3.9 \times 10^{-3}$ ($1.02\pm0.83 \times 10^{-3}$)	$0.14\text{-}1.1 \times 10^{-3}$ ($0.43\pm0.27 \times 10^{-3}$)	E
Irish Sea				0.36-0.68 (0.44±0.13)		(0.06)	(0.034)	(0.081)	F
East Pacific California Coast 1974	0.11-0.15 (0.13±0.02)	0.11-0.18 (0.14±0.02)			0.14-0.30 (0.19±0.06)	0.044-0.09 (0.067±0.013)	0.78-5.8 (2.7±2.22)	<0.03-2.7 (1.3±0.8)	G
Continental Atmospheres									
Ireland 51°40'N 09°45'W July-Aug. 1970		(0.1±0.04) (0.019±0.03)							H
Great Britain				$0.38\text{-}1.13 \times 10^{-3}$ ($0.66\pm0.48 \times 10^{-3}$)		$0.88\text{-}2.0 \times 10^{-3}$ ($1.29\pm0.37 \times 10^{-3}$)	$0.34\text{-}4.8 \times 10^{-3}$ ($1.82\pm1.5 \times 10^{-3}$)	$1.1\text{-}7.7 \times 10^{-3}$ ($3.23\pm2.2 \times 10^{-3}$)	E
La Jolla, California		0.01-2.2 (0.37±.56)							I

Continental Atmospheres

Location, Date	CCl_2F_2	CCl_3F	CH_3I	$CHCl_3$	CH_3CCl_3	CCl_4	$CHCl=CCl_2$	$CCl_2=CCl_2$	Reference
San Diego, California		0.01-1.8 (0.29±0.24)							I
Los Angeles Basin Sept.-Oct. 1972		0.11->1.25 (0.72±0.41)			0.01-2.32 (0.46±0.25)	0.10-1.63 (0.27±0.12)		<0.01-4.2 (1.49±0.74)	J
San Bernadino Mt. Oct. 6, 1972		(0.08)			(0.05)	(0.18)		(0.09)	J
New Brunswick, New Jersey		(0.37)	(0.08)		(0.27)	(0.17)		(0.12)	K
Bentrum, Germany Feb. 21, 1974		(0.60)			(2.83)	—		(1.35)	L
Gr. Feldberg, Germany, Feb. 21, 1974		(0.04)			(0.11)	(0.045)		(0.15)	L
Los Angeles Basin July 25, 1972	0.5-0.9 (0.72±0.07)	0.37-1.1 (0.56±0.14)							M
Feb. 1-14, 1972	<0.1-1.4 (0.44±0.23)	0.06-1.47 (0.25±0.15)							
Ireland June-July, 1974	(0.102±0.03)	(0.08±0.05)		(0.027±0.008)	(0.065±0.017)	(0.11±0.01)	(0.015±0.012)	(0.028±0.009)	C
Washington, D.C. July 9, 1974 July 11, 1974 July 12, 1974		0.38 0.20 0.16							D
U.S.A. excluding La Jolla, March 1974-April 1975	0.33-4.8 (1.2±0.82)	0.16-4.9 (0.89±0.86)		0.3-2.6 (0.88±0.52)	0.2-1.3 (0.48±0.2)	0.05-0.35 (0.11±0.045)	0.55-2.6 (1.49±0.67)	0.09-7.3 (1.38±1.26)	G
Europe March 1974-Sept. 1975	0.41-11.4 (3.9±3.1)	0.16->1.1 (1.1±0.73)		0.06-3.0 (0.97±0.87)	0.04	0.02->1.5 (0.2±0.26)	0.75-5.8 (3.5±1.8)	0.03-3.2 (1.6±1.4)	G

Table 14-4 — Continued

Location, Date	CCl_2F_2	CCl_3F	CH_3I	$CHCl_3$	CH_3CCl_3	CCl_4	$CHCl=CCl_2$	$CCl_2=CCl_2$	Reference
Continental Atmospheres									
U.S.A. La Jolla, Calif. Apr 1974-Jan 1976	0.18-2.5 (0.73±0.59)	01.-0.56 (0.24±0.1)		<0.06-2.8 (0.53±0.38)	<0.06-1.1 (0.39±0.24)	0.06-0.18 (0.12±0.03)	<0.5-5.4 (1.49±1.3)	<0.03-3.1 (0.61±0.56)	G
Liverpool/Manchester Suburban area		0.82		0.56-1.5	<0.017-1.0	0.15-2.9	0.17-3.4	<0.014-1.4	O
Moel Famau, Flintshire				<0.019-0.075	0.34-0.67	0.10-0.44	0.17-1.5	<0.014-0.34	O
Rannoch Meir, Argyllshire		2.0-2.1		0.019-0.094	0.17-0.25	0.58-0.73	0.43-1.4	0.041-0.14	O
Forest of Dean, Monmouthshire		0.39		1.7	0.47	0.058	0.85	0.41	O
Seagirt, N.J. June 18-19, 1974	0.25-0.38 (0.32)	0.12-0.55 (0.25)	<0.001-0.016 (0.007)	<0.01-0.06 (0.04)	0.044-0.2 (0.10)	0.12-0.36 (0.23)	<0.05-2.8 (0.26)	0.10-0.88 (0.32)	P
New York, N.Y. June 27-28, 1974	0.5-3.8 (1.5)	0.50-3.8 (1.44)	0.003-0.032 (0.01)	<0.01-0.48 (0.16)	0.10-1.6 (0.61)	0.16-0.37 (0.27)	0.11-1.1 (0.71)	1.0-9.75 (4.5)	P
Sandy Hook, N.J. July 2-5, 1974	0.4-3.2 (1.2)	0.10-0.75 (0.28)	0.003-0.16 (0.04)	<0.01-0.63 (0.03)	0.03-0.33 (0.15)	0.085-0.66 (0.22)	<0.05-0.80 (0.34)	0.15-1.4 (0.39)	P
Delaware, Del. July 8-10, 1974	0.4-1.3 (0.68)	0.095-0.21 (0.13)	0.002-0.018 (0.01)	<0.01 —	0.03-0.3 (0.10)	0.06-0.14 (0.09)	0.05-0.56 (0.35)	<0.02-0.51 (0.24)	P
Baltimore, Md. July 11-12, 1974	0.25-0.8 (0.40)	0.10-0.68 (0.24)	<0.001-0.013 (0.008)	<0.01 —	0.044-0.21 (0.12)	0.06-0.16 (0.12)	<0.05	<0.02-0.29 (0.18)	P
Wilmington, Ohio July 16-26, 1974	0.08-0.18 (0.11)	0.078-0.38 (0.14)	<0.001-0.016 (0.007)	<0.01-9.8 (0.34)	0.030-0.35 (0.097)	0.096-0.39 (0.20)	<0.05-0.63 (0.19)	<0.02-0.69 (0.15)	P
White Face Mt., N.Y. Sept. 16-19, 1974	0.085-0.17 (0.10)	0.10-0.18 (0.13)	<0.001-0.003 (0.003)	<0.01-0.25 (0.009)	0.032-0.13 (0.067)	0.20-0.33 (0.26)	<0.05-0.35 (0.10)	<0.02-0.19 (0.07)	P
Bayonne, N.J. March-Dec. 1973	0.06-47 (6.27)	0.046-8.8 (1.34)	<0.001-38 (0.32)	<0.01-15 (1.03)	0.075-14.4 (1.59)	0.05-18 (1.63)	<0.05-8.8 (0.92)	0.30-8.2 (1.63)	P

Location, Date	CCl_2F_2	CCl_3F	CH_3I	$CHCl_3$	CH_3CCl_3	CCl_4	$CHCl=CCl_2$	$CCl_2=CCl_2$	Reference
Continental Atmospheres									
Washington State Nov. 8-Dec. 2, 1974	0.23-0.43	0.12-0.40							Q
Pullman, Wash. Dec. 1974-Feb. 1975	(0.23±0.01)	(0.13±0.008)	(<0.005)	(0.02±0.01)	(0.1±0.015)	(0.12±0.015)	(<0.005)	(0.020±0.010)	R

References: (A) Lovelock *et al.*, 1973. (B) Wilkniss *et al.*, 1973. (C) Lovelock, 1974. (D) Wilkniss *et al.*, 1975. (E) Murray and Riley, 1973. (F) Goldberg, 1975. (G) This work. (H) Lovelock, 1974. (I) Su and Goldberg, 1973. (J) Simmonds *et al.*, 1974. (K) Lillian and Singh, 1974. (L) Bergent and Betz, 1974. (M) Hester *et al.*, 1974. (N) Riley, Private communication, 1971. (O) Lillian *et al.*, 1975. (P) Grimsrud and Rasmussen, 1975. (Q) Grimsrud and Rasmussen, 1975.

Table 14-5
Halocarbon Concentrations in the Hydrosphere
(10^{-12} g/ml)

Location	CCl_2F_2	CCl_3F	CH_3I	$CHCl_3$	CH_3CCl_3	CCl_4	$CHClCCl_2$	CCl_2CCl_2
Marine								
Open ocean East Pacific, mixed layer, 0 – 100 m.		0.93±0.55	0.78±0.41	14.8±5.3	6.2±4.3	0.51±0.28	8.6±4.3	2.0±0.94
Scripps Inst. of Oceanography Pier water, Surface 1-28-75 to 7-8-75								
(a) Including rainy season		0.58±0.14	1.85±0.32	11.8±5.8	10.5±6.8	0.67±0.17	9.4±7.2	10.4±6.7
(b) Excluding rainy season		0.63±0.1	2.1±0.3	9.3±3.6	6.2±1.3	0.72±0.06	3.3±0.5	5.5±1.2
Rain								
La Jolla, Calif.		1.6±0.8	1.3±0.9	17±13	8.1±3.8	2.8±2.2	5±2.6	5.7±4.0
Runcorn, U.K.[1]				200	90	300	150	150
Snow								
Southern California		0.25	4.6	20	6.2	0.33	30	2.3
Central California		0.26	3.8	2.8	0.6	0.36	<1.5	1.4
Alaska	250±60	2600±650	3.1±1.5	94±25	27±4.5	2.2±0.4	39±8	16±3.3

Location	CCl_2F_2	CCl_3F	CH_3I	$CHCl_3$	CH_3CCl_3	CCl_4	$CHClCCl_2$	CCl_2CCl_2
Ice								
Commercial machine		0.59	0.0	80	3.9	1.4	20	3.8
Reservoir								
Lake[b]								
Untreated water		1.45±0.9	1.4±0.6	11±4	3.6±1.5	1.4±0.09	38-65 / 5.1±4.6	140-420 / 2.91±1.2
Treated water		1.1		6×10^3		} 3.8	1	2
Fresh water U.K.[a]				2×10^3		} 3×10^2		
Drinking water								
Scripps		4.9	11.8	5×10^4		15.3	7.9	8.8
Tap (Switzerland)[b]							105	2100
Hot Springs								
Southern California		0.037±0.02	0.0	81±22	<0.1	<0.01	53±16	0.0
Central California		0.025±0.02	0.0	<0.6	<0.1	∿0.03	0.0	0.0
Other Marine Values								
Atlantic[c]		0.047±0.044	0.85±1.56	8.3±1.8		0.41±0.12		
Northeast Atlantic[d]					} 0.25	0.17±0.04	8.2±2.0	0.48±0.22
Liverpool Bay[a]							0.3	0.12

[a] Pearson and McConnell, 1975.
[b] Grob and Grob, 1974.
[c] Lovelock et al., 1973.
[d] Murray and Riley, 1973.

Table 14-6
Environmental Sinks for the Halocarbons

	Area 10^{16} cm²	Volume 10^{24} cm³	CCl_2F_2 10^{-12} ml/ml	CCl_2F_2 10^{-13} g/ml	CCl_2F_2 10^{12} g	CCl_3F 10^{-12} ml/ml	CCl_3F 10^{-13} g/ml	CCl_3F 10^{12} g	CH_3CCl_3 10^{-12} ml/ml	CH_3CCl_3 10^{-13} g/ml	CH_3CCl_3 10^{12} g
(A) Average Concentration											
1. Oceanic atmosphere			160	5.4		100	6.2		100	5.7	
2. Continental atmosphere			500	38		500	31		500	29	
3. Surface seawater							6.0			62	
4. Fresh waters							15			36	
5. Surface snow (Alaska)				2,500			26,000			270	
6. Fresh Snow											
(B) Dimensions of sinks											
1. Atmosphere		4.33									
2. World ocean	360										
3. Polar ice	16										
4. Terrestrial water	134	0.00023									
(C) Production of trichlorofluoromethane 1950 - 1973, integrated, estimated (grams)					4			2.6			2.6
(D) Environmental levels											
1. Atmosphere: (A-1) × 4.33 × 10^{24} ml					4.2			2.7			2.5
2. Surface ocean: (A-3) × 360 × 10^{16} cm² × 10^4 cm								0.022			0.22
3. Terrestrial waters: (A-4) × 2.3 × 10^{20} cm³								0.00036			0.0008
4. Stratosphere								0.2[a]			—
5. Polar Ice (10 cm/year accumulation rate for ten years yield a depth of 1 meter) (A-5) × 16 × 10^{16} cm² × 10^2 cm					0.004			0.04			0.0004

[a]From Krey and Lagmarsino (1975).

References

Bergent, K.H. and V. Betz (1974). Erfahrungen bei der quantitative Analyse Von flüchtrgen organischen Mikroverunreinigungen in Luft. *Chromatographia, 7,* 681–687.

Burreson, B.J., R.E. Moore and P. Roller (1975). Halogorms in the essential oil of the alga Asparagopsis Taxiformis (Rhodophytoa). *Tetrahedron Letters,* 1975, 473–476.

Goldberg, E.D. (1975). Marine Pollution. In *Chemical Oceanography* (J.P. Riley and G. Skirrow, eds.), 2nd Edition, Volume III, Academic Press, N.Y pp. 81.

Grimsrud, E.P. and R.A. Rasmussen (1975). The analysis of chlorofluorocarbons in the troposphere by gas chromatography-mass spectrometry. *Atmospheric Environment, 9,* 1010.

Grimsrud, E.P. and R.A. Rasmussen (1975). Survey and analysis of halocarbons in the atmosphere by gas chromatography-mass spectrometry. *Atmospheric Environment, 9,* 1014.

Grob, K. and G. Grob (1974). Organic substances in potable water and in its precursor, Part II. Application in the area of Zurich. *J. of Chromatography, 90,* 303.

Hester, N.E., E.R. Stephens and O.C. Taylor (1974). Fluorocarbons in the Los Angeles Basin. *J. Air Poll. Control Assn., 24,* 591–595.

Howard, P.H. and A. Hancett (1975). Chlorofluorocarbon sources of environment contamination. *Science, 189,* 217–219.

Krey, P.W. and R.J. Lagomarsino (1975). Stratospheric concentrations of SF_6 and CCl_3F. *Health and Safety Laboratory Environmental Quarterly,* 1 July 1975. Energy Research and Development Administration, New York. 1-97-1-123.

Lillian, D. and H.B. Singh (1974). Absolute determination of atmospheric halocarbons by gas phase coulometry. *Anal. Chem., 46,* 1060–1063.

Lillian D., H.B. Singh, A. Appleby, L. Lobban, R. Arnts, R. Gumpert, R. Hague, J. Toomey, J. Kazazis, M. Antell, D. Hansen and Barry Scott (1975). Atmospheric fates of halogenated compounds. *Environ. Sci. and Tech., 9,* 1042.

Liss, P.S. and P.G. Slater (1974). Flux of gases across the air-sea interface. *Nature, 247,* 181–184.

Lovelock, J.E. (1974). Atmospheric halocarbons and stratospheric ozone. *Nature, 252,* 293–294.

Lovelock, J.E., R.J. Maggs and R.J. Wade (1973). Halogenated hydrocarbons in and over the Atlantic. *Nature, 241,* 194–196.

Molina, M.J. and F.S. Rowland (1974). Stratospheric sink for chlorfluoromethanes: Chlorine atomic-catalysed destruction of ozone. *Nature, 249,* 810–812.

Murray, A.J. and J.P. Riley (1973). Occurrence of some chlorinated aliphatic hydrocarbons in the environment. *Nature, 242,* 37–38.

NAS (1975). Assessing potential ocean pollutants. U.S. National Academy of Sciences, Washington, D.C., 438 pp.

Pearson, C.R. and G. McConnell (1975). Chlorinated C_1 and C_2 hydrocarbons in the marine environment. *Proc. Roy. Soc. Lond. 189B,* 305–332.

OECD (1975). Fluorocarbons. An assessment of worldwide production, use and environmental issues. Paris, 50 pp.

Seltzer, Richard J. (1975). Reactions grow to trichloroethylene alter. *Chem. Eng. News,* May 19, 41–43.

Simmonds, P.G., S.L. Kerrin, J.E. Lovelock and F.H. Shair (1974). Distribution of atmospheric halocarbons in the air over the Los Angeles Basin. *Atmospheric Environ., 8,* 209–216.

Su, Chih-wu (1976). Low molecular weight halocarbons. In *Strategies for Monitoring Marine Pollutants* (Edward D. Goldberg, ed.), Interscience, N.Y. (in press).

Su, C. and E.D. Goldberg (1973). Chlorofluorocarbons in the atmosphere. *Nature, 245,* 27.

Wilkniss, P.E., R.A. Lamontagne, R.E. Larson, J.W. Swinnerton, C.R. Dickson and T. Thompson (1973). Atmospheric trace gases in the southern hemisphere. *Nature, 245,* 45–47.

Wilkniss, P.E., J.W. Swinnerton, R.A. Lamontagne and D.J. Bressan (1975). Trichlorofluoromethane in the troposphere, distribution and increase, 1971–1974. *Science, 187,* 832–834.

15

Concentrations and Fluxes of Phthalates, DDTs and PCBs to the Gulf of Mexico

C. S. Giam, H. S. Chan and G. S. Neff

From the large production volume and wide uses of the phthalic acid esters (PAEs), mainly as plasticizers, it appeared that PAEs were likely contaminants of the marine environment. (Tables 15-1 and 15-2 are summaries of their production volumes and uses.) The discovery of two of the most common phthalates, di-(2-ethylhexyl) phthalate (DEHP) and dibutyl phthalate (DBP) in a number of terrestrial and aquatic samples (*Environ. Health Persp.*, 1973; Mathur, 1974; Mayer, 1972; Williams, 1973; Zitko, 1973) along with reports of toxicity to aquatic organisms further indicated that the phthalates were probable marine pollutants. On the basis of these presumptions, a program was instituted in our laboratory to analyze marine samples for the presence of the phthalates. The data below is the first systematic study of the distribution of PAEs in the marine environment.

The major tasks of this program were (1) to establish methodology, including the necessary ultra-low background reduction techniques for the accurate assessment of phthalate levels in biota, sediment, water and air, (2) to analyze selected samples from the Gulf of Mexico for the phthalates and (3) to assess, on

Table 15-1
Phthalate — Plasticizers Uses (1970)

Plasticizer Uses of Phthalates	Usage (in million Kg.)
Building and Construction	177
House Furnishings	91
Automobiles	54
Apparel	32
Food Packaging	11
Medical Products	9
Nonplasticizer Uses of Phthalates	23

Table 15-2
Production of Phthalates

	Production (in million Kg)	
Phthalates	1961	1971
DEHP	63	177
DBP	7	10
Other Miscellaneous Phthalates	102	255
Total	172	442

the basis of the analyses, production figures and use data, the probable rates and routes of input of the phthalates into the marine environment, specifically the Gulf of Mexico. The achievement of the first of these tasks has been reported elsewhere (Giam, Chan and Neff, 1975a, b; Giam et al., 1976); this report is concerned mainly with the analytical results and their significance. The phthalates emphasized in this study were DEHP and DBP as they had been found in other environmental studies and because their production volumes were sufficiently high to suggest that measurable levels would be present if phthalates were contaminating the marine environment. Since we had also been conducting surveys of the levels of DDTs and PCBs in the Gulf (Giam et al., 1972, 1973, 1974), and since these chlorinated pollutants are copollutants with the phthalates (PAEs), the concentrations of these compounds are also included to aid in discerning if patterns of pollution distribution of manmade contaminants exist in the Gulf. (Various aspects concerning PCBs and DDTs distribution in the marine environment, including those of other workers, will not be discussed here.)

Concentrations of PAEs, PCBs and DDTs in the Gulf of Mexico

Biota, surface (0 – 10 cm) sediment, and water samples were obtained from more than 35 stations in the Mississippi River Delta and the Gulf of Mexico. The mean values of the analytical results are presented in Table 15-3. More details on location and type of samples may be found in Chan (1975) and Kakareka (1974). Several significant observations can be made from this data. One is that the distribution of concentrations of the phthalates in the analyzed components of the marine environment appear to be quite different from that of the chlorinated hydrocarbons. While the DDTs and PCBs are most concentrated in biota, the phthalates are more concentrated in sediment than biota samples. Since DEHP, DDTs and PCBs are hydrophobic compounds with a high affinity for surfaces, the concentration differences are probably due to dissimilar routes and rates of uptake, metabolism and excretion rather than selectivity in sediment absorption. Another observation is that the levels of DDTs and PCBs in

Table 15-3
Mean Concentrations of Organic Pollutants in Gulf of Mexico

Number of Stations	Samples	DEHP	DBP	PCBs	t-DDT
44	Biota (71-72)	ND[a]	ND	56.0	49.00 (ng/g)
24	Biota (73-74)	5.0	0.1	26.0	2.30 (ng/g)
36	Sediment (73-75)	9.0	3.0	0.3	0.20 (ng/g)
34	Water (73-74)	112.0	87.0	1.8	0.60 (ng/l)
8	Air (73-75)	0.4	0.3	0.4	0.04 (ng/m³)

[a]ND — No Data.

biota (in the Gulf of Mexico) declined slightly relative to our previous 1971-72 survey; these results suggest but do not prove conclusively that the recent restrictions of DDTs and open uses of PCBs are yielding slight changes in the environmental distribution of these compounds.

Distribution of phthalates, DDTs and PCBs in the Gulf of Mexico

The patterns of distribution of these contaminants vary widely depending on location, type of sample and class of pollutant. For convenience, the concentrations for each pollutant (or group of pollutant) have been grouped in three ranges, low, medium and high. A detailed presentation of this data is presented in Figures 15-1 to 15-5. The mean concentrations in different media (from different areas of the Gulf) are summarized in Table 15-3.

Sediment

For phthalates, PCB and t-DDT[a] distributions, the trends are similar (Figures 15-1 to 15-4) and indicate that the Mississippi River is one source of these pollutants. The mean levels in the Mississippi River Delta region are 116 ng/g DEHP, 16 ng/g DBP, 18 ng/g PCBs and 5 ng/g t-DDT while in the areas beyond the Delta, the levels dropped to 9 ng/g DEHP, 3 ng/g DBP, 0.3 ng/g PCBs and 0.2 ng/g t-DDT. This is consistent with the fact that deposition rates drop off significantly beyond the Delta (see Dr. B.J. Presley's discussion of trace metals distribution).

---

[a]t-DDT is used to refer to total DDT, or the sum of DDT and DDE concentrations.

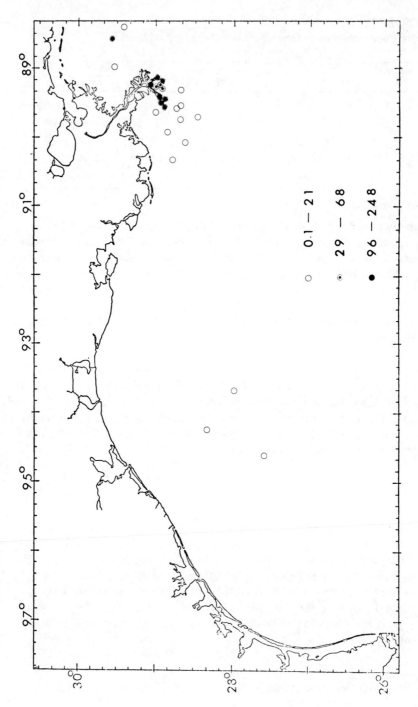

Figure 15-1. DEHP in Sediment, ng/g.

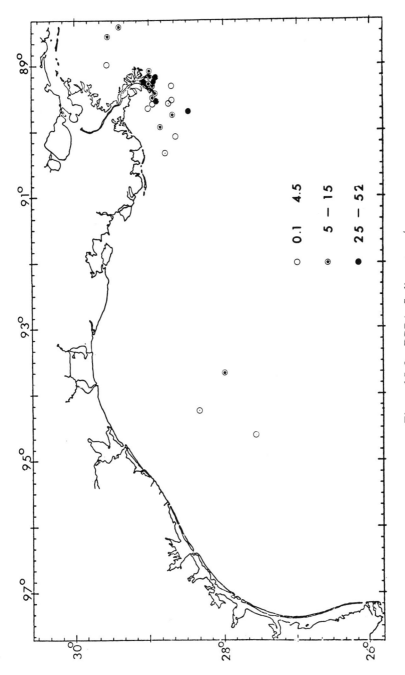

Figure 15-2.   DBP in Sediment, ng/g.

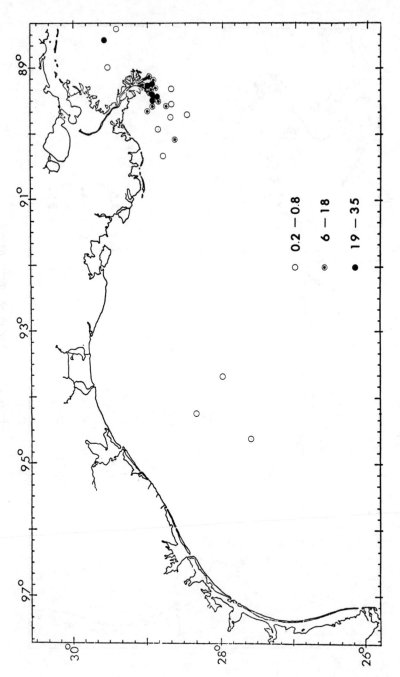

Figure 15-3. PCB's in Sediment, ng/g.

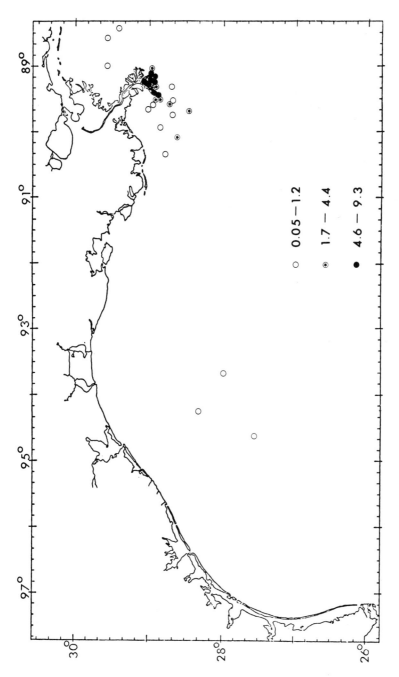

Figure 15-4. Total DDT in Sediment, ng/g.

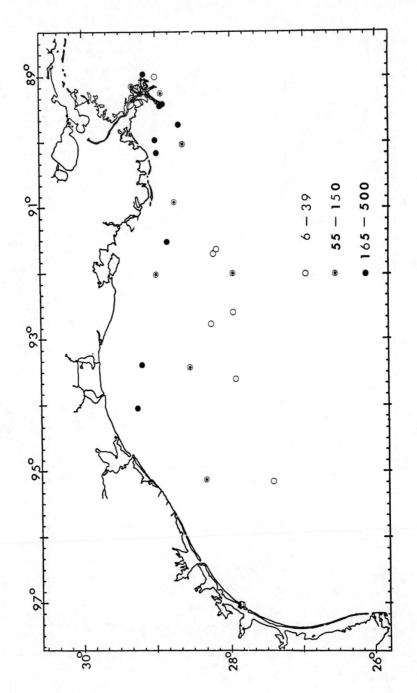

Figure 15-5.   DEHP in Water, ng/L.

Water

Figure 15-5 shows the distribution of DEHP in the northwestern part of the Gulf of Mexico. In general, concentrations of DEHP and DBP were higher near the coastal regions than in waters in the open Gulf. The mean concentrations of DEHP and DBP were higher in this region (112 ng/1 DEHP and 87 ng/1 DBP), than in the lower Mississippi River waters (63 ng/1 DEHP and 14 ng/1 DBP). This observation suggests that besides the Mississippi River, there are other sources of phthalates pollution into the Gulf of Mexico, such as open-ocean dumping, atmospheric transport or industrial discharges. This will be further investigated.

In contrast to the phthalates, higher levels of PCBs and t-DDT were found in the Mississippi River waters (2.5 ng/1 PCBs and 2.3 ng/1 t-DDT) than in the Gulf of Mexico waters (1.8 ng/1 PCBs and 0.6 ng/1 t-DDT). The concentrations in the Gulf waters correlate well with those recently determined in the Sargasso Sea (1.1 ng/1 PCBs and 0.06 ng/1 DDT) (Bidleman and Olney, 1974).

Biota

Very low levels of phthalate esters are present in the biota samples analyzed. About 68% of the biota analyzed fall within the range of <0.1 - 4 ng/g, 20% within 8-12 ng/g and 12% within 20-135 ng/g. DEHP was the only phthalate detected with a mean of about 5 ng/g. DBP, DEP and DMP, if present, were at less than detection limit of 0.1 ng/g. The biota samples had higher levels of PCBs (25 ng/g) and t-DDT (11 ng/g) than of phthalates. The levels of these chlorinated hydrocarbons appear to be decreasing relative to the 1971 survey (Giam *et al.*, 1972); the mean residue levels obtained in 1971 were about 60 ng/g for PCBs and 50 ng/g for t-DDT. There is also a significant decrease in the p,p'-DDT/p,p'-DDE ratio between 1971 (p,p'DDT/p,p'-DDE = 2.8) and 1973-1974 (p,p'-DDT/p,p'-DDE = 0.16).

The higher levels of PCBs than phthalates in biota but lower in water and sediment, and the lower levels of phthalates in liver than in muscle samples suggest that phthalates may be readily metabolized in open-ocean biota samples.

Atmosphere

Preliminary atmospheric measurements of phthalates over the Gulf of Mexico showed that the mean concentrations of DEHP, DBP, PCBs and t-DDT were 0.4 ng/m^3, 0.3 ng/m^3, 0.4 ng/m^3 and 0.04 ng/m^3 respectively (Table 15-4). The levels of PCBs and t-DDT were similar to those found in Sargasso Sea Atmosphere (0.5 ng/m^3 PCB; 0.03 ng/m^3 t-DDT) (Bidleman and Olney, 1974).

Table 15-4

Phthalate Esters and Chlorinated Hydrocarbons in the Air from the Gulf of Mexico

Date Obtained	Location	(ng/m^3)			
		DDT	PCB	DBP	DEHP
2/75	28° 36.7' 96° 45.2'	0.05	0.19	0.08	0.4
2/75	28° 54.1' 28° 53.3' 89° 29.4' 89° 31.8'	0.03	0.91	0.7	<0.9
2/75	28° 59' 29° 3.4' 89° 13.8' 89° 12.3'	0.09	0.27	1.5	1.4
2/75	29° 13' 89° 18.5'	<0.03	0.84	<0.7	<0.9
2/75	29° 13' 91° 47.5' Galveston	0.1	0.4	0.1	0.9
3/75	27° 51' 91° 4'	0.02	0.6	–	2.3
2/74	27° 55.7' 93° 36'	<0.02	<0.4	<0.2	<0.4
2/74	27° 58.8' 93° 39.5'	<0.04	<0.9	<0.4	<0.9

Accumulation and Fluxes of PAEs to the Gulf of Mexico

The extensive contribution of the Mississippi River to the organic pollutant levels in the Gulf of Mexico is illustrated by the higher levels of all pollutants in Delta area sediments relative to the other areas sampled. In the case of PCBs and DDTs, high water levels in addition to high sediment levels in the Delta region implicate the Mississippi as the prime source of these pollutants in the Gulf. The water levels of the phthalates, however, indicated that the Mississippi may not be the only major source of these compounds in the Gulf.

Based on the concentrations presented earlier, and the reasonable assumption that DEHP is present uniformly in the top 50 m of waters, and top 3 cm of sediments in the U.S. Continental Shelf in the Gulf, the current amount of DEHP in the Gulf was calculated at 3.5×10^6 Kg. A current flux to the Gulf of 0.2×10^6 Kg/yr of DEHP can be estimated from this figure and the rates of production of DEHP from its initial time of high production to the present (Chan, 1975). In this study, the contribution of the Mississippi was found to be only 0.04×10^6 Kg/yr. Thus, approximately 0.16×10^6 Kg/yr or four times more DEHP is introduced into the Gulf annually from sources other than the Mississippi. Disposal data on plastics indicate that 0.03×10^6 Kg/yr may enter the Gulf from this source. Similar calculations for DBP indicate a flux of 0.09×10^6

Kg/yr, with only 0.01 × 10⁶ Kg/yr contributed by the Mississippi. The origins of the remaining inputs to the Gulf have not yet been determined, but atmospheric transport may be a significant factor.

Acknowledgments

We wish to thank NSF-IDOE for the support of this work through Grant GX 37349. Also, we wish to thank Drs. B.J. Presley and W.M. Sackett, Department of Oceanography, Texas A&M University, for their assistance in sampling and some calculations of fluxes.

References

Bidleman, T.F. and C.E. Olney (1974). Chlorinated hydrocarbons in the Sargasso Sea atmosphere and surface water. *Science, 183,* 516-517.

Chan, H.S. (1975). A study of the transfer processes of phthalate esters to the marine environment. Ph.D. Dissertation, Texas A&M University.

Environ. Health Persp., 2 (1972), *3* (1973) and references therein.

Giam, C.S., H.S. Chan and G.S. Neff (1975). Sensitive method for determination of phthalate ester plasticizers in open-ocean biota samples. *Anal. Chem., 47*(14), 2225.

Giam, C.S., H.S. Chan and G.S. Neff (1975). Rapid and inexpensive method for detection of polychlorinated biphenyls and phthalates in air. *Anal. Chem., 47*(14), 2319.

Giam, C.S., H.S. Chan, T.F. Hammargren, G.S. Neff, and D.L. Stalling (1976). Confirmation of phthalate esters from environmental samples by derivatization. *Anal. Chem., 48*(1), 78.

Giam, C.S., R.L. Richardson, D. Taylor and M.K. Wong (1974). DDT, DDE and PCBs in the tissues of reef dwelling groupers (serranidae) in the Gulf of Mexico and the Grand Bahamas. *Bull. Environ. Contam. Toxicol., 11,* 189.

Giam, C.S., M.K. Wong, A.R. Hanks, W.M. Sackett and R.L. Richardson (1973). Chlorinated hydrocarbons in plankton from the Gulf of Mexico and Northern Caribbean. *Bull. Environ. Contam. Toxicol., 9,* 376.

Giam, C.S., A.R. Hanks, R.L. Richardson, W.M. Sackett and M.K. Wong (1972). DDT, DDE and polychlorinated biphenyls in biota from the Gulf of Mexico and Caribbean Sea—1971. *Pestic. Monit. J., 6,* 139.

Kakareka, J.P. (1974). *A Study of organic pollutant transfer processes in the estuarine environment.* M.Sc. Thesis, Texas A&M University.

Mathur, S.P. (1974). Phthalate esters in the environment: Pollutants or natural products? *J. Enviorn. Quality, 3*(3), 189 and references therein.

Mayer, F.L., Jr., D.L. Stalling and J.L. Johnson (1972). Phthalate esters as environmental contaminants. *Nature, 238,* 411.

Williams, D.T. (1973). Dibutyl- and di(2-ethylhexyl) phthalate in fish. *J. Agr. Food Chem., 21,* 1128.

Zitko, V. (1973). Determination of phthalates in biological samples. *Intern. J. Environ. Anal. Chem., 2,* 241.

List of Contributors

T. F. Bidleman
Department of Chemistry
University of South Carolina
Columbia, South Carolina 29208

W. M. Broenkow
Moss Landing Marine Laboratories and
San Jose State University
Moss Landing, California 95039

J. M. Brooks
Department of Oceanography
Texas A&M University
College Station, Texas 77843

K. W. Bruland
Marine Sciences
University of California
Santa Cruz, California 95064

M. Burnett
Division of Geological and Planetary Sciences
California Institute of Technology
Pasadena, California 91125

J. N. Butler
Division of Engineering and Applied Physics
Harvard University
Cambridge, Massachusetts 02138

J. Cadwallader
Bermuda Biological Station for Research
St. George's West, Bermuda

H. S. Chan
Department of Chemistry
Texas A&M University
College Station, Texas 77843

R. A. Duce
Graduate School of Oceanography
University of Rhode Island
Kingston, Rhode Island 02881

387

W. M. Dunstan
Skidaway Institute of Oceanography
P. O. Box 13687
Savannah, Georgia 31406

J. L. Fasching
Department of Chemistry
University of Rhode Island
Kingston, Rhode Island 02881

W. F. Fitzgerald
Department of Marine Sciences and
Marine Sciences Institute
University of Connecticut
Groton, Connecticut 06340

I. S. Fletcher
Graduate School of Oceanography
University of Rhode Island
Kingston, Rhode Island 02881

W. S. Gardner
Skidaway Institute of Oceanography
P. O. Box 13687
Savannah, Georgia 31406

J. Geiselman
Woods Hole Oceanographic Institution Graduate Program
Massachusetts Institute of Technology
Cambridge, Massachusetts 02139

C. S. Giam
Department of Chemistry
Texas A&M University
College Station, Texas 77843

E. D. Goldberg
Scripps Institution of Oceanography
La Jolla, California 92037

J. L. Heffter
Air Resources Laboratory
National Oceanic and Atmospheric Administration
Silver Spring, Maryland 20910

E. J. Hoffman
Graduate School of Oceanography
University of Rhode Island
Kingston, Rhode Island 02881

G. L. Hoffman
Graduate School of Oceanography
University of Rhode Island
Kingston, Rhode Island 02881

B. W. de Lappe
Bodega Marine Laboratory
University of California
Bodega Bay, California 94923

J. H. Martin
Moss Landing Marine Laboratories
Moss Landing, California 95039
and California State University
San Francisco, California 94132

J. M. Miller
Air Resources Laboratory
National Oceanic and Atmospheric Administration
Silver Spring, Maryland 20910

B. F. Morris
Bermuda Biological Station for Research
St. George's West, Bermuda

G. S. Neff
Department of Chemistry
Texas A&M University
College Station, Texas 77843

C. E. Olney
Department of Food and Resource Chemistry
University of Rhode Island
Kingston, Rhode Island 02881

G. A. Paffenhofer
Skidaway Institute of Oceanography
P. O. Box 13687
Savannah, Georgia 31406

C. Patterson
Division of Geological and Planetary Sciences
California Institute of Technology
Pasadena, California 91125

S. R. Piotrowicz
Graduate School of Oceanography
University of Rhode Island
Kingston, Rhode Island 02881

B. J. Presley
Department of Oceanography
Texas A&M University
College Station, Texas 77843

B. J. Ray
Graduate School of Oceanography
University of Rhode Island
Kingston, Rhode Island 02881

C. P. Rice
Department of Food and Resource Chemistry
University of Rhode Island
Kingston, Rhode Island 02881

R. W. Risebrough
Bodega Marine Laboratory
University of California
Bodega Bay, California 94923

B. Schaule
Division of Geological and Planetary Sciences
California Institute of Technology
Pasadena, California 91125

D. Settle
Division of Geological and Planetary Sciences
California Institute of Technology
Pasadena, California 91125

T. D. Sleeter
Bermuda Biological Station for Research
St. George's West, Bermuda and
Division of Engineering and Applied Physics
Harvard University
Cambridge, Massachusetts 02138

C. Su
Scripps Institution of Oceanography
La Jolla, California 92037

J. H. Trefry
Department of Oceanography
Texas A&M University
College Station, Texas 77843

W. Walker II
Bodega Marine Laboratory
University of California
Bodega Bay, California 94923

G. T. Wallace
Graduate School of Oceanography
University of Rhode Island
Kingston, Rhode Island 02881

P. R. Walsh
Graduate School of Oceanography
University of Rhode Island
Kingston, Rhode Island 02881

H. L. Windom
Skidaway Institute of Oceanography
P. O. Box 13687
Savannah, Georgia 31406